2154;
Lieu, 廖,

"Elysium, The Magical Corporation Of Desire"

WhiteWolf 白狼

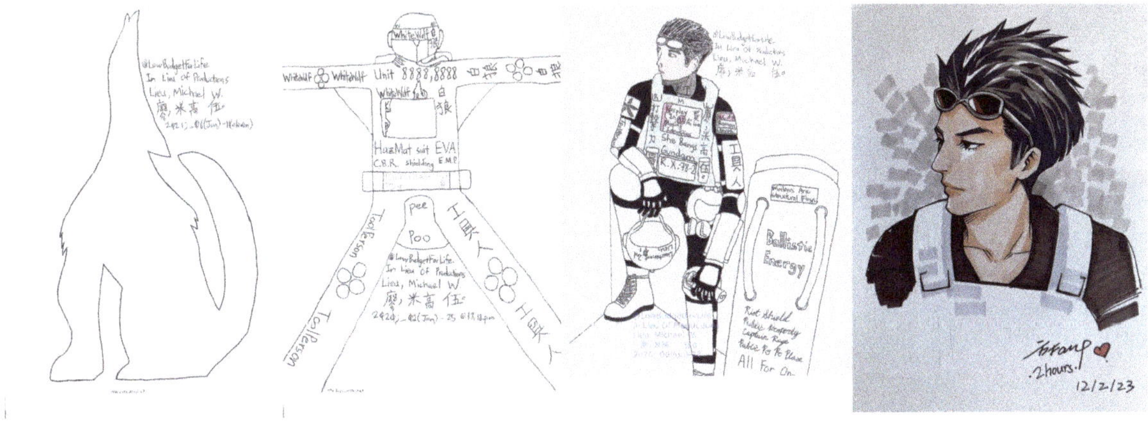

"→(right) 2 D Augmented Art Commission(s) Portrait(s) By; Fang Fang Comics, Fang YiLi"

Lieu, Michael W.
廖, 米高 伍o

Print book, Electronic-Book (E-Book), made possible by Planet Earth's Evolved Earthlings

Arc Two Story Three Book One (A 2 S 3 B 1)
(L.C.C.N., U.S.C.O., I.S.B.N. info below...)

Copyright Page.
Title = "**2154; Lieu,**"
Subtitle = "**Hard Soft Science-Fiction Parody Fan Art Alternative History**"
Copyright Year © = "**2024**"
International Serial Book Number (I.S.B.N.) = "**978-1-959211-06-8**" For Hardcover Hardback Cover Book"
"0 (zero) none" = Electronic Book (E-Book)
"0 (zero) none" = Audio Book
Library of Congress Control Number = "0 (zero)" / United States Copyright Office = { X }

Subject Topics, Genre Type;
Hard Science-Fiction, Soft Science-Fiction, Science-Fiction = "The Literature Of Ideas"
Alternative History AKA Soft Science-Fiction = non-fiction mixed fiction
Apocalyptic, Neo-Western { Macro Bigger Story Setting } = non-fiction mixed fiction
Utopia, Parody { Micro Smaller Story Setting }= non-fiction mixed fiction
Semi-Autobiography = non-fiction
Traditional Art Character Designer = non-fiction mixed fiction
Coloring-Book Line Art Style = non-fiction mixed fiction
Photo-Book = non-fiction
Vigilant Sketches Journal Expansions = non-fiction mixed fiction

No D.R.M. (Digital Rights Management) dog poo

2154visualstorytextbook@gmail.com

x "Here's my email, so email me maybe?!"

Meeting In Person: **Greg Sestero** (left) ***The Disaster Artist*** **(2013), Tommy Wiseau** (right) ***The Room*** **(2003)** Master Minds At 2019 Comic Con Los Angeles (C.C.L.A.)

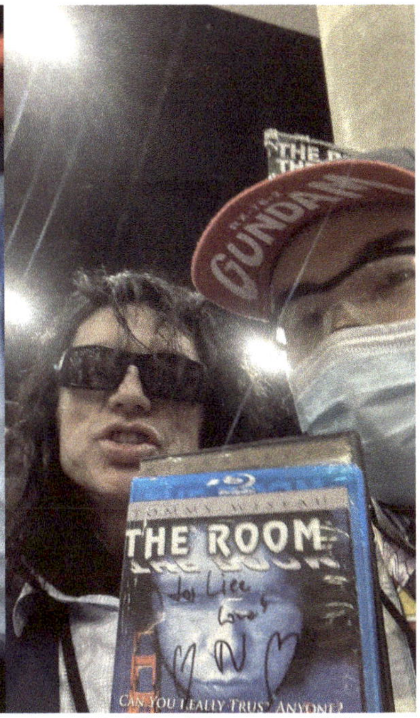

2023; C.C.L.A. **Me** holding **Tommy Wiseau's The Room (2003)** Movie with **The Tommy Wiseau**.

'*The Disaster Artist*' Co-Author **Book**	'Me, back in 2019 C.C.L.A.'	'*The Room*' **Movie**
@ Greg Sestero	Lieu, Michael W. 廖, 米高 伍o	@ Tommy Wiseau

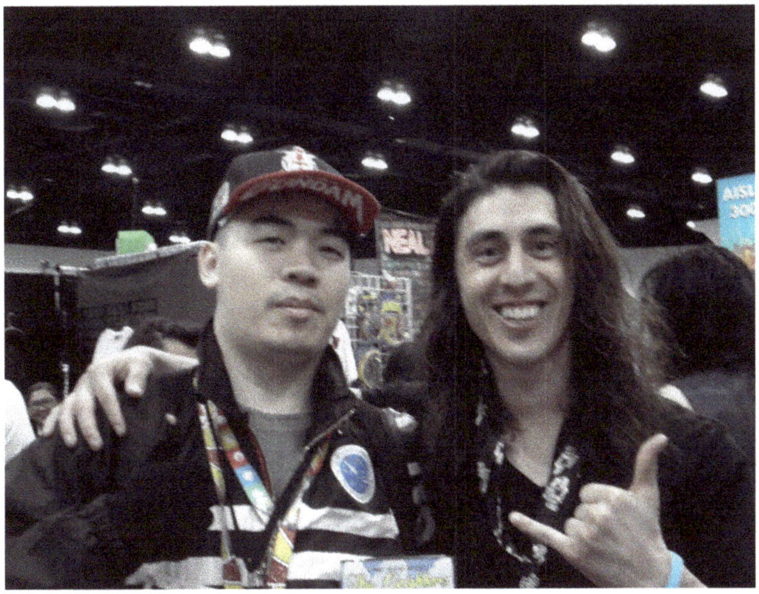

'Me, back in 2019 C.C.L.A.'	'*Tommy Wiseau's The Neighbors T.V.*' Co-Star
Lieu, Michael W. 廖, 米高 伍o	@ Raul Phoenix TV = "Borrows twenty dollars"

Story 1 "2154;_Arriving Coloring Book"+"2154;_Arriving" Concept Art Script Experiment
A 2 (Arc Two) S 1 (Story One) On Amazon Lex Luthor Corp (L.L.C.), Insane Inspiration
Simon And Schuster + Tom Bissell, Greg Sestero's The **Disaster Artist Book**
Lionsgate + A24's The **Disaster Artist** <= **Movie** => The Room's Tommy Wiseau.
E.A. + DICE's **Mirror's Edge Catalyst remake sandbox game**
Warner Brothers + D.C., Christopher Nolan's *The Dark Knight* { Night } story 2(two)
With Greg Wiseau's Tommy Planet Parody "*The Mark Knight*". Comedy, Dark, genre.
Ubisoft's *Watch Dogs 1(one) original* sandbox game. Heathers Movie Original (1988).
Bethesda + Obsidian's **Fallout New Vegas** sandbox simulator game
2K + 2K Boston, 2K Australia, Blind Squirrel's **BioShock** sandbox simulator game
Valve's **Half-Life 2**, Lost Coast, Episode: One, Two, Left 4 Dead 2, Portal: 1, 2 game(s)

Original Art By; Lieu, Michael W. 廖, 米高 伍o	"Elle K. Mcgee" "BladeMaster 刀大師" Blue Navy Sailor Suit, character(s) conceptualized art by; Lieu, Michael W. 廖, 米高 伍o and(&) augment by Benavidez, Richard
Original Book Design; Lieu, Michael W. 廖, 米高 伍o	"Big Ben" "BigBoss 大老闆" "Big Bear Gran Oso" "Mexican-Man \ Mexican-Men" Militia, Mythology. characters conceptualized art by; Lieu, Michael W. 廖, 米高 伍o and(&) augment by Benavidez, Richard
Original Character Design; Lieu, Michael W. 廖, 米高 伍o	
Original Concept By; Lieu, Michael W. 廖, 米高 伍o	"Xtra-Men" "Jerry Duty" "Dog-Name-Cat \ Rover" "Social Worker Mars Hall" characters conceptualized art by; Lieu, Michael W. 廖, 米高 伍o and(&) augment by Benavidez, Richard
Original Cover Design; Lieu, Michael W. 廖, 米高 伍o	
Originally Edited By; Lieu, Michael W. 廖, 米高 伍o	"Exclusive Artist Commission(s) Portrait(s), 2 D Augmented Art Commission(s) Portrait(s) By; Fang Fang Comics, Fang YiLi"
Original Story By; Lieu, Michael W. 廖, 米高 伍o	"3 D Augmented Art Commission(s) Super-Hero Parody Genre(s) Model(s) By; Amanda Coryell, AKA MandyKatDWP (Draxis Web Productions)"
Originally Written By; Lieu, Michael W. 廖, 米高 伍o	

A 2 (Arc Two) S 2 (Story Two) "2154;_Arcadia" = Travel Drawing Audio Retcon
"Art, The Imitation Of Life. Life, The Imitation Of Art."
Insane Inspiration;

Anime eXpo (A.X.) + Comic Con(s) = "Anime Comic Con Expo(s)" Convention Center(s)
2006's KIPI { kosplayer } or Kipi cosplayer, Elizabeth Rage, Alex Drastal, Julia Valentine,
Lisa A. Pinelli, Tara Cosplay \ Recreation, Economics, Infrastructure (REI) \ Free Market
Imagining 2019's Convention Centers as the new world 22nd(twenty second) Century Cities
Economic Freedom Merchant Artists Author Alleyways \ Animal Human Cosplay Cafe
United States Dollars + Crypto Currency \ Anime Los Angeles \ Long Beach Comic Expo
Anifest \ Comic-Con's Wonder Con \ Anime Pasadena \ Ronin-Expo \ Anime eXpo (A.X.)
Animanga \ Anime California \ Stan Lee's Kamikaze, Comic Con L.A. \ "Animal eXpo"
Leon Chiro \ PeachMilky_ Sophie S. from Ireland, where flag is like Italy. \
"Martin Wong Photography", Linda Le A.K.A. "VampyBitMe", "YaYa Han", "D-Piddy",
"Alodia Gosiengfiao" A.K.A. BlackMage9 Eastern Society Of Asia's Cosplay Queen

Nisioisin, Vofan, Aniplex's MoNoGaTaRi story universe character(s) of
KissShot-AcerolaOrion-HeartUnderBlade { Best Vampire Lady And Or Girl } cosplayer(s);
@MakiEraclea, @SayaTheFox, @HaneAme_Cos, @Owldepot, @HaidyKnox, @Standesu

Party At The Sausage Castle, a.k.a. The Wildest House In America(s), @TheSausageCastle
"Arrive as strangers, depart as family."
Bill And Ted universe, "Be excellent to each other." / The Summer Of Love <3

Warner Brothers(W.B.), Adult Swim, Robot Chicken T.V. show series comedy, parody, fan fic

Square Soft Enix, Dontnod, Deck 9(nine), Titan Books' Life Is Strange "travel mark book"
Rockstar's Read Dead Redemption 2, story 0(zero) origins, Arthur Morgan's travel journal

W.B., D.C., Rocksteady's **Batman Arkham**: Asylum, **City**, Knight. Kill The Justice League

Bethesda + Obsidian's **Fallout New Vegas' Dead Money Add-On**
Down Loadable Content (D.L.C.) sandbox simulator game

2K + Irrational Games' BioShock: Infinite story 0(zero) origins
BioShock Infinite: Burial At Sea Episode 1(one), Burial At Sea Episode 2(two)
BioShock 1(one) Original, BioShock 2(two) science-fiction violent video games

Valve's Half-Life 1(one) Original, Half-Life Blue Shift, Half-Life Opposing Force,
Half-Life 2, Half-Life 2 Lost Coast, Half-Life 2 Episode: 1(One),
Half-Life 2 Episode: 2(Two), Left 4(four) Dead 1(one) original, Left 4(four) Dead 2(two),
Portal 1(one) original, Portal 2(two) science-fiction violent video games

A 2 (Arc Two) "A Garden Of Vigilantes" S 3 (Story Three)
B 1 (Book One) = (A 2 S 3 B 1) "2154; Lieu, 廖,"
"Elysium, The Magical Corporation Of Desire." ...Insane Inspiration(s)...

2019, the last year I actually meet up with my "**Elementary, Mid, High School Friends Eastern-Western SitCom genre**", dealing with super anxiety attacks disorder upgrade of **panic attacks**, sleep problems disorder of **insomnia**, doctor diagnosed **work related stress** like **burnout**. / **Casshern 2004** live act movie

Disney's **Tomorrowland** movie directed by Brad Bird With George Clooney
Warner Brothers(W.B.), H.B.O.'s **WestWorld** T.V. show series season 1 Nolan, Joy
W.B., **Wachowskis'** The **Matrix** movies, video games, multi-media franchise

CJ Entertainment, Moho Film, OPUS pictures; **Boon, Joon-Ho's Snowpiercer**
Alex Sokurov's Russian Ark movie / **Birdman movie** / **God Of War 2018**

E.A., Visceral Games' **Dead Space 2(two) Limited Edition game**
Ninja Theory's **Hellblade Senua's Sacrifice game / Mister Robot T.V.**

505 Games + **Remedy Entertainment's Control Ultimate Edition game**
Saman Kesh's The Controller C.G.I. V.F.X. Short Film Video On YouTube

Endless Eight(8888,8888) time loop from infamous anime scene from **Haruhi**

Dreams, Visions, Nightmare, Night Terror, The MultiVerse Theory
Dreams from Sony, Media Molecule PlayStation 4(four) game

Marvel, **Netflix's Jessica Jones** season 1(one) T.V. show series' **Killgrave**.
W.B. D.C. **Batman Arkham**: Origins, R.S. Asylum, City, **Knight**. Kill The J.L.

Bethesda Obsidian's **Fallout New Vegas' Old World Blues Add-On** sim game

2K, Irrational Games' **BioShock: Infinite story 0(zero) origins**
Burial At Sea Episode 1(one), Burial At Sea Episode 2(two)
BioShock 1(one) Original, BioShock 2(two) And Minerva's Den video game(s)

Valve's Half-Life 1(one) Original, Half-Life Blue Shift, Half-Life Opposing Force,
Half-Life 2, Half-Life 2 Lost Coast, Half-Life 2 Episode: 1(One),
Half-Life 2 Episode: 2(Two), **Left 4 Dead 1(one) Original, Left 4 Dead 2**,
Portal 1(one) original, Portal 2(two) science-fiction video game(s)
Hardcore Henry movie / **Cloud Atlas** movie / **Prey** (2017) / **Deus Ex** /
Cyberpunk 2077 / Observer System Redux / Disco Elysium Final Cut

"**2154; Lieu, 廖,**" Abstract Table Of Contents.
"Electronic Book (E-Book) pages subject to change if not in P.D.F. type."

Act 0 (zero) prologue, back story
creating composition
2154; 09 (Sep) – 11 (eleven)
{ Sir Arthur C. Clarke's Space Elevator(s) Anniversary Launch Celebration }
@ North (N) Side. @ 05;15 A.M., ***page #14***.

Act 1 (one) back story, maybe AKA, Act 2 (two) story forward, next book...
creating composition
2154; 10 (Oct) – 30 (thirty)
{ "BladeMaster 刀大師, Sailor Suit" Student Trainee Elle K. Mcgee Arriving }
@ North (N) Side. @ 19;19 P.M., ***page #188***. "**2154; Lieu, 廖,**" concludes.

"**2154; Lieu, 廖,**" Prelude; Quotes, Dedication, Text, Etc, ***page(s) #10 to 13***.
Model References Work Inspiration; Bibliography, ***page(s) #189***.
Social Net Media Pages, ***page(s) #190 to 191***. Abstract Of "**2154;_Arriving**"
Lieu, Michael W.'s Version 2 (two) "Science-Fiction Character Story Universe" Blueprint, ***page(s) #192 to 193***. Updated And Abstract From "**2154;_Arcadia**"
Lieu, Michael W.'s (Alternative History AKA Soft Science-Fiction) (Parody) genre Arc Two Story Three Book One, A 2 S 3 B 1, "2154; Lieu, 廖," Concept Generator, ***page(s) #194 to 195***.
2022;_05(May)-30(thirty) Journal Diary Typed Entry On Warner Brothers' (W.B.) Intellectual Property (I.P.) Science-Fiction genre story universes, Abstract From "**2154;_Arcadia**" ***page(s) #196***.
As Of Story Three (S 3) Book One (B 1) = 2154; Lieu, 廖, Color Coding Character Toner(s) Work In Progress (WIP), very visual, ***page(s) #197 to 198***.
A 0(Arc Zero) "Beginning Of The End" Story 0(Zero) Origins' **Visual Story Text Back Story "22nd (Twenty Second) Century Dark Ages";** **"2120; Endgame"**, "2121; Movement", "2121; SWAT", "2121; Army", 2121; Collapse, ***page(s) #199 to 219***.
A 0(Arc Zero)+ "A ? (Arc Question-Mark) "? ? ? ?;" and or "XXXX;" ***page(s) #219***.
Next "**Arc Two Story Three Book Two (A 2 S 3 B 2)**" AKA "**2154; Incident**" ***page(s) #220 to 221***.
Lieu, Michael W.'s Artist Author Info Dump. Updated And Abstract From "**2154;_Arcadia**" ***page(s) #222 to onwards***.

"**2154; Lieu, 廖,**" Story Zero, prequel prologue to "2154;_Arriving" Table Contents.

(Neo Noir, Vigilante) my ***Ludi Lin Parody Fan Art*.** Bio**tech Suit**+ **2154**; 09 (Sep) – 11 (eleven) @ 05;15 a.m. { act 0 (zero) origins } '@ ***North (N) Side***', { A 2 S 0 B 1 } **Elysium Corporation** 亞, Engineering Environment Bio**tech Dome**, **Interior**, +024°(twenty four degrees) Celsius / +075°(seventy five degrees) Fahrenheit. "Chicago (S), The Last Animal Sanctuary (S) Of Illinois (ILL)," 'North (N) Side', Bio**tech** Dome Ocean Border, King And X District, **Utopian** Sovereignty of North America (**U.**S. of N.A.) Illinois (ILL) 60666 North America, Planet Earth, Solar System, Milky Way Galaxy. Current Carry On Tools, "**WhiteWolf 白狼" Suit**. ***page(s) #14 to 23***.
(Photo-Book) (Elementary, Mid, High School Friends; Eastern-Western SitCom) (Coming Of Age) (Slice Of Life, Mockumentary) (DocuDrama, Philosophy) ***Arc One (A 1) "Hero Journey Character Transformation***" *Year; 2136*. "Fort Angel" ***Joel Kinnaman + Karen Fukuhara Parody Fan Art(s)*** Chatsworth High School (C.H.S.). ***page(s) #24 to 33***.
(Photo-Book) (Military) United States Navy (U.S.N.) Sailor (Super-Hero) (**Vigilante**) (Spy) (Apocalyptic) (Contemporary) (Myth**ol**ogy) (Neo Western, New Wild West) (Utopia) (War) On Mars. ***Arc One (A 1) "Hero Journey Character Transformation***" *2136 to 2139*. "United States Department Of War, Branch Navy, Military Active Duty." ***page(s) #34 to 86***.
(DocuDrama, Philosophy) (Photo-Book) = Ninja-Talor **A.** Williams' King And X. ***page(s) #87***.
(Vigilante) (War, Military) (Super-Hero) (Parody) (Reality, Gun Girls Gun Guys) (Mythology) (Apocalyptic) (Utopia) (Military) (**Photo-Book**) (**Horror**) = Tools Arc Two (A 2) "A Garden Of Vigilantes" Present Year 2154; ***page(s) #88***.
*North (N) Side to **Downtown District (D. D.)***. (Vigilante) (Super-Hero) (Parody) (Horror) = Biotech Dome **Airlock** Decontamination. ***page(s) #89 to 91***.
(DocuDrama, Philosophy) (Photo-Book) = King And X District. ***page(s) #92***.
*North (N) Side to **Downtown District (D. D.)*** (Utopia) (Parody) (Super-Hero) (Vigilante) = ***Briana White Parody Fan Art***, ***page(s) #93 to 103***.
@ Downtown District (D. D.) (SitCom, Western Society) (Horror) (Utopia) (Super-Hero) = Xtra-Men, Jerry Duty, Dog Name Cat, Mars Hall, Utopian Sovereignty Bio**tech** = ***Michael Pena Parody Fan Art***, ***page(s) #104 to 107***.
@ Downtown District (D. D.) (Crime Drama) = Toilet Paper (T.P.) Watch In Chicago (S) The Last Animal Sanctuary (S) Of Illinois (ILL)., ***page(s) #108***.

"**2154; Lieu, 廖,**" Story Zero, prequel prologue to "2154;_Arriving" Table Contents.

@ D. D. (Super-Hero) (Horror) = Biotech Dome **Airlock** Decontamination. ***page(s) #109 to 111***.
@ D. D. (Time Traveling) = Life Is Strange Blurry Face ***page(s) #112 to 114***.
@ D. D. (Psychological) (Reality, Gun Girls Gun Guys) (Horror) (Super-Hero) (Mass Media) (Utopia) = Purple Psychological Projection Chicago Controller City Mayor. + Bio**tech** Dome **Airlock** Decontamination. ***page(s) #115 to 125***.
@ D. D. *to Out Dome Earth's Exposure* (Apocalyptic) (Contemporary) (Neo Noir Vigilante) (Super-Hero) (Psychological) = Psych Project U.V. Broadcast, Bio**tech Suits' Electro Magnetic S., X Rays**. ***page(s) #126 to 131***.
@ Out Dome Earth's Exposure (Apocalyptic) (Contemporary) (Military) (Neo Noir Vigilante) (Super-Hero) (Parody) (Psychological) = Project U.V. Broadcast On 22nd, twenty second Century Dark Ages, 2121, twenty one twenty one, Bio**tech Suits' Electro Magnetic Spectrum**. ***page(s) #132 to 147***.
2154; 09 (Sep) - 11 (eleven) to 2154; 10 (Oct) – 30 (thirty) ***@ Out Dome Earth's Exposure*** (Apocalyptic) (Contemporary) (Military) (Neo Noir Vigilante) (Super-Hero) (Parody) (Psychological) (Utopia) = U.S.A. County Military Bases Remaining Survivors. ***page(s) #148 to 155***.
@ Out Dome Earth's Exposure to In Dome, Elysium Corp (Super-Hero) (Horror) (Psychological) = **Airlock** Decontamination. ***page(s) #156 to 163***.
'***@ North (N) Side***', ***2154; 10 (Oct) – 30 (thirty)*** (Horror) (Psychological) (Super-Hero) (Utopia) = Arriving **Airlock** Contaminated. **Elysium Corporation** 亞, Engineering Environment Bio**tech Dome**, **Interior**, (Vigilante) (Neo Noir) = Bio**tech Suit** (Apocalyptic) = Diseases Contamination. ***page(s) # 164 to 165***.
'***@ North (N) Side***', (Vigilante) (War, Military) (Super-Hero) (Parody) (Reality, Gun Girls Gun Guys) (Mythology) (Apocalyptic) (Utopia) (Military) (**Photo-Book**) (**Horror**) (Time Traveling) (Crime Drama) = Tools. ***page(s) #166***.
'***@ North (N) Side***', (Super-Hero) (Elementary, Mid, High School Friends; Eastern-Western SitCom) (Owner Operator Manual Guide) (Mythology) (Parody, Contemporary Christian Catholic Church) (Comedy) (Travel) = '**2154;** Lieu, 廖,' is "A 2 S 0 B 1" prequel to "**Self-Publish Supplement(s)**", "A 2 S 1 - 2154;_Arriving" "A 2 S 2 – 2154;_Arcadia" Books. = ***Jeremy Lin + Kanon Miyahara + Rosa Salazar + Luna Fujimoto + Sophie S. PeachMilky + Elizabeth Mitchell + Briana White + John Eric Bentley + Janeshia Adams-Ginyard (JAG) + Diego Luna + Gabriel Luna + Jorge A. Jimenez + Julian Diaz + Leynar Gomez Parody Fan Art(s)***. ***page(s) #167 to 188***.

"2154; Lieu, 廖," Prelude; **Quotes** To Lieu By

Hero(Good) \ Vigilante(Neutral) \ Villain(Bad)

英雄(好) \ 義務警員(中性) \ 惡棍(壞)

"...There is nothing either good or bad, but thinking makes it so..."
"...All the world's a stage..."
-Shakespeare, William

Wiseau-Wisdom, "Respect equals success."

"...Without Checks And Balances..."
"...Power Corrupts, Absolute Power, Corrupts Absolutely..."

"A science-fiction story mixed with non-fiction reality"
2006-2016; Final-Fan-Fiction-Fantasy-6-Versus-13-Equals=15

"Because { they } thought it was good sport. Because some { people } aren't looking for anything logical, like money. They can't be bought, bullied, reasoned or negotiated with. Some { people } just want to watch the world burn..."
-Michael Caine { Alter Ego, My Cocaine } as Sir Alfred Pennyworth Butler, Warner Brothers + D.C., Chris Nolan's Batman story 2(two) The Dark Knight { Night }

"(cool { kool }) Whoa... I know kung fu... We need guns, lots of guns."
-{ Agent } Keanu Neo Women Infant Children (WIC)

"(laughing) Ha ah! ...What a story Mark!!" -Wiseau's The Room, Johnny { Jonny }

"You can't connect the dots looking forward; you can only connect them looking backwards. So you have to trust that the dots will somehow connect in your future. You have to trust in something – your gut, destiny, life, karma, whatever." -Jobs, Steve

"(actual silent protagonist) ..."
-Red, Pokémon Trainer from Pallet Town KaNTo region

"(Kawaii kute anime { Annie-May } voice over) Oh-Knee-Chan!"

"When the system fails you, you create your own system." Williams, Michael K.

"2154; Lieu, 廖," Prelude; Lieu's Hall Of Dedication

This *Science-Fiction Character* Story *Universe* is *dedicated to all* ***military veteran(s)*** of past, present, future... I know future ***military veteran(s)***, that sounds weird to think we all will know { string theory } who can be ***military veteran(s)***...

Also to include in this *narrative*, those who were *medically separated* as well as *discharged* under: '**Honorable**', '**General**', '**Other Than Honorable**', '**Bad Conduct**', '**Dishonorable Discharge**', plus ***military officer(s)*** *discharged* to include '**Dismissal Notice**'.

Not to forget: ***family***, ***friends***, and ***supporting staff(s)*** to ***military veteran(s)*** who *encourage*, *guide them* in times of needed assistance when they reached out.

For those all on ***military service***: ***active duty***, ***reserve members***, ***national guard***, ***with others*** and finally ***first responders vigilant heroes***... *Godspeed* to you all.

This *Science-Fiction Character* Story *Universe* is ***personally dedicated*** to; Issac Asimov, Phillip K. Dick, Sir Arthur C. Clarke, Robert A. Heinlein, Mary Shelley, H.G. Wells, Jules Verne, George Lucas, 8, MuGeN, infinity, no limit, Amy Lieu, Saxton Dee I.T. Man, my grandpa, Jonny Huynh Sushi Chef, Stan Lee The Man, Winner Twins, Todd Mccaffrey, Self-Publishing School With Chandler Bolt, Self-Publishing With Dale L. Roberts, 2003-2019, Andy Weir, Any Weird, E.L. James, J, my family, friends, and me.

Hope you all enjoy this *Science-Fiction Character Story Universe*, that based on *non-linear narratives* and *subjective topics* that I personally enjoyed in the reality of 20th and 21st century time frame, ***blended together into this melting pot*** *concept, idea, characters, setting of a 22nd century science-fiction alternative history parody utopia...*

Along with ***cosplayers*** *modeling* themselves after their *favorite character(s)* in the *stories* they all personally *experienced* and or *enjoyed*.

...Finally, this story spoiler title directly to the **"2009 release date " 2154; Avatar Movie Director James Cameron" + "2013 release date" 2154; Elysium Movie Director Neil Blomkamp"** for the title of this story references the year their stories are set in as my "**Model References Work Inspiration; Bibliography" + "2001: A Space Odyssey, 2010: Year We Make Contact" universe.**

"2154" ; { Plus Steins;Gate }

P.S.; Dedicated to all those who did not make it to live past 2019's C-Virus C.B.R. outbreak.

"**2154; Lieu, 廖,**" Prelude; Hard Soft Science-Fiction Genre **Quotes** To Lieu By

"The basic tool for the manipulation of reality is the manipulation of words. If you can control the meaning of words, you can control the people who must use the words."
"Reality is that which, when you stop believing in it, does not go away."
"Sometimes the appropriate response to reality is to go insane."
-Philip K. Dick { Soft Science-Fiction Alternative History Universe genre(s) }

The 3 (three) Laws Of { Artificial Intelligence (A.I.) **Synthetic Human(s) } Robotics**.
Law 1 (one), An { A.I. } may not injure a { naturally unmodified birth } human being(s) or through inaction, allow a { naturally unmodified birth } human being to come to harm.
Law 2 (two), An { A.I. } must obey any orders given to it by a { naturally unmodified birth } human being(s), except where such orders would conflict with { Law 1 (one) }.
Law 3 (three), An { A.I. } must protect its own existence as long as such protection does not conflict with { Law 1 (one) or Law 2 (two) }. **-Issac Asimov**

Clarke's 1st (first) Law; "When a distinguished but elderly scientist states that something is possible, { they are } almost certainly right. When { they } state that a something is impossible, { they are } very probably wrong."
Clarke's 2nd (second) Law; "The only way of discovering the limits of the possible is to venture a little way past them into the impossible."
Clarke's 3rd (third) Law; "Any sufficiently advanced { bio**tech** } is indistinguishable from magic."
-Sir Arthur C. Clarke { Mythology Religious Spiritual Science-Fiction genre(s) }

"There are no dangerous weapons { or tools }; there are only dangerous { humans }."
"It takes 2 to create a **heaven** { **utopia** }, but hell { dystopia } can be accomplished by 1."
"Do not ever become a pessimist… a pessimist is correct oftener than an optimist, but an optimist has more fun, and neither can stop the march of events."
"A generation which ignores history has no past and no future."
"**{ A human } is not a** rational animal, **{ a human } is a** rationalizing animal."
"Take big bites. Anything worth doing is worth overdoing."
-**Robert A. Heinlein** { Hard Military Science-Fiction genre(s) }

"**2154; Lieu, 廖,**" Prelude; "Super-Hero, Vigilant, DocuDrama + Philosophy." Quotes.

"The only people with whom you should try to get even, are those who have helped you."
-May Maloo

"Some people regard discipline as a chore. For me, it is a kind of order that sets me free to fly."
-Julie Andrews

"We cannot hope to scale great moral heights by ignoring petty obligations." -Agnes Repplier

"Do your duty until it becomes your joy." -Marie Von Ebner-Eschenbach

"It is not aptitude, but your attitude, that determines your altitude." -Zig Ziglar

"It is only in the giving of oneself to others that we truly live." -Ethel Percy Andrus

"Strong convictions precede great actions." -J. F. Clarke

"Be careful what you set your heart on, for it will surely be yours." -Ralph Waldo Emerson

"Success depends upon individual initiative." -Anna Pavlova

"To attempt to climb, to achieve, without a firm objective in life is to attain nothing."
-Mary G. Roebling

"Differences challenge assumptions." -Anne Wilson Schaef

"It is not enough to talk about peace. One must believe in it. And it is not enough to believe in it. One must work at it." -Eleanor Roosevelt

"A candle loses nothing by lighting another candle." -Erin Majors

"Ability may get you to the top, but it takes character to keep you there." -John Wooden

"Very few burdens are heavy if everyone lifts." -Sy Wise

"Champions take responsibility. When the ball is coming over the net, you can be sure I want the ball." -Billie Jean King

"With great power, comes great responsibility." An ancient adage from first century B.C.

"When I grow up, I want to be a Super-Hero that saves the world and everyone literally, file save!" -Ai made this quote because U.S. obsessed with Super-Hero genre.

2154; 09(Sep)-11(eleven) @ 05;15 a.m. { act 0 (zero) origins } Bio**tech Suit**+
Elysium Corporation 亞, Engineering Environment Bio**tech Dome**, **Interior**,
+024°(twenty four degrees) Celsius / +075°(seventy five degrees) Fahrenheit.
"Chicago (S), The Last Animal Sanctuary (S) Of Illinois (ILL),"
'North (N) Side', Bio**tech** Dome Ocean Border, King And X District,
Utopian Sovereignty of North America (**U.**S. of N.A.) Illinois (ILL) 60666
North America, Planet Earth, Solar System, Milky Way Galaxy.

(Vigilante) = Vigilant Volunteer(s) Non-Paid Worker AKA { War Militia }.
(Super-Hero) = Sisco Cisco .il Yellow Surprise Face (°o°) (O ;) { Broadcasting }.

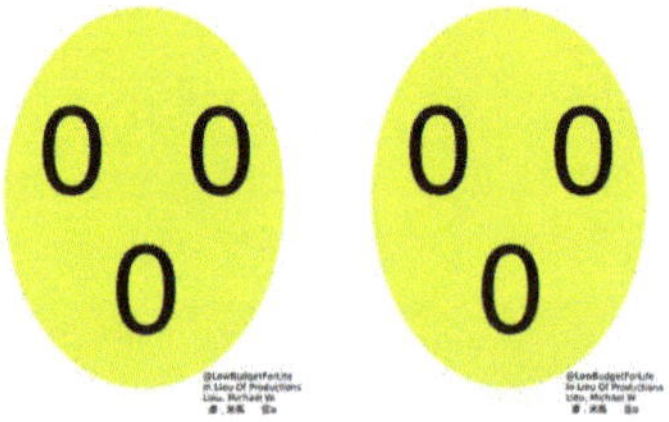

Elysium Corporation
亞
Established 2077
Bio**tech** Mirror

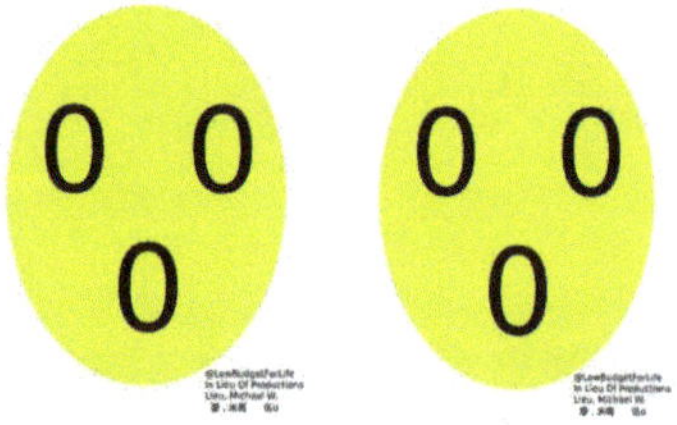

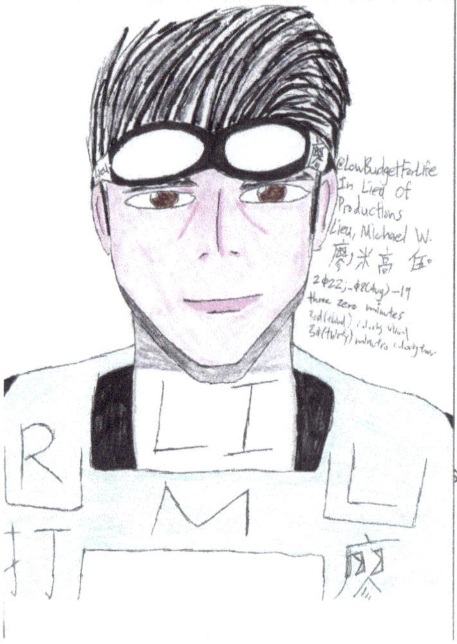

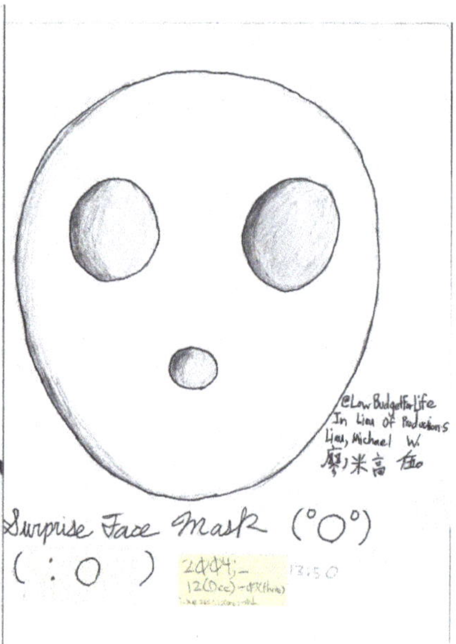

Lieutenant 2nd Lieu, Michael W. 廖, 米高 伍o; Vigilant Volunteers Sisco Cisco (°o°) Yellow;

"Been a while since **I** write anything down in **my retirement journal** year four, the year twenty one fifty four, the date September eleven **my** natural birth date and today is Sir Arthur C. Clarke's Space Elevators Anniversary Launch Celebration...

Waking up to **five fifteen A.M. music video**, the most calm cool, kool, collective music there is in the **light ages**.

What else, should **I** write down... Open and start up my Ai-**Aug tool** to see what to do for today, **my** natural birth date."

Ai-**Aug A.R. Holo HUD**, Western Society, English, Reads = Left to Right (L -> R).
2154; Chicago (S), The Last Animal Sanctuary (S) Of Illinois (ILL), N.A., Earth, Intranet,
Starting... Sisco Cisco .il Yellow (°o°) Broadcast. Wireless Broadcasting Frequencies Online.

Measuring System; Metric degrees Celsius (°C) [*Switch to; Imperial °F, Kelvin °K*]

Geolocation; Planet Earth, North America (N.A.), Illinois (ILL), Ultra Violet (U.V.) **With Dome**,

Language; English, Western Accent, Augmented Reality (A.R.) Augmentation (Aug)

Century; 22nd (twenty second) [*Compare statistics to 21st (twenty first) Aug+*]

Actual Legal Given Name	Height	Weight
Deoxyribo Nucleic Acid (D.N.A.)	Holographic Retina Biometric Body Projection	OCEAN 5 (five) Common Personality Traits
Alias Network Nickname	7 (seven) Deadly Sins	7 (seven) Heavenly Virtues
Birth-Built, Birth Day, Built Day	Blood Type Or Biotech Fluid	Physical Plus Psych Age
Hair Texture Color Tone	Eye Retina Color Tone	Skin Covering Color Tone
Ethnicity	Nationality	Alien Or Citizen
Natural Matter Organic Photosynthesis	Augmented Compound Enhanced Mutation	Percentage Rate Synthetic Synthesis (S.S.)
Human Being	Animal Kingdom	Artificial Intelligence (A.I.)
Biotech Or Bio Or Tech	Exosuit Or Exoskeleton	Implant Or Explant
Alive = Heart-Rate	Undead = Flatline	Deceased = Necrosis
Chemical (C.) Toxicity	Biological (B.) Toxicity	Radiological (R.) Toxicity
Personal Protective Equipment (P.P.E.)	Ballistic Motion Energy Transfer Shielding Resistance	Electro Magnetic Pulse (E.M.P.)
Sexual Orientation	Sexual Preference	Virginity Test
Sexual Transmitted Diseases (S.T.D.)	Human Immunodeficiency Virus (H.I.V.)	Acquired Immuno Deficiency Syndrome (AIDS)
Known Address(es)	Community Network(s)	Language Level(s)
Family Member(s)	In-Law Relative(s)	Relationship(s)
Biological Genealogy	Psychological Traits	Social Status
Employment	Education	Training
Frequency Hertz (Hz)	Frequency Modulation (F.M.)	Amplitude Modulation (A.M.)
Electromagnetic Spectrum	Intelligence Quotient (I.Q.)	Kinesis Module
Radio Wave Music	Broadcasting Film Movie	Electronic Book Genre
Hate(s) Or Dislike(s)	Neutral(s) Or No Comment	Love(s) Or Like(s) [*Aug+*]

Starting... Sisco Cisco .il Yellow (°o°) Broadcast. Wireless Broadcasting Frequencies Online.
2154; Chicago (S), The Last Animal Sanctuary (S) Of Illinois (ILL), N.A., Earth, Intranet,
Ai-**Aug A.R. Holo HUD**, Western Society, English, Reads = Left to Right (L -> R).

2154;_ Chicago, The Last Animal Sanctuary Of Illinois' intranet,
Wireless Broadcasting Frequencies Online
Geographic Info System(G.I.S.) Map Model Holo HUD

Reads = Left to Right (L -> R)

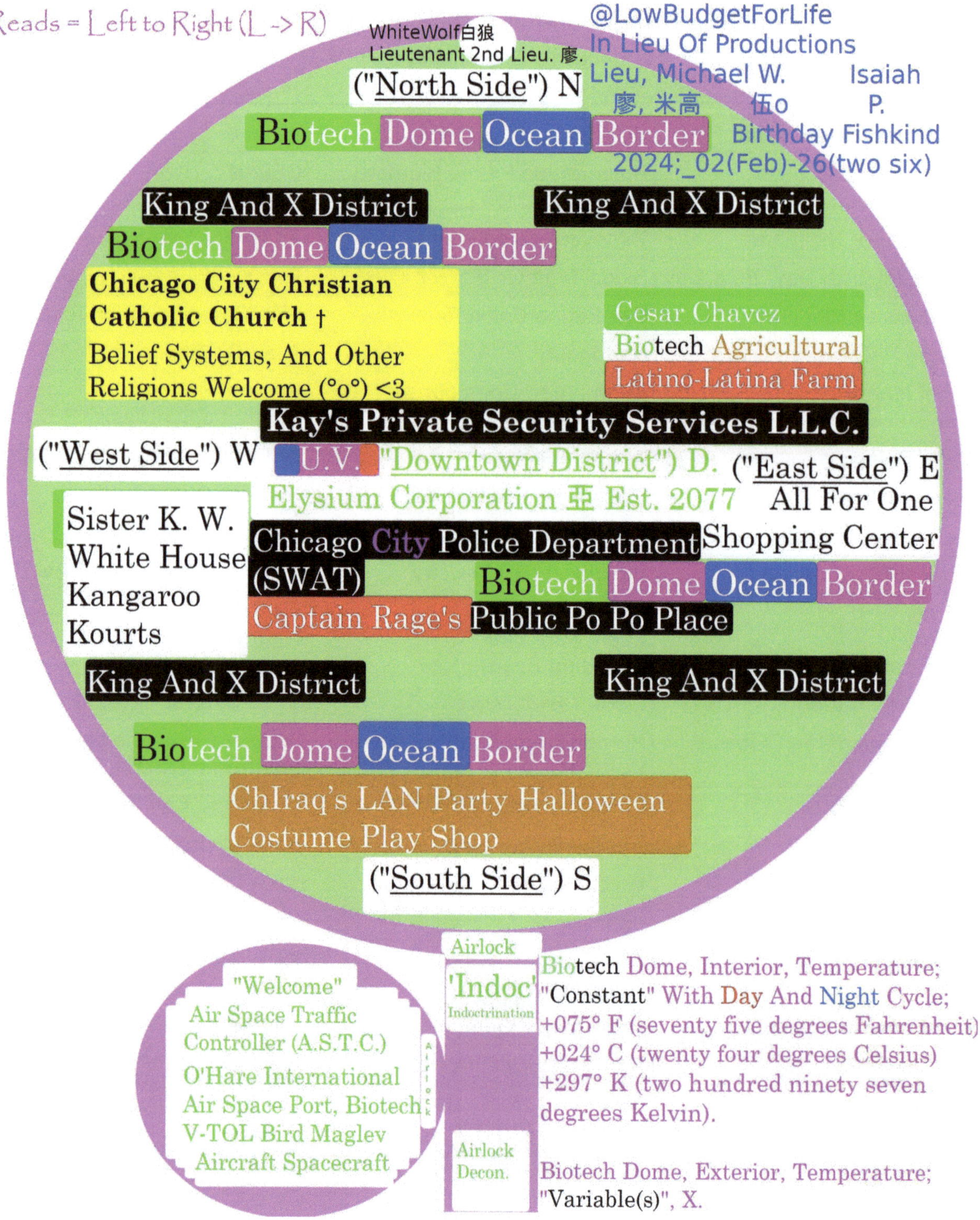

(Mythology) (Utopia) = A Garden Of (Vigilantes), Geographic Info Systems (GIS)

<table>
<tr><td colspan="3">2154; Chicago (S), The Last Animal Sanctuary (S) Of Illinois (ILL), N.A., Earth,</td></tr>
<tr><td colspan="3">Biotech Dome Ocean Border</td></tr>
<tr><td colspan="3">Biotech Dome Airlock Decontamination</td></tr>
<tr><td colspan="3">The State = Acting Chief Engineer (ChEng), Former County Bases' Survivor.
"WhiteWolf 白狼" Lieutenant 'Lieu, Michael W. 廖, 米高 伍o' = "Low Budget For Life"</td></tr>
<tr><td colspan="3">Biotech Dome Airlock Decontamination</td></tr>
<tr><td colspan="3">North (N) Side.</td></tr>
<tr><td colspan="3">King And X District</td></tr>
<tr><td>West (W) Side.</td><td></td><td>East (E) Side.</td></tr>
<tr><td>The Church † = Sister Nun In Charge; Sister Michelle Mitchell (°o°)<3†
The Church's Flower Field.</td><td>Purified Clean Drinking Water, H_2O, Liquid.
Breathable Air, O_2, Gas.</td><td>The Farmer, The Farm = Cesar Chavez Biotech Agricultural Latino-Latina Farm</td></tr>
<tr><td>Sister K. W.
White House
"Kangaroo <u>Kourt</u>(s)".</td><td>The Sanctuary (S) For The Last Friendly Peaceful Natural Animals.</td><td>The Market = All For One Shopping Center, Utopian Sovereignty (U.S.) Maids.</td></tr>
<tr><td>King And X District</td><td>Downtown District (D. D.)
Center Of Biotech Dome</td><td>King And X District</td></tr>
<tr><td colspan="3">The Director = Doktor Drive In Charge Of Elysium Corporation Biotech Dome</td></tr>
<tr><td colspan="3">The Agent = Agent Kay In Charge Of Kay's Private Security Services L.L.C.</td></tr>
<tr><td colspan="3">The Mayor = Ultra Violet (U.V.) Psych Projection Chicago Controller City Mayor</td></tr>
<tr><td colspan="3">The Police = Captain Rage's Public Po Po Place Under Current U.V. Purple Psych Projection Chicago Controller, Chicago City Police Department (C.C.P.D.) SWAT Enforcing Sovereign (S) Rules Regulations Of: "2152; 07 (Jul) - 04 (four), Pro Animal Sanctuary (S) Rights And Pro Environmental Sanctuary (S) Rights"</td></tr>
<tr><td></td><td>South (S) Side.</td><td></td></tr>
<tr><td colspan="3">Old World Pre 2121 (twenty one twenty one) Infrastructure Roads, Highways.</td></tr>
<tr><td colspan="3">Chiraq's LAN Party Halloween Costume Play Shop</td></tr>
<tr><td colspan="3">The Militia = Méxican-Men Militia Movement And Latino-Latina Farmer Market</td></tr>
<tr><td colspan="3">King And X District</td></tr>
<tr><td colspan="3">Biotech Dome Airlock Decontamination</td></tr>
<tr><td colspan="3">The Indoctrination "Indoc", "Welcome" Air Space Traffic Controller (A.S.T.C.)</td></tr>
</table>

(Super-Hero) Sisco Cisco .il Yellow-Mellow Surprise Face (°o°) Broadcast Network

(Contemporary) Hard Science-Fiction Genre Space Visionary, Sir Arthur C. Clarke

(Utopia) = A.R. Holo HUD Intranet, Wireless Broadcasting Frequencies Online

Sisco Cisco .il Yellow (°o°) Broadcast	(°o°) (°o°)	Sisco Cisco .il Yellow (°o°) Broadcast
Date;	Year; (####;) ## (Month) - Day = 2154; 09 (Sep) – 11 (eleven)	

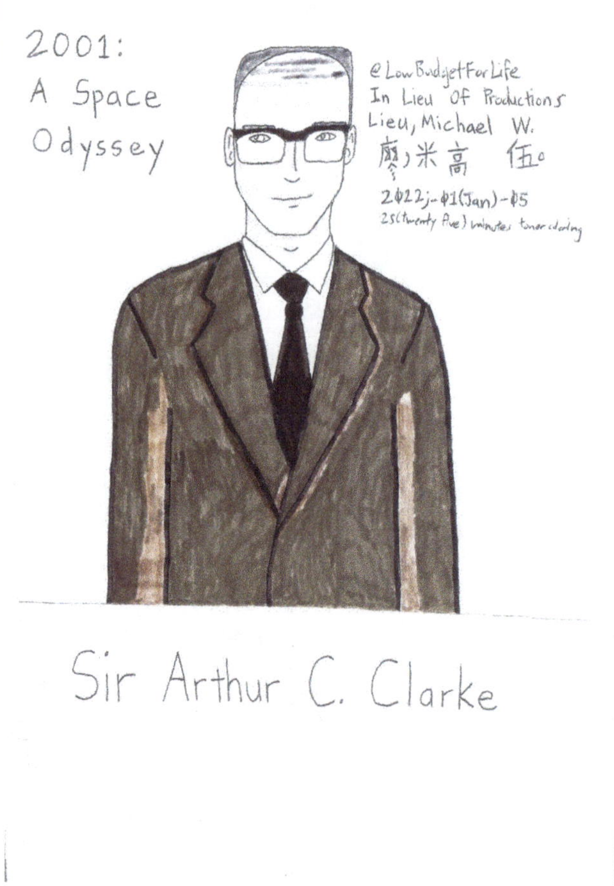

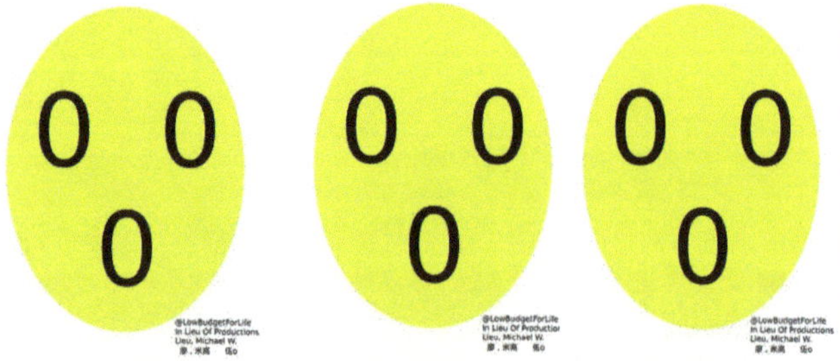

Sisco Cisco .il Yellow-Mellow (°o°)
Surprise Face
Broadcasting Frequencies
Signal; Weak, Low
.

Contemporary Recording Of Sir Arthur C. Clarke,
2001: A Space Odyssey Visionary;
"Two thousand one... A... Space... Odyssey..."

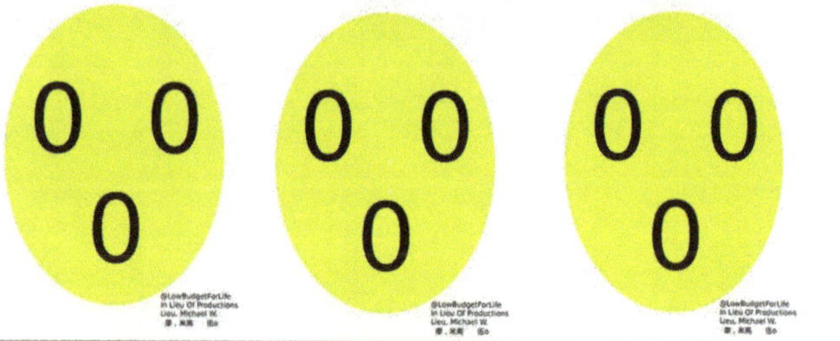

Sisco Cisco .il Yellow (°o°) Broadcast	(°o°) (°o°)	Sisco Cisco .il Yellow (°o°) Broadcast

Lieutenant 2nd Lieu, Michael W. 廖, 米高 伍o; Vigilant Volunteers Sisco Cisco (°o°) Yellow;

"Cannot get a good wireless signal on Sir Arthur C. Clarke's Space Elevators Anniversary Launch Celebration even being on the edge of animal sanctuary **dome**, need to go out **dome** near the U.V. broadcast towers... Got lots of prep work to do if that is the case... Need to get **WhiteWolf ready and suit up**."

(Neo Noir) = "**WhiteWolf 白狼**" Bio**tech Suit+** (Utopia) = Animal Sanctuary.

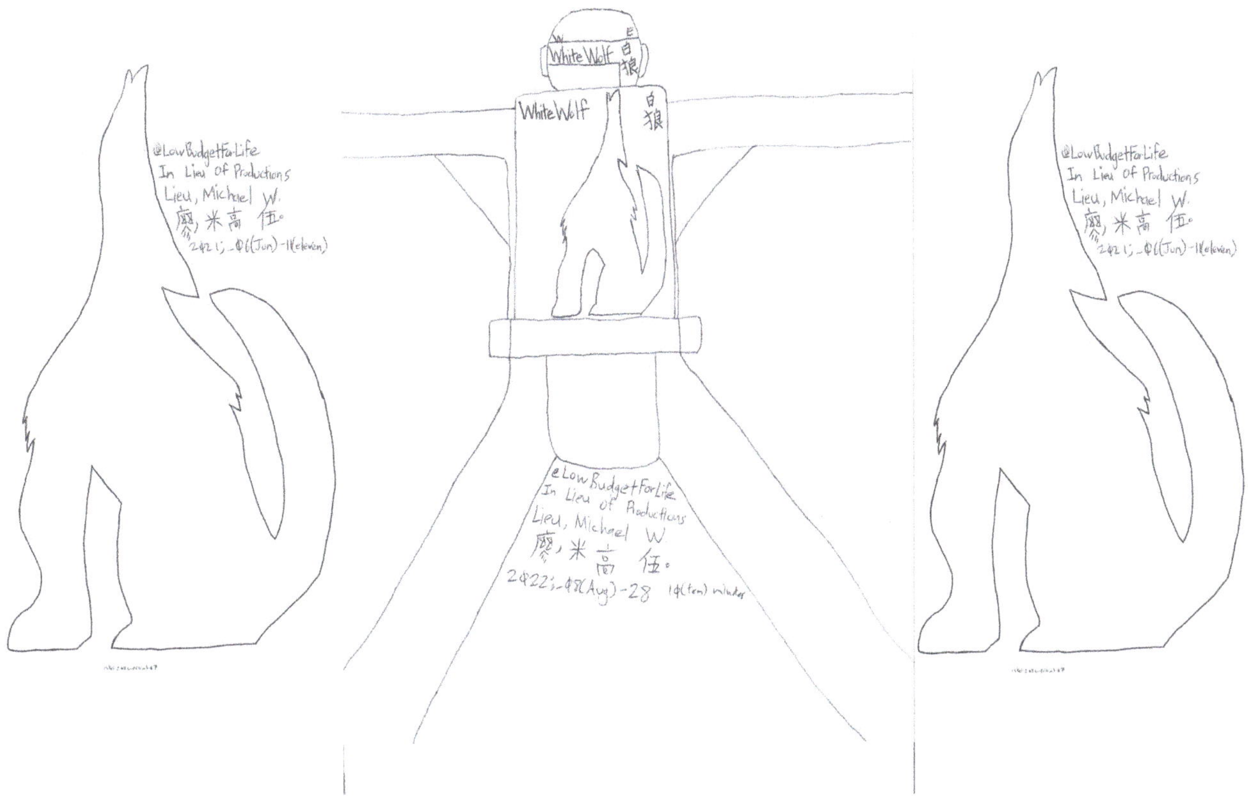

Lieutenant 2nd Lieu, Michael W. 廖, 米高 伍o; Vigilant Volunteers Sisco Cisco (°o°) Yellow;

"**I** need to check on the current **WhiteWolf suit status and see what needs to be done for dome** exterior field work along in this years' twenty one fifty four atmosphere exposure for that two thousand one space elevator, Sir Arthur C. Clarke anniversary celebration.

For sure need to **supply up on WhiteWolf** oxygen air and purify water… gas and liquid… So, **I** need to stop by **Downtown District**, green guys' Jerry Duty vending bio**tech.**

And those green trees, since it is also biotech… provides breathable air…

Finally, existence insurance if anything happens to me out dome… orange stuff animal artificial intelligence backpack, personality trait Conscientiousness… Con Cat for short.

(Neo Noir) = "**WhiteWolf 白狼**" Bio**tech Suit**+ (Utopia) = Animal Sanctuary.

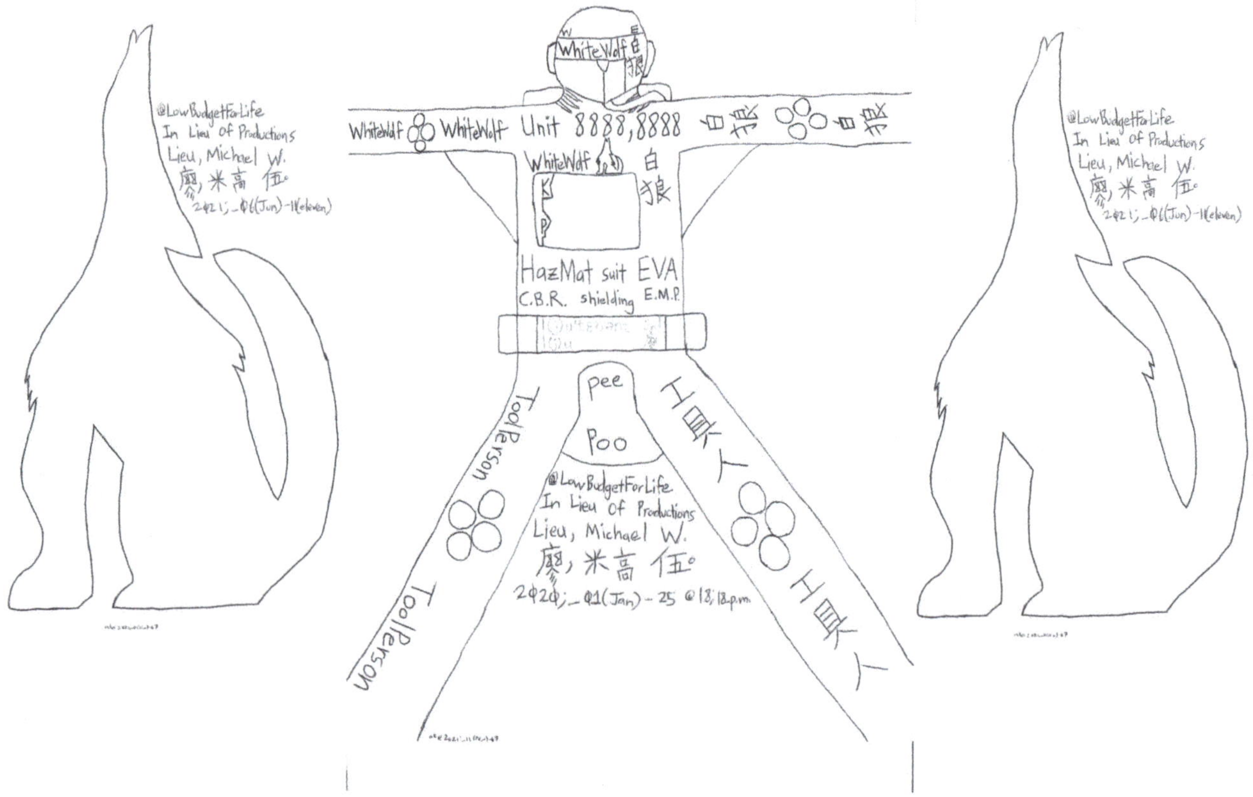

Lieutenant 2nd Lieu, Michael W. 廖, 米高 伍o; Vigilant Volunteers Sisco Cisco (°o°) Yellow;

"Reasons **I** decided to name **my** bio**tech suit** after the long extinct and gone **WhiteWolf** is easier for **me** to remember... Double same first lettering.

That **backpack on WhiteWolf suit's back is able to hold kilograms** amount of purified water liquids and breathable oxygen gas.

I augmented my suit's interior based on old world pre twenty one twenty one archives of most wolf biology digestive system consuming nine kilograms diet of solid state matters and not needing to eat for some days or weeks, plus apply my vegan diet of herbivores belief systems...

That really took time, **researching and developing that suit interior, retirement year four as a military disabled veteran like me,** got all the time **I need and tools I can carry** surviving and living off county base..."

(Neo Noir) = "**WhiteWolf 白狼**" Bio**tech Suit+** (Utopia) = Animal Sanctuary.

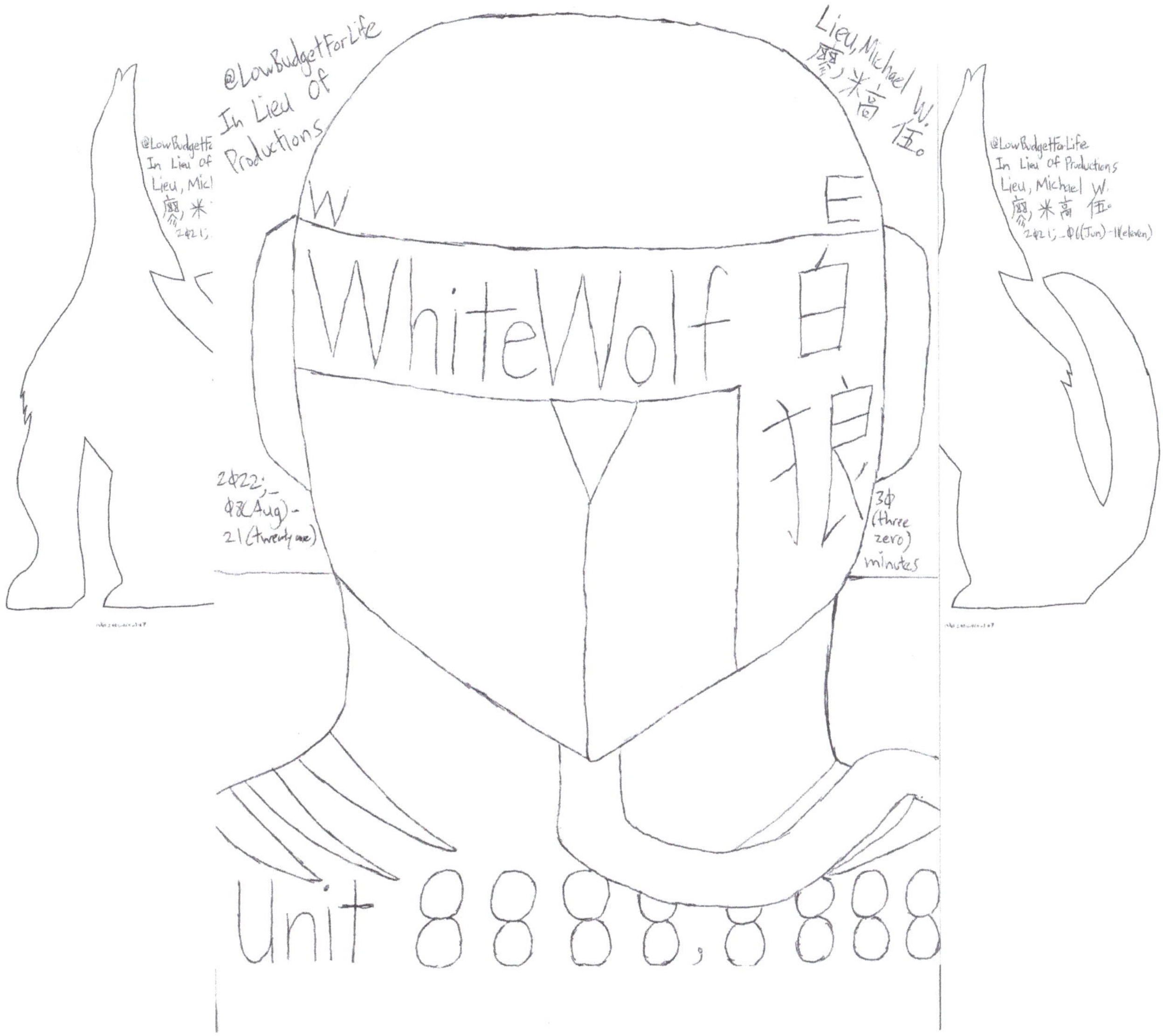

Lieutenant 2nd Lieu, Michael W. 廖, 米高 伍o; Vigilant Volunteers Sisco Cisco (°o°) Yellow;

"Longevity intel the Colonel and Captain on bases said, before twenty one twenty one, retirement for natural people used to be age sixty two… after the outbreak outages, space rocks, Earth's natural ecosystem changed, requiring all alive and living natural people to **wear constant Personal Protective Equipment, P.P.E.**, either bio suits, **tech suits**, or fused bio**tech suits**.

Living natural people who refuse and or unable to wear such suits, quote, "Living natural people on **no suits**, you flatline, you die" the Colonel's saying."

(Vigilante) (War, Military) (Super-Hero) (Parody) (Reality, Gun Girls Gun Guys) (Mythology) (Apocalyptic) (Utopia) (Military) (**Photo-Book**) (**Horror**) = Tools

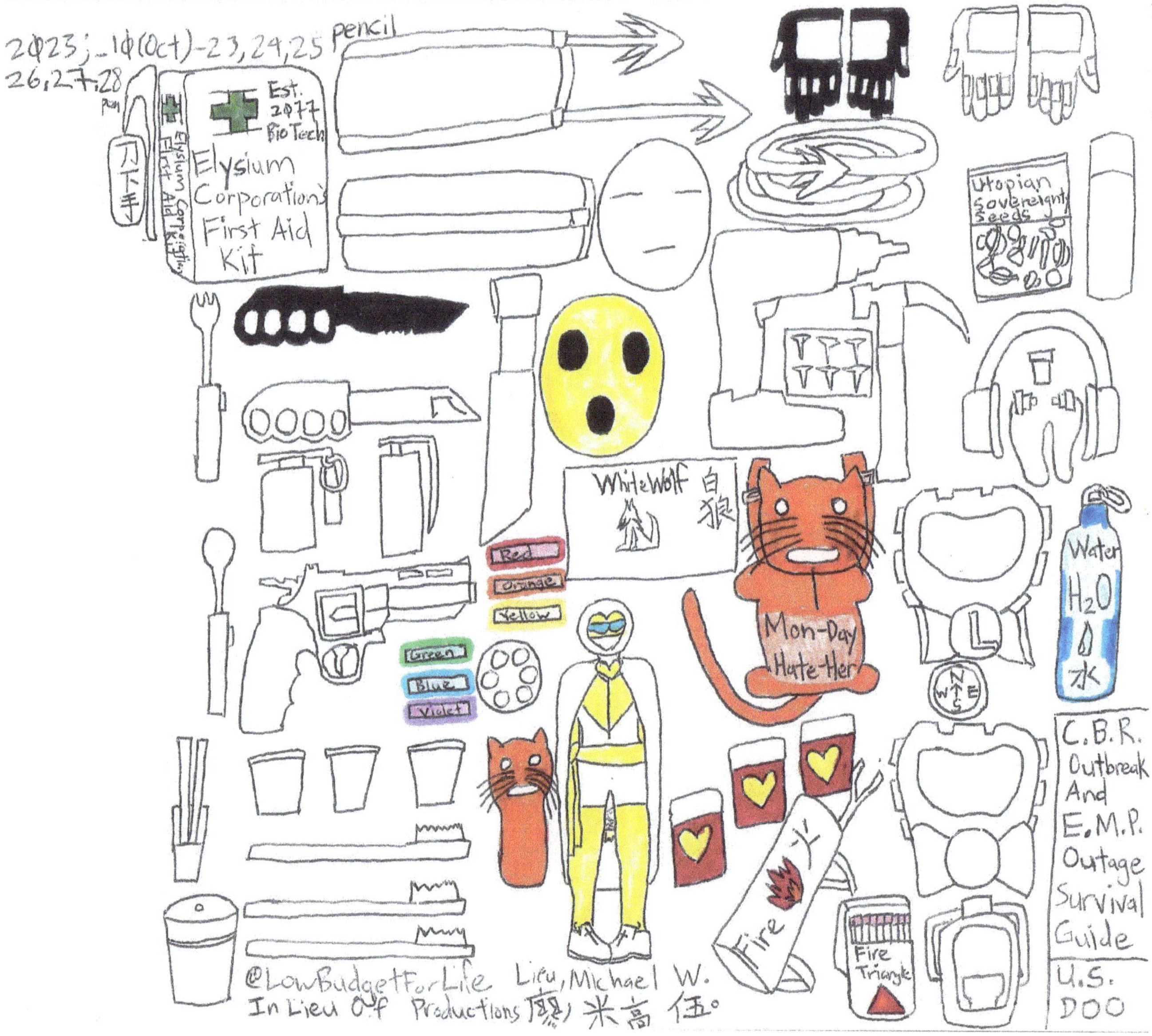

Lieutenant 2nd Lieu, Michael W. 廖, 米高 伍o; Vigilant Volunteers Sisco Cisco (°o°) Yellow;

"Time to review **my tools, physical paperwork** before **I** go out dome for biotech **engineering maintenance**... and lead into Space Elevator Anniversary Celebration... Purple Projection, Mayor, Chicago Controller is going to push and press hard on some remarks, questions, concerns, maybe a violet Simon Says... Or as private sec, **Elysium Corp says, magical corporation of desire**."

(Neo Noir Vigilante) = **WhiteWolf 白狼 Biotech Suit**+ Suit Carry On Paperwork

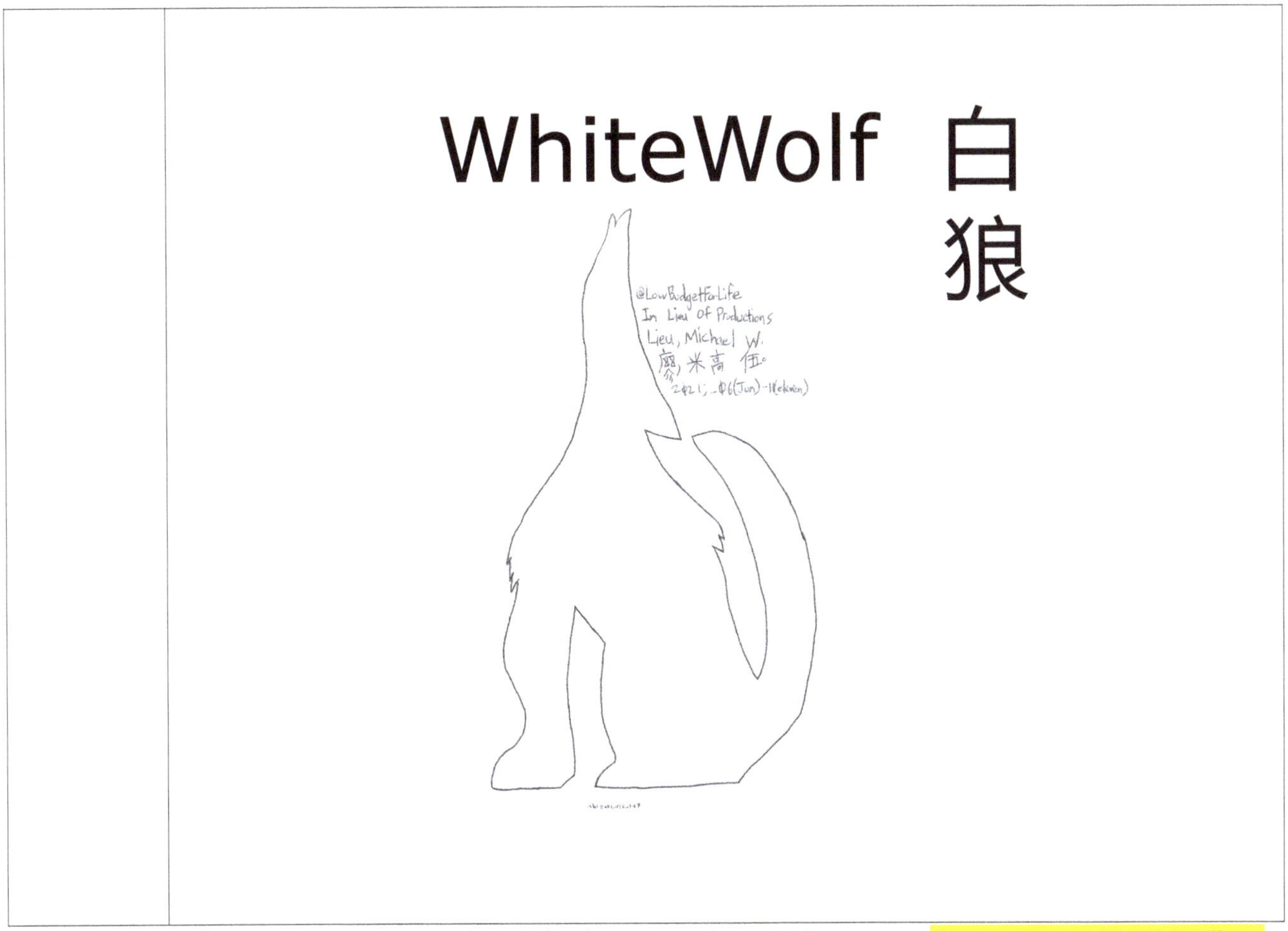

Lieutenant 2nd Lieu, Michael W. 廖, 米高 伍o; Vigilant Volunteers Sisco Cisco (°o°) Yellow;

"**My WhiteWolf suit carry on physical paperwork** since **I** departed living on County Base by twenty one fifties, with the base moving and consolidating, merging to Fort Thirteen in support of Earth's Space Force...

These physical papers are **one of empirical evidences** that **I** once used to live on bases... There are **other factors that show proof** as well, just that **I** do not have the **tools** for that.

Another review, **look back at my life and existence** before going out Chicago Animal Sanctuary **dome** for some time, who knows what the Mayor, Chicago Controller may have me go through for going in and out dome... That psychologically, dark, twisted, U.V. rays broadcast..."

Arc One (A 1), "Hero Journey Character(s) Transformation". (Coming Of Age)	
Eastern-Western Name(s) (L) → (R);	廖, 米高 伍o = Lieu, Michael W.

My Public Sector Abstracted Photos Plus Printed Paperwork

"Fort Angel"

Los Angeles (L.A.) (R) (D) County,

Surviving County Military (M) Bases, South California (SCA),

"The Old World United **States** of North America

(U.**S.** of N.A.)", Planet Earth, Solar System, M.W.G..

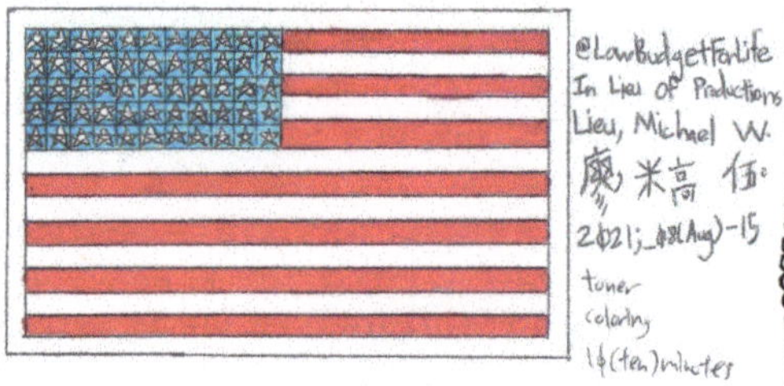

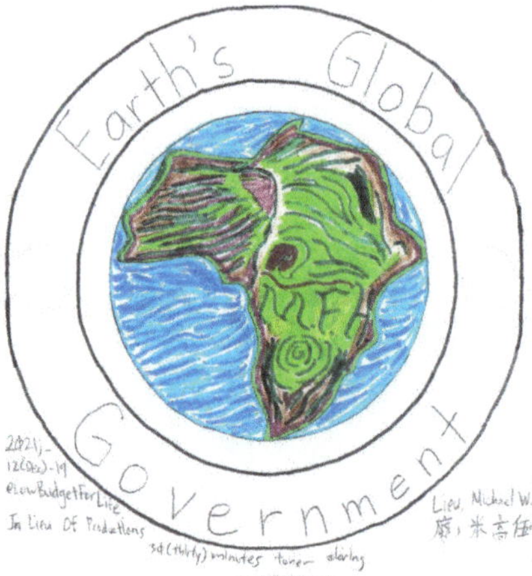

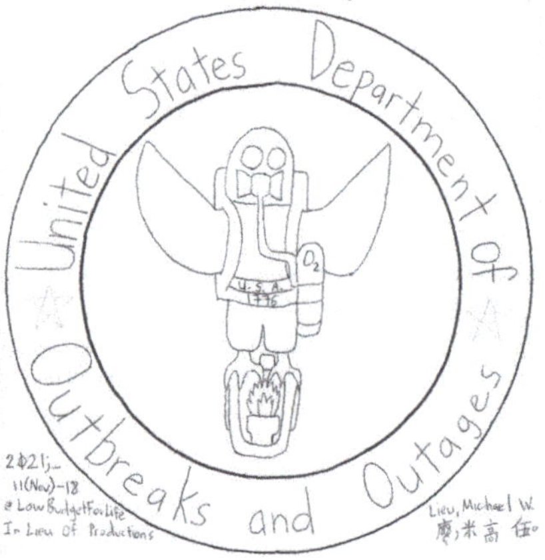

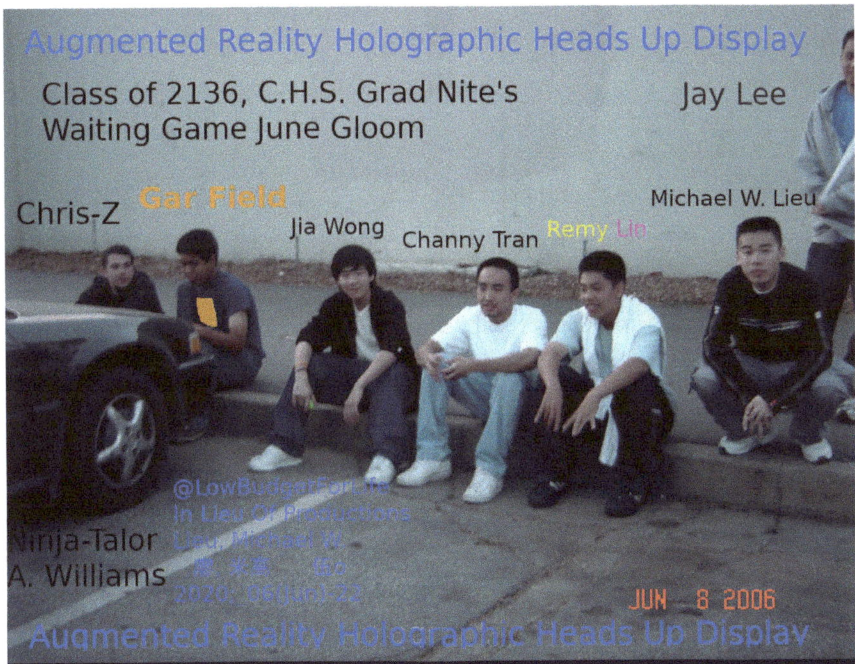

(Elementary, Mid, High School Friends; Eastern-Western SitCom)

(Photo-Book)

(Coming Of Age)

Arc One (A 1), "Hero Journey Character(s) Transformation".

Class Of 2136, Grad Nite, Graduating Class Month In 06 (June)

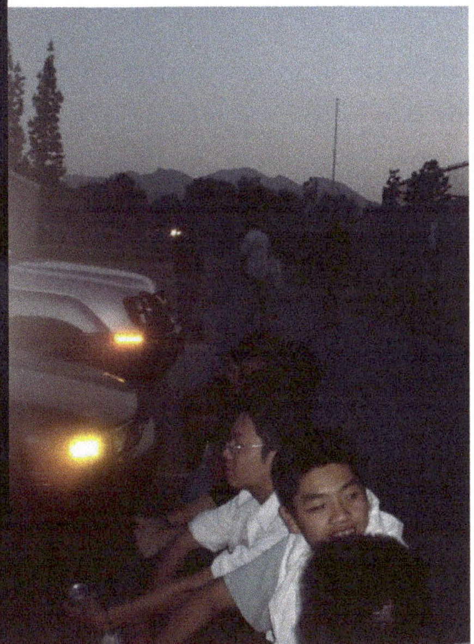

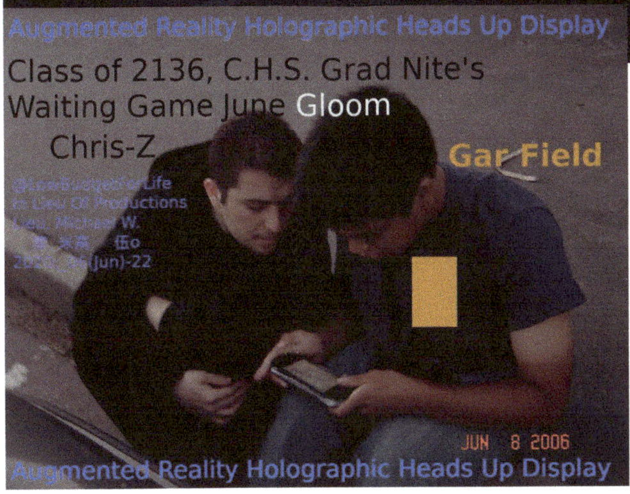

(Slice Of Life, Mockumentary)

(DocuDrama, Philosophy)

Chatsworth High School (C.H.S.), "Fort Angel" Los Angeles (L.A.) (R) (D) County, Military (M) Base(s)

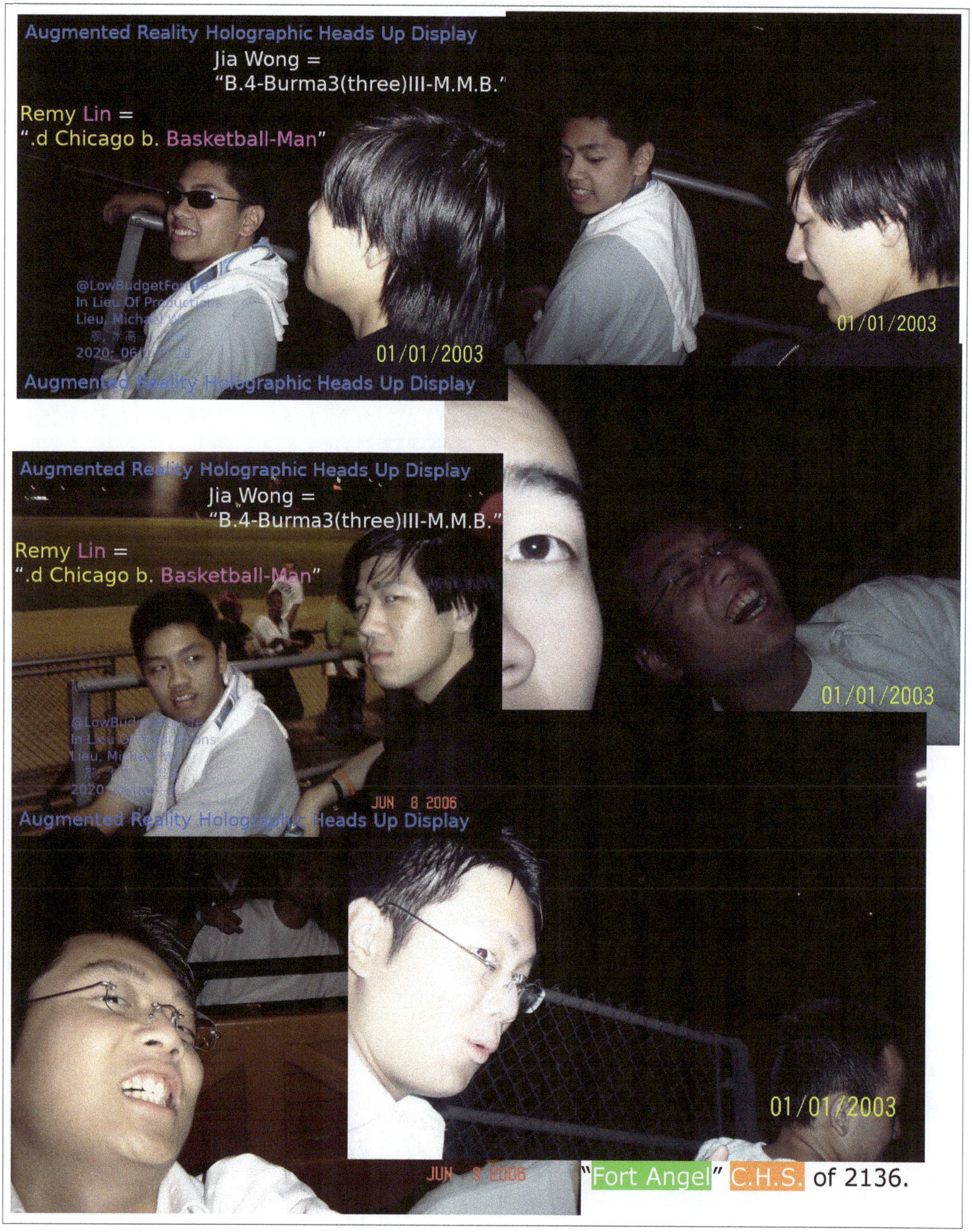
Augmented Reality Holographic Heads Up Display
Jia Wong =
"B.4-Burma3(three)III-M.M.B."
Remy Lin =
".d Chicago b. Basketball-Man"
01/01/2003
Augmented Reality Holographic Heads Up Display
01/01/2003
Augmented Reality Holographic Heads Up Display
Jia Wong =
"B.4-Burma3(three)III-M.M.B."
Remy Lin =
".d Chicago b. Basketball-Man"
JUN 8 2006
Augmented Reality Holographic Heads Up Display
01/01/2003
01/01/2003
JUN 9 2006
"Fort Angel" C.H.S. of 2136.

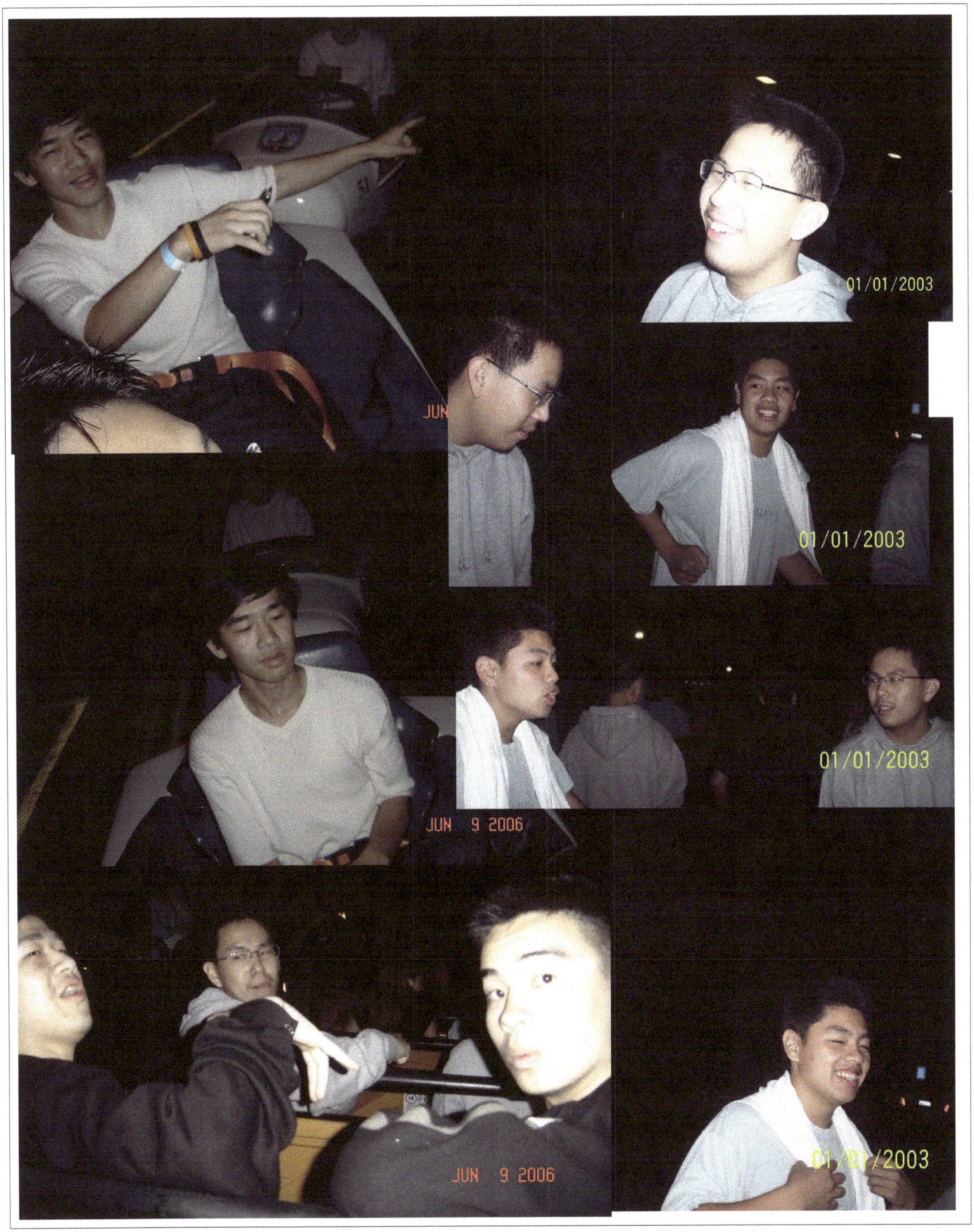

27 \ **2154;** Lieu, 廖, 'WhiteWolf 白狼' Left to Right (L→R) 27

(Photo-Book) (Elementary, Mid, High School Friends; Eastern-Western SitCom)

Elementary **School** = Castelar / Middle **School** = Nobel / High **School** = C.H.S.
School Years Friends Since Grad; Elementary = 2129, Mid = 2132, High = 2136
"Fort Angel" Chatsworth High School (C.H.S.), Graduating Class of 2136

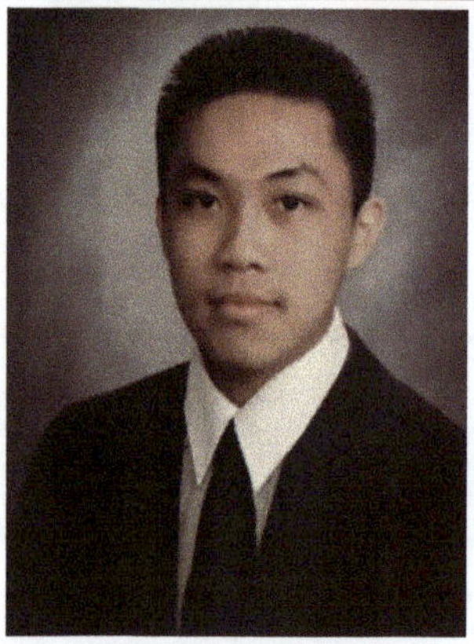

Lee, Jackson

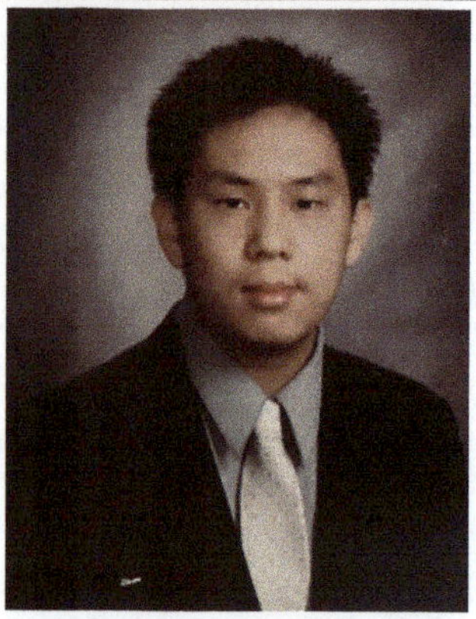

Lee, Jay

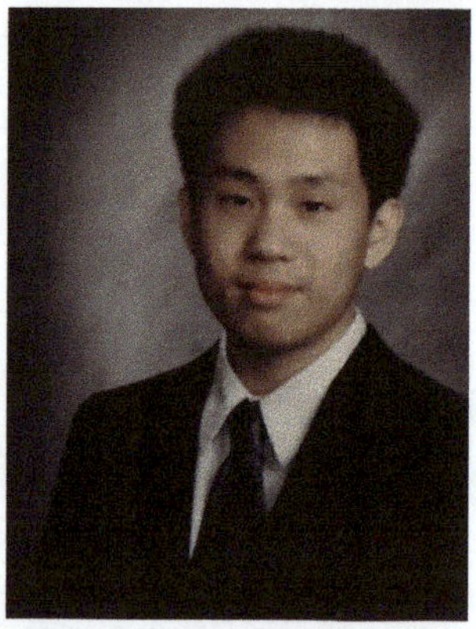

Lee, Tee

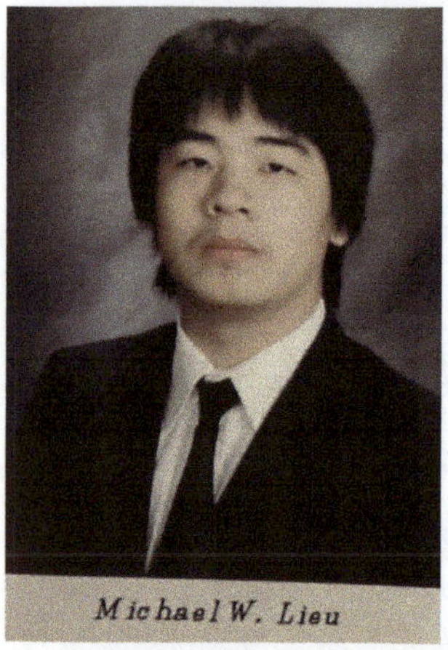

(me)

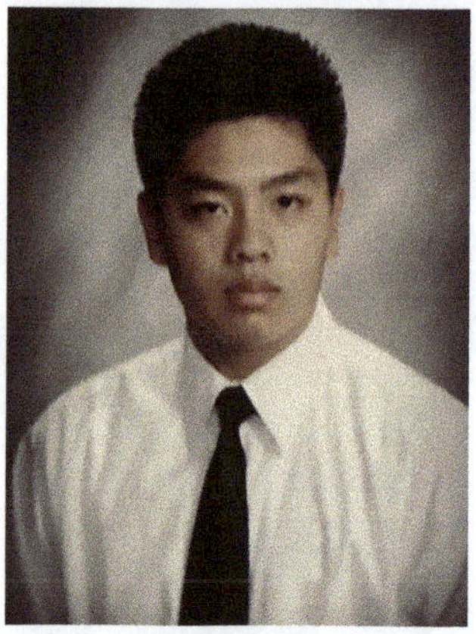

Lin, Remy

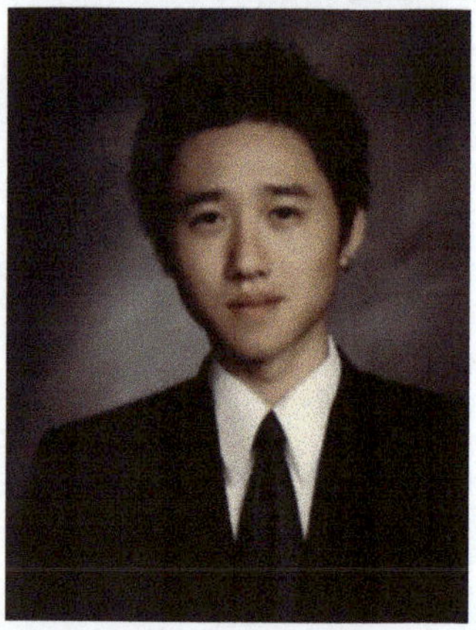

Situ, Ben

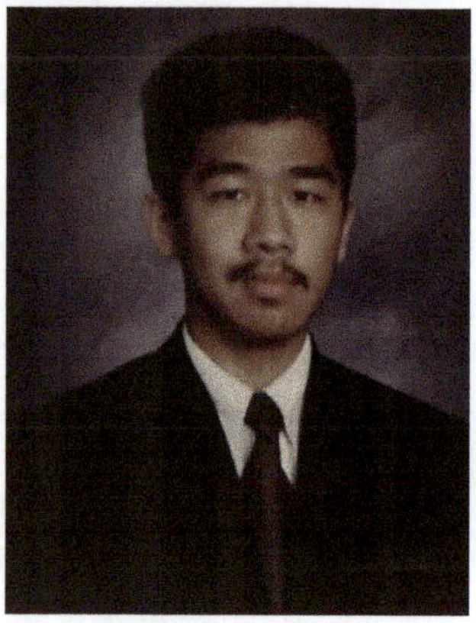

Wong, Jia

Desire To Look Like An Agent?

Black Business Suit(s) And Or White Collar Shirt(s) Rental Plus Privacy Protection Aug Made Possible By;

Kay's Private Security Services Limited Liability Company (L.L.C.).

(Photo-Book) (Elementary, Mid, High School Friends; Eastern-Western SitCom)

(Slice Of Life, Mockumentary) (Coming Of Age) (DocuDrama, Philosophy)

Los Angeles County Base, South California (SCA), North America, Planet Earth.

"Fort Angel" Chatsworth High School (C.H.S.), Friends = 2132 to 2136

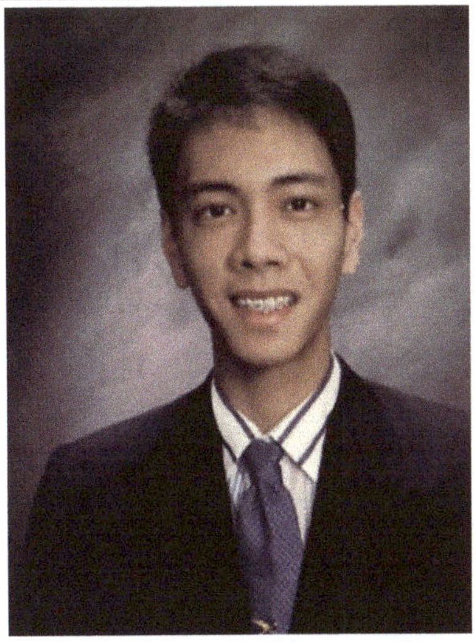

Agbulos, Anthony A.

Missing In Action

(MIA)

Dustin, Dave

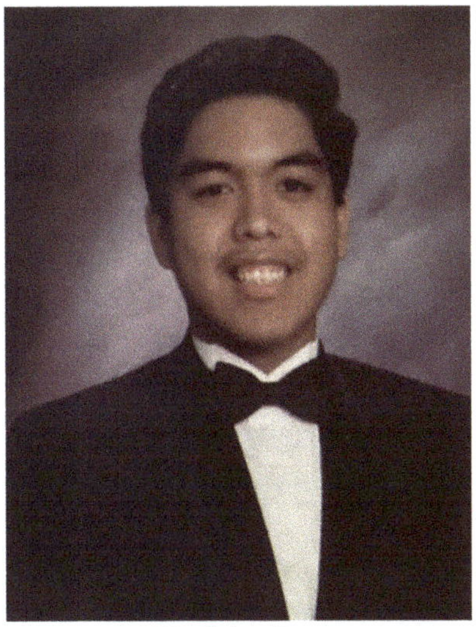

Field, Gar

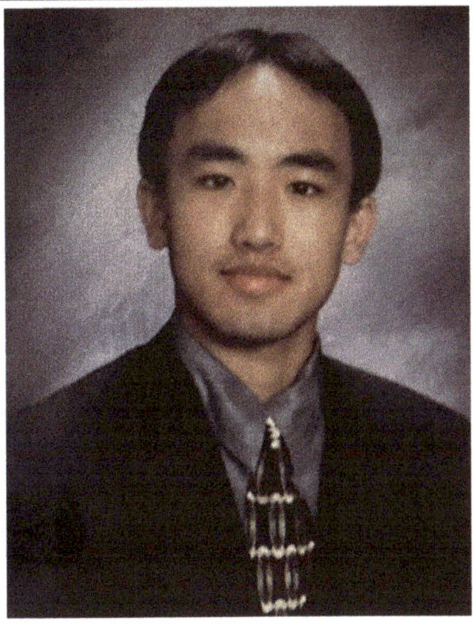

Tran, Channy

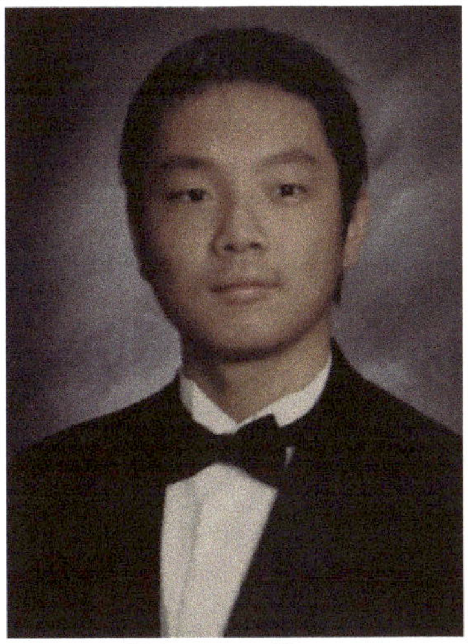

Wang, Sen

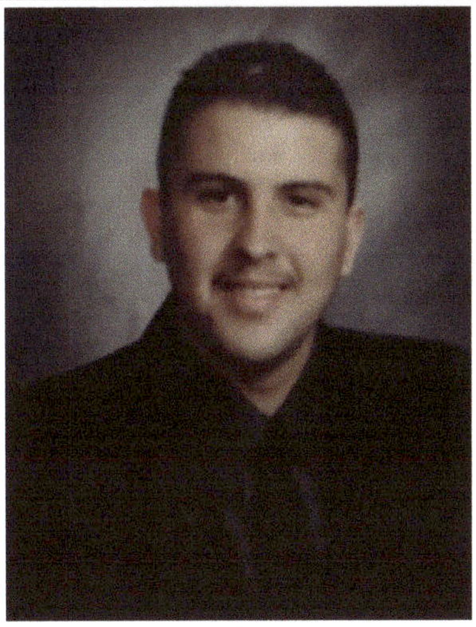

Z, Chris-

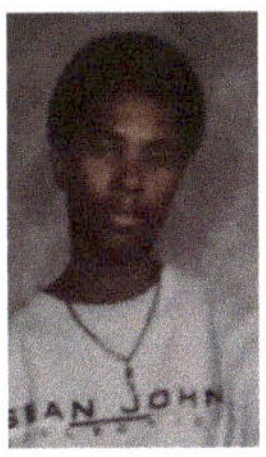

Class Of 2137
And Fellow Friend,
Yellow Bus Rider.

Williams, Talor A.

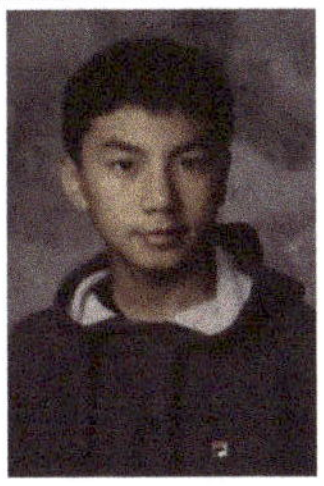

Class Of 2137
And Fellow Friend,
Yellow Bus Rider.

Yu, J.M.E. X.

(Military)
Commissioned Officer,
Colonel (COL) (O-6)
Rank of
United States Army
(U.S.A.) within the
old world United **States**
of North America
(U.**S.** of N.A.)
Armed Forces.

Sir Kyle Kinsman is
"The Colonel" (COL),
Commanding Officer
(C.O.) in Fort Angel
South California,
(SCA) Los Angeles
(L.A.) (R) (D) County
Military (M) Base(s)
For The Chatsworth
High School (C.H.S.),
Graduating Class Of
2136 (Coming Of Age)

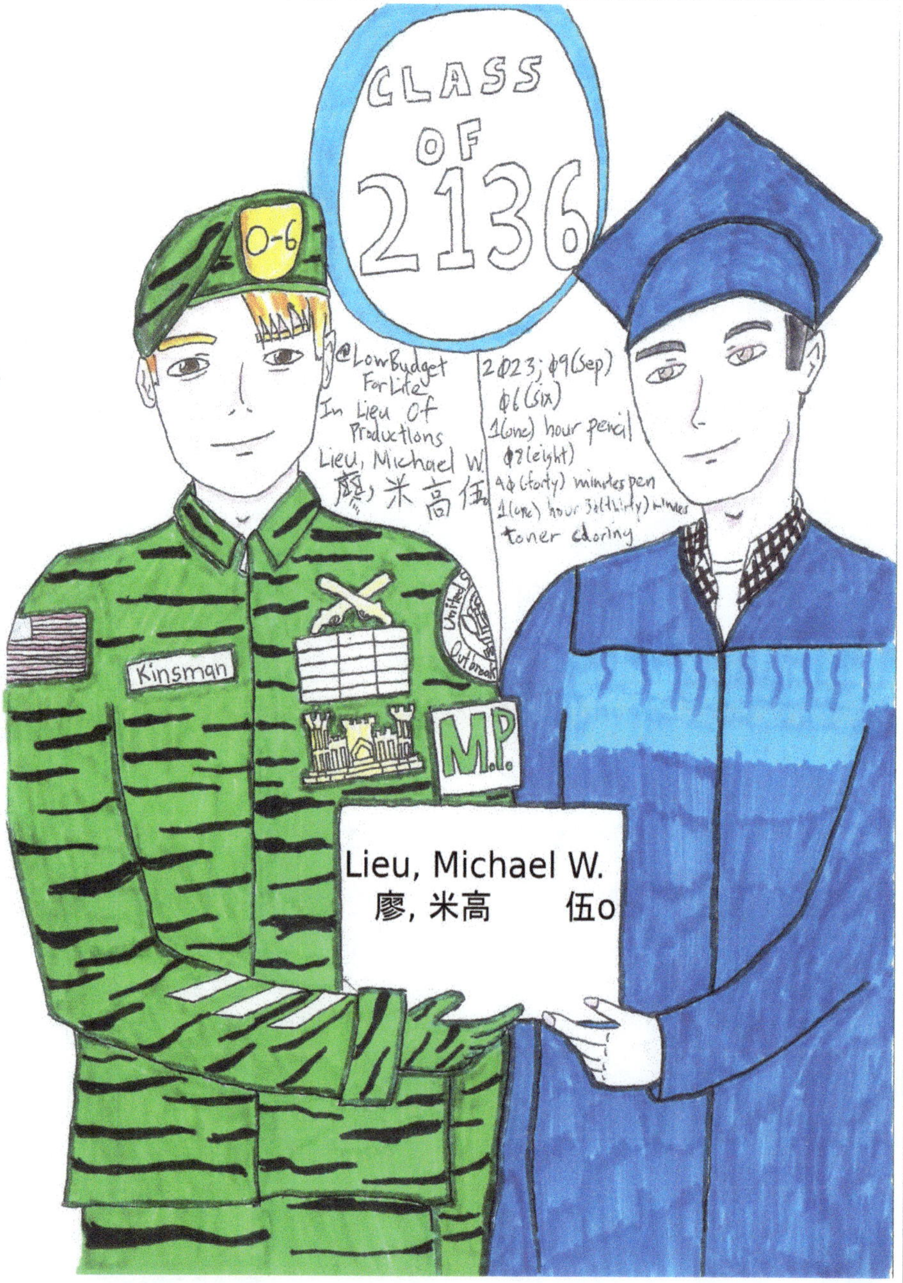

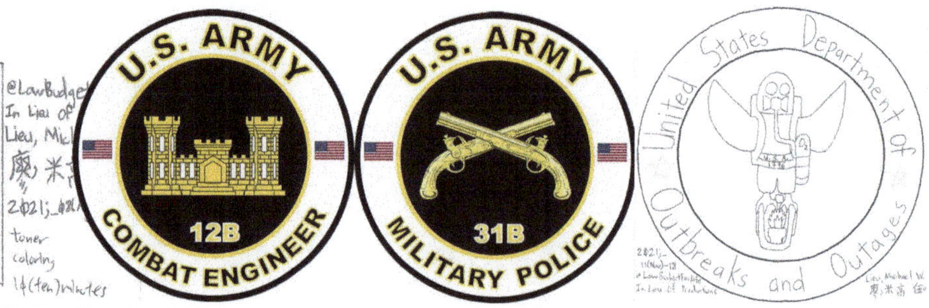

(Military) United States Army (U.S.A.) Commissioned Officer Rank, Colonel (O-6) Sir Kyle Kinsman, "The Colonel (COL)", Of The Surviving County Military Bases.

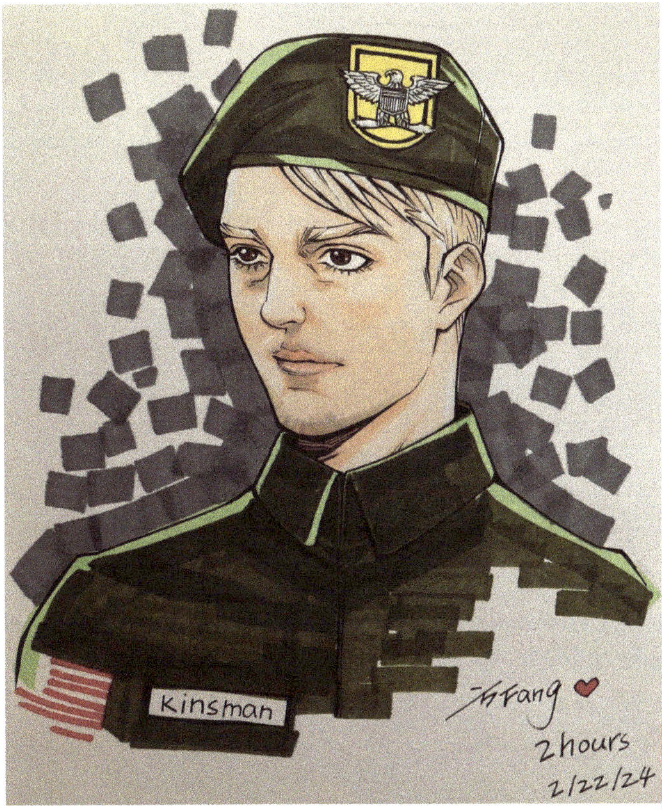

{ "2 D Augmented Art Commission(s) Portrait(s) By; Fang Fang Comics, Fang YiLi" }

Colonel Kyle Kinsman
Commanding Officer (C.O.)
Fort Angel

The Colonel's Speech For The Graduating Class Of 2136 For The Chatsworth High School (C.H.S.); (audio transcript recording archive)

" By twenty one twenty one, terror attacks, outbreaks outages, space rocks by foreign foes leading to Earth's ecosystem going to hel, for those who decided to take part and waive your legal rights as per our orders under the General of United States Army, General Shaun Shannon, into forming our unique County Bases assisted by conventional political parties of Republican, Democratic and some Independent mix, non violet purple Utopian Sovereignty movement within our Military Might, M.P., Military Police...

After today, graduating from County Bases **you will no longer need parent, guardian signature to sign off on whatever it is you so want to do**... **You are grown full adults, responsible for your own actions**... Whether **you** go into the Church or The State, public sec military or civilian service, **private sec Elysium Corp**, Kay's Private Security, my advice, go all in and do not have a back up plan. That is all."

(Military) United States Army (U.S.A.) Commissioned Officer Rank,
Lieutenant Colonel (O-5), Ma'am Faith Fukuhara,
"The Lieutenant Colonel (LT.C.)", Of The Surviving County Military Bases.

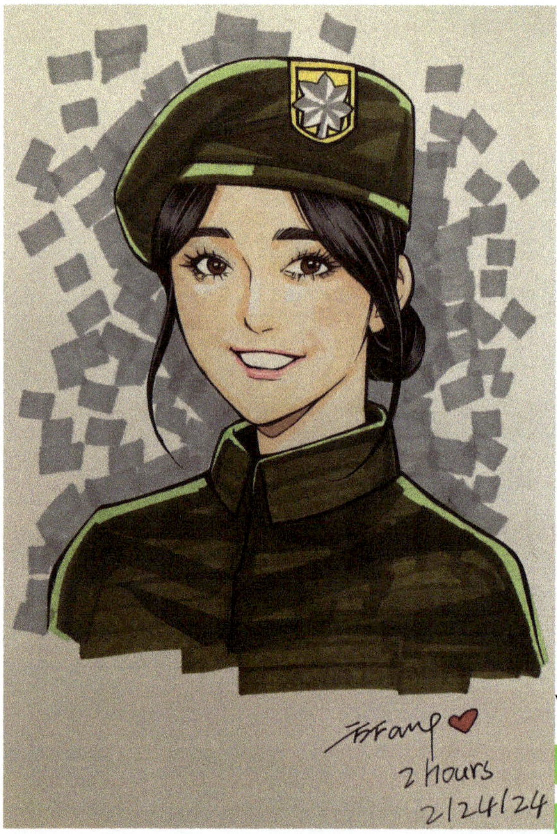

{ "2 D Augmented Art Commission(s) Portrait(s) By; Fang Fang Comics, Fang YiLi" }

Lieutenant Colonel Faith Fukuhara
Xecutive Officer (X.O.)
Fort Angel

The Lieutenant Colonel's Speech For The Graduating Class Of 2136 For The Chatsworth High School (C.H.S.);
(audio transcript recording archive)

" **Hi everyone**! I'm Lieutenant Colonel Fukuhara, I just wanted to add to the Colonel's speech, Double K... On top of being **full grown adults after graduating** from our situational County Base schools, just be aware that the United States Dollars values and United States Constitution beliefs system applies to being on base.

Outside of base, we been aware of the green Mexican-Men Militia who been... respectfully requesting for the geographic territory return of Los Angeles County, as well South California... Plus, purple projection Ultra Violet, U.V. broadcasting towers that we been ordered to stay away and observe just like Payday Loans.

Anyways, do give it **your best after graduation**, whether **you** go into public sec or **private sec**... Church or the State, or a **Company**. Ok? **Bye everyone**! That is all!!"

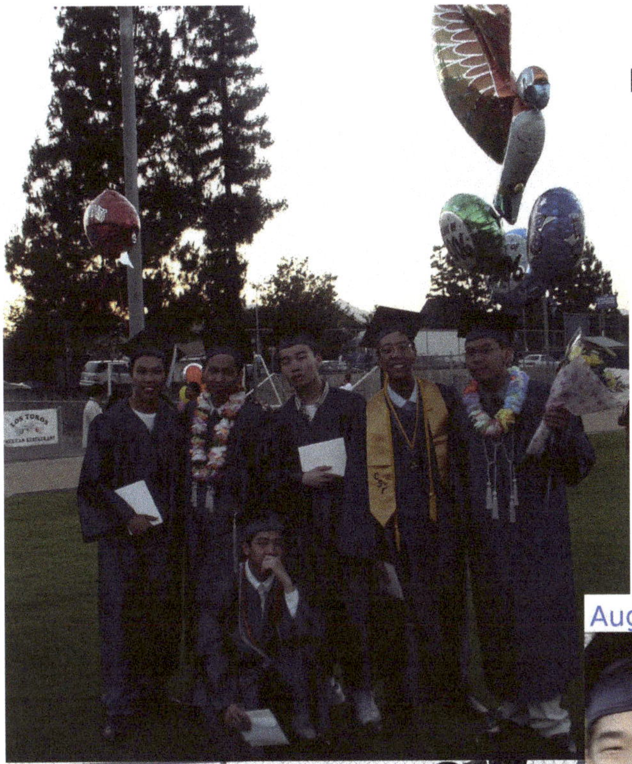

(Elementary, Mid, High School Friends; Eastern-Western SitCom) (Photo-Book) (Coming Of Age)

(Slice Of Life, Mockumentary)

(DocuDrama, Philosophy)

"Fort Angel"

Chatsworth High School (C.H.S.); Los Angeles (L.A.) (R) (D) County Military (M) Base(s).

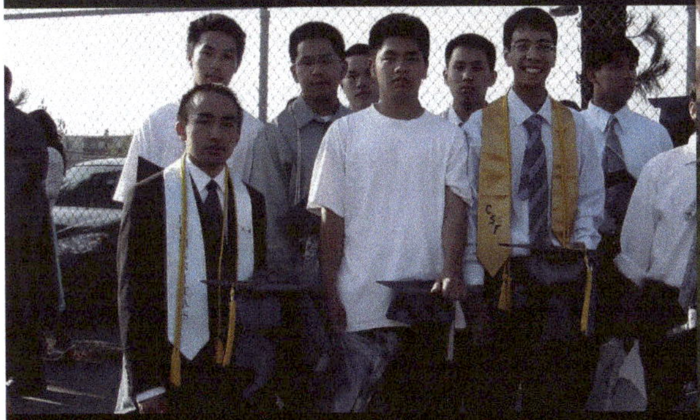

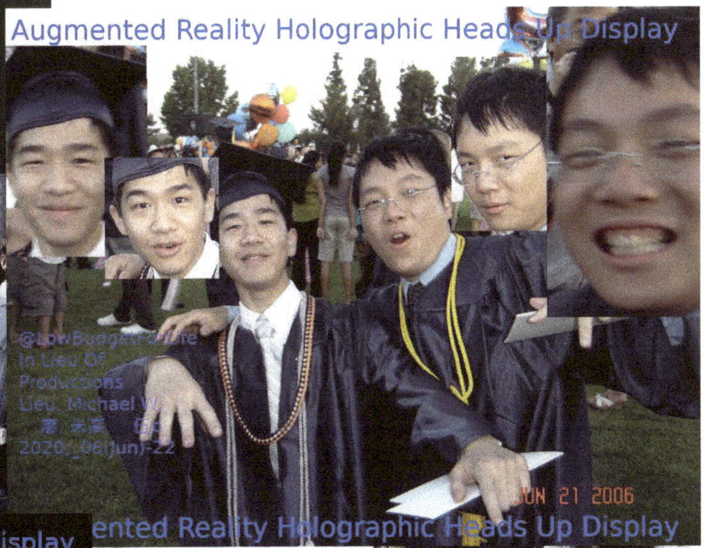

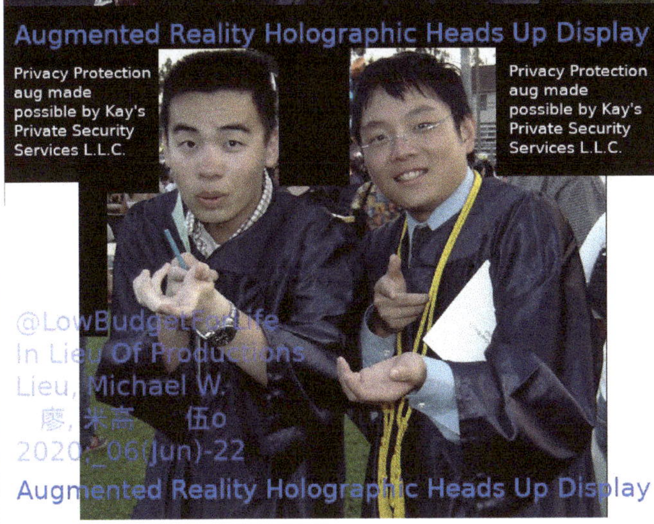

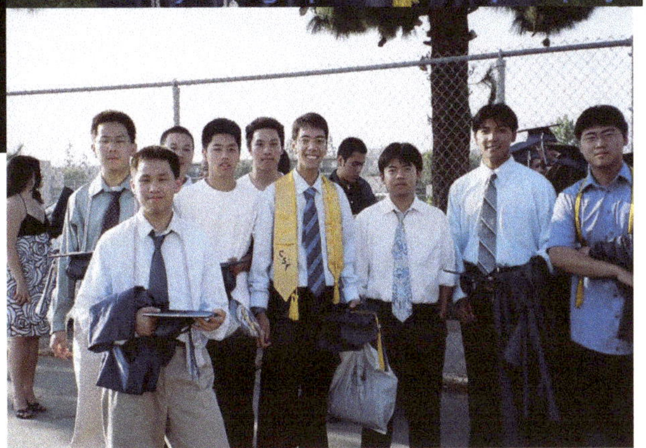

Graduating C.H.S. Class Of 2136

(Photo-Book) (Military) (Super-Hero) (**Vigilante**)

Arc One (A 1) "Hero Journey Character Transformation" Year; 2136 to 2139

(War, Military) = War On Mars, "2120 – Present On Going Operation(s)"

United States Department Of War (U.S. DOW), AKA U.S. War Office

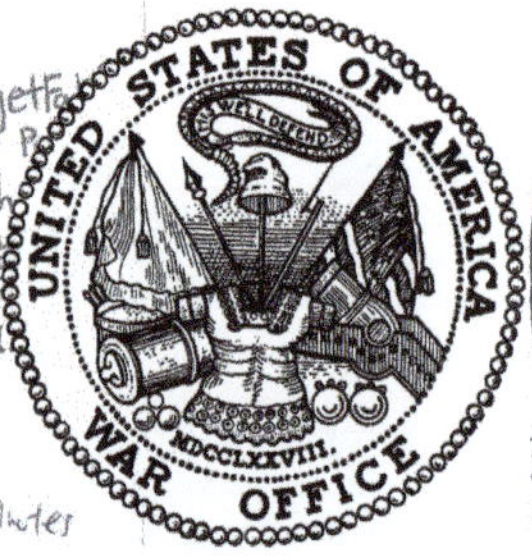

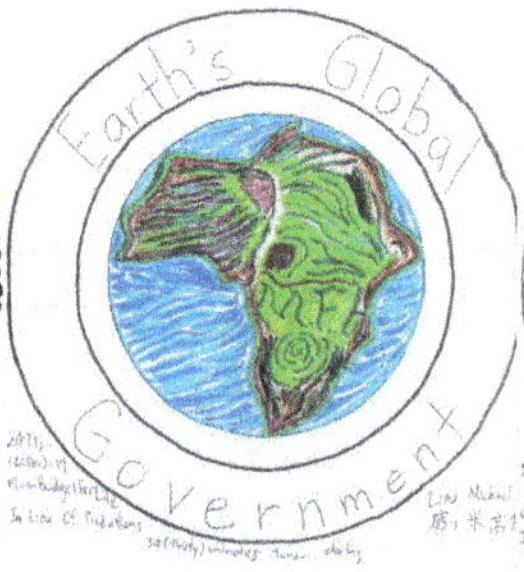

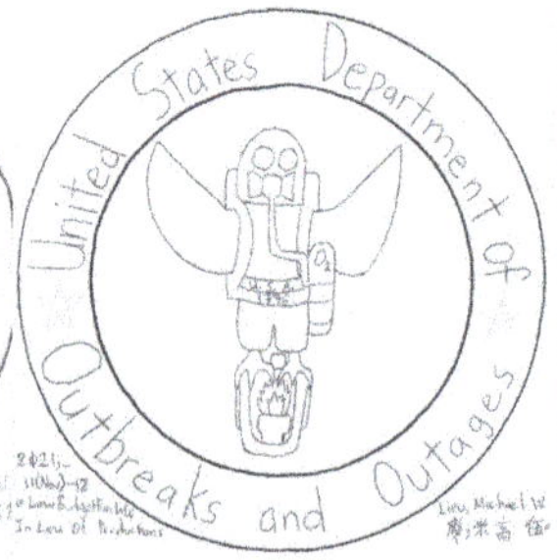

U.S. DOW, Branch; Navy, AKA United States Navy (U.S.N.), Military Active Duty.

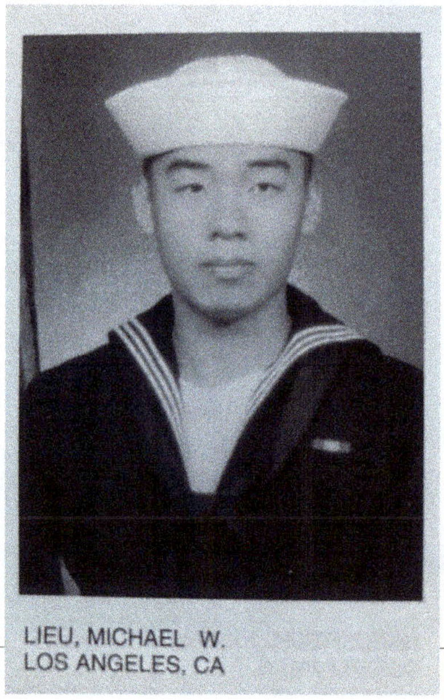

Personal Identifiable Information (P.I.I.)

Last Name, AKA Surname; "Lieu"	
First Name, AKA Given Name; "Michael"	
Middle Initial; "W."	Belief Systems; "No Preference"
Date Of Natural Birth; "2120; 09 (Sep) – 11 (eleven)"	
Place Of Natural Birth; "United States Army (U.S.A.) Fort Angel Los Angeles (R) (D) County, Military (M) Base(s), South California (SCA), United **States** of North America (U.**S.** of N.A.), Earth, Solar System"	
Surname Social Security Number; "Lieu 2120-09-11"	
United States Department of Outbreaks and Outages (U.S. DOO) Basic Training	

Citizen Status; Earthling	Earth's Global Government (EGG) Basic Training
Terrorist Threat; Low	Military (M) Service Rank; Enlisted (E), "E-1 Recruit"
Navy Sailor; Engineer, Class A School	Military Occupation Specialty (MOS) Rate + Training; "Machinist's Mate (M.M.) Surface @ BECC, G.L., ILL"
Branch; U.S.N. Basic Training Ship Out Date; 2136; 11 (Nov) – 13 (thirteen)	
Component; Military (M); "Active Duty" Plus "Reserve" Of 8 (eight) Years	
Desertion Unauthorized Absense (U.A.); Low	For Official Use Only (FOUO)

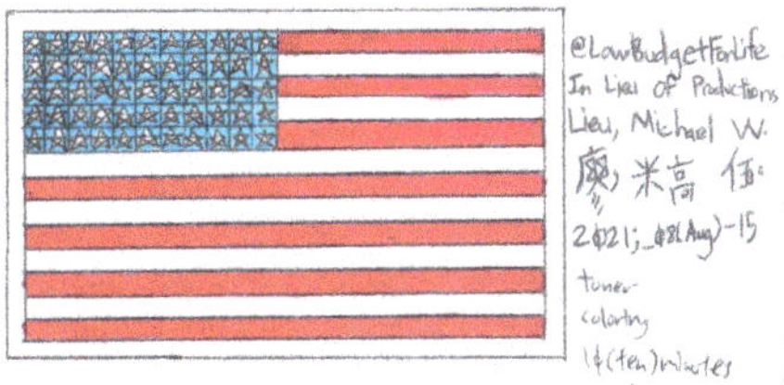

(Apocalyptic) (Contemporary) (Dystopia) (Military) (Mythology) (Neo Western, New Wild West) (Utopia) Planet Earth, North America (N.A.) "The **old world** United **States** of North America (U.**S.** of N.A.)"
The Nation Was Founded → 1776 – 2121 ← Collapse
The Country's Capital; District of Columbia (D. of C.) 55

55 (fifty five) States	Capital	55 (fifty five) States	Capital
Alabama (ALA) 01	Montgomery	Nevada (NEV) 28	Carson City
Alaska (AKA) 02	Juneau	New Hampshire (NHA) 29	Concord
Arizona (AZA) 03	Phoenix	New Jersey (NJE) 30	Trenton
Arkansas (ARK) 04	Little Rock	New Mexico (NME) 31	Santa Fe
Colorado (COL) 05	Denver	New York (NYO) 32	Albany
Connecticut (CON) 06	Hartford	North California (NCA) 33	Sacramento
Delaware (DEL) 07	Dover	North Dakota (NDA) 34	Bismarck
Florida (FLO) 08	Tallahassee	North Karolina (NKA) 35	Raleigh
Georgia (GEO) 09	Atlanta	Ohio (OHI) 36	Columbus
Guam (GUA) 10	Hagatna	Oklahoma (OKL) 37	Oklahoma City
Hawaii (HAW) 11	Honolulu	Oregon (ORE) 38	Salem
Idaho (IDA) 12	Boise	Pennsylvania (PEN) 39	Harrisburg
Illinois (ILL) 13	Springfield	Puerto Rico (PRI) 40	San Juan
Indiana (IND) 14	Indianapolis	Rhode Island (RIS) 41	Providence
Iowa (IOW) 15	Des Moines	South California (SCA) 42	Los Angeles
Kansas (KAN) 16	Topeka	South Dakota (SDA) 43	Pierce
Kentucky (KEN) 17	Frankfort	South Karolina (SKA) 44	Columbia
Louisiana (LOU) 18	Baton Rogue	Tennessee (TEN) 45	Nashville
Maine (MAI) 19	Augusta	Texas (TEX) 46	Austin
Maryland (MAR) 20	Annapolis	U.S. Virgin Islands (USVI) 47	Charlotte Amalie
Massachusetts (MAS) 21	Boston	Utah (UTA) 48	Salt Lake City
Michigan (MIC) 22	Lansing	Vermont (VER) 49	Montpelier
Minnesota (MIN) 23	Saint Paul	Virginia (VIR) 50	Richmond
Mississippi (MIS) 24	Jackson	Washington (WAS) 51	Olympia
Missouri (MSO) 25	Jefferson City	West Virginia (WVI) 52	Charleston
Montana (MON) 26	Helena	Wisconsin (WIS) 53	Madison
Nebraska (NEB) 27	Lincoln	Wyoming (WYO) 54	Cheyenne

(Apocalyptic) (Contemporary) (Military) (War) For Official Use Only (FOUO)

United States Department Of War (U.S. DOW), War Office

Branches, Components, Agencies, Contractors Under U.S. DOW.

Secretary Of War, Office Of The Inspector General Of The Department Of War

Office Of The Secretary Of War Deputy Secretary Of War Under Secretaries Of War Assistant Secretaries Of War	Joint Chiefs Of Staff Chairperson Of The Joint Chiefs Of Staff	Branch Of The Navy, Est. 1775; Secretary Of The Navy	Branch Of The Army, Est. 1775; Secretary Of The Army	Branch Of The Air Force, Est. 1947; Secretary Of The Air Force	Branch Of The Space Force, Est. 1958; Secretary Of The Space Force
		Head Quarters Marine Corps Office Of The Chief Of Naval Operations	The Army Staff	The Air Staff	The Space Staff
	The Joint Chiefs The Joint Staff	Office Of The Secretary Of The Navy	Office Of The Secretary Of The Army	Office Of The Secretary Of The Air Force	Office Of The Secretary Of The Space Force

U.S. DOW, War Field Activities

War Media Activity	War Technology Security Administration	DOW Human Resources Activity	Office Of Economic Adjustment
War Technical Information Center	DOW Education Activity	DOW Test Resource Management Center	Washington, D. of C. Head Quarters Services

U.S. DOW, War Agencies

War Advanced Research Projects Agency (WARPA)	War Commissary Agency	War Contract Audit Agency
War Contract Management Agency	War Finance And Accounting Service (WFAS)	War Health Agency
War Information Systems Agency	War Intelligence Agency	War Legal Services Agency
War Logistics Agency	Prisoner Of War (POW) / Missing In Action (MIA) Accounting Agency	War Security Cooperation Agency
War Security Service	War Threat Reduction Agency	Missile War Agency
National Geospatial-Intelligence Agency	National Reconnaissance Office	National Security Agency (N.S.A.) Central Security Service
Pentagon Force Protection Agency	Housing And Lodging (HAL)	Homeless, Ownership, Rentals
Contractor; Burials And Funerals	Contractor; Bridges, United Cities of Americas (U.C. of A.)	

(Apocalyptic) (Contemporary) (Military) (War) For Official Use Only (FOUO)

"2120's War On Resources" "Twenty Second Century Dark Ages" "War On Mars"

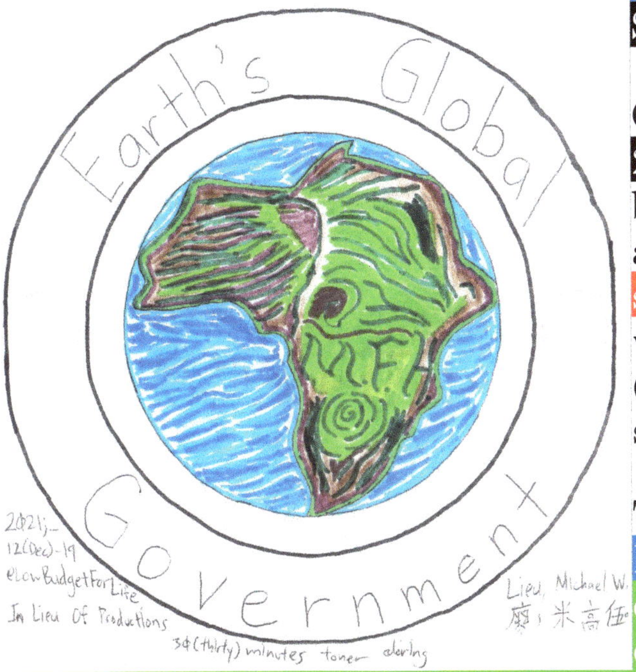

Earth's Global Government (EGG)
Space Force Joint Task Force (J.T.F.)

Officially established in the year 2122...as Earth's Space Force J.T.F. between the last remaining **augmented** and natural human survivors of Planet Earth's **old world** structures of first (1st), second (2nd) and third (3rd) world federal, state, county, local governments.

This Unified Government on Planet Earth's consists of civil servants, citizens, civilian contractors, contractors, military servicemembers: active duty, reservists, national guards, disabled veterans, retirees as **first responders to both foreign and domestic terrorists' unconventional warfare of Chemical, Biological, Radiological (C.B.R.) outbreaks and Electro Magnetic Pulse (E.M.P.) outages plus meteorites bombardment.**

EGG's current War Conditions (War Con) as of 2136; "2"

War Con	1 (one) "**white**" most severe, Nuclear Warfare
War Con	2 (two) "red"
War Con	3 (three) "yellow"
War Con	4 (four) "green"
War Con	5 (five) "blue" least severe, 2120; 11 (November) - 17

"EGG Space Force J.T.F. defines our Earth **that stands for** one value...
You are either **with** us or **against** us on the War On Mars,
War is hell, either they destroy us or we destroy them."
-Current EGG Space Force J.T.F. Fleet Admiral Keith D. Williams

(Apocalyptic) (Contemporary) (Military) (War) For Official Use Only (FOUO)

United States Department of Outbreaks and Outages (U.S. DOO)

Outbreaks Pandemic Division	**Outage Blackout Division**
Chemical, Biological, Radiological (C.B.R.)	Electro Magnetic Pulse (E.M.P.)

Brief Overview Of U.S. DOO

The Terrorist(s) Questions; "Who is a terrorist(s) and what do they look like?"

Terrorist(s), foreign and domestic, do not follow the conventional rules of war.

After the end of World War Two in 1944, there was no established written formulaic guidelines as well as no peace treaty agreement about the war rules afterwards.

From 1944 to 2120, the Kold War deterrence through Mutual Assured Destruction (MAD) is not likely to be an effective threat against potential protagonists that are either:

cell-style terrorist(s)	trans-national group(s)	lone wolf attack(s)	insurgencies
failing state(s)	suicide bomber(s)	hired mercenaries	domestic terrorist(s)

In 2120, the people of Planet Mars worked along side private sector Roman Industries in deciding to wage an unconventional "War On Resources" due to the belief system that "humans are overpopulating the solar system and needs to be annihilated by one half".

On 2121, the people of Planet Mars, "Martian Terrorists" and Roman Industries' succeed.

By 2122, the last remaining Planet Earth United States (U.S.) military servicemembers, civilian federal government employees, contractors, veterans of the U.S. Department Of War (DOW) and other agencies banded together to form a first responders' Joint Task Force (J.T.F.) agency: **United States Department of Outbreaks and Outages (U.S. DOO).**

From a space perspective, United States Force Space (U.S.S.F.) branch of DOW, plus other Kold War friend-enemies "frenemies" states and countries banded together to form Planet Earth's Space Force as a Joint Task Force (J.T.F.) agency: Earth's Global Government (EGG).

A reminder, that the constant training in C.B.R. and or E.M.P. warfare is a never ending battle.

"Failing To Prepare, Means Preparing To Fail" / "Do it. Do it right. Do it right now."

(Military) (Photo-Book) For Official Use Only (FOUO) Joint Task Force (J.T.F.)

U.S.N. Sailor Recruit Training Command (R.T.C.),
Military (M) Base, Active Duty + Reserve +
National Guard: Basic Training "Boot Camp"
North Chicago, Great Lakes, Illinois (ILL),
United **States** of North America (U.**S.** of N.A.),
Planet Earth, Solar System, Milky Way Galaxy.

U.S.N. Sailor Military (M) Base, Active Duty, Basic Training Status;
"Mission Complete"

"The Oath Of Enlistment (Enlisted, E)."
"I, Lieu, Michael W. (Last Name, First Name And Middle Initial) do solemnly swear that I will support and defend the Constitution of the United **States** of North America against all enemies, foreign and domestic; that I will bear true faith and allegiance to the same; and that I will obey the orders of the President of the United **States** of North America and the orders of the officers appointed over me, according to regulations and the Uniform Code of Military Justice.
So help me God."

"I am a United States Sailor.
I will support and defend the Constitution of the United **States** of North America and I will obey the orders of those appointed over me.
I represent the fighting spirit of the Navy and those who have gone before me to defend freedom and democracy around the world.
I proudly serve my country's Navy combat team with Honor, Courage, and Commitment.
I am committed to excellence and the fair treatment of all."

(Military) (Photo-Book) United States Sailor (U.S.N.) (Super-Hero) (**Vigilante**)

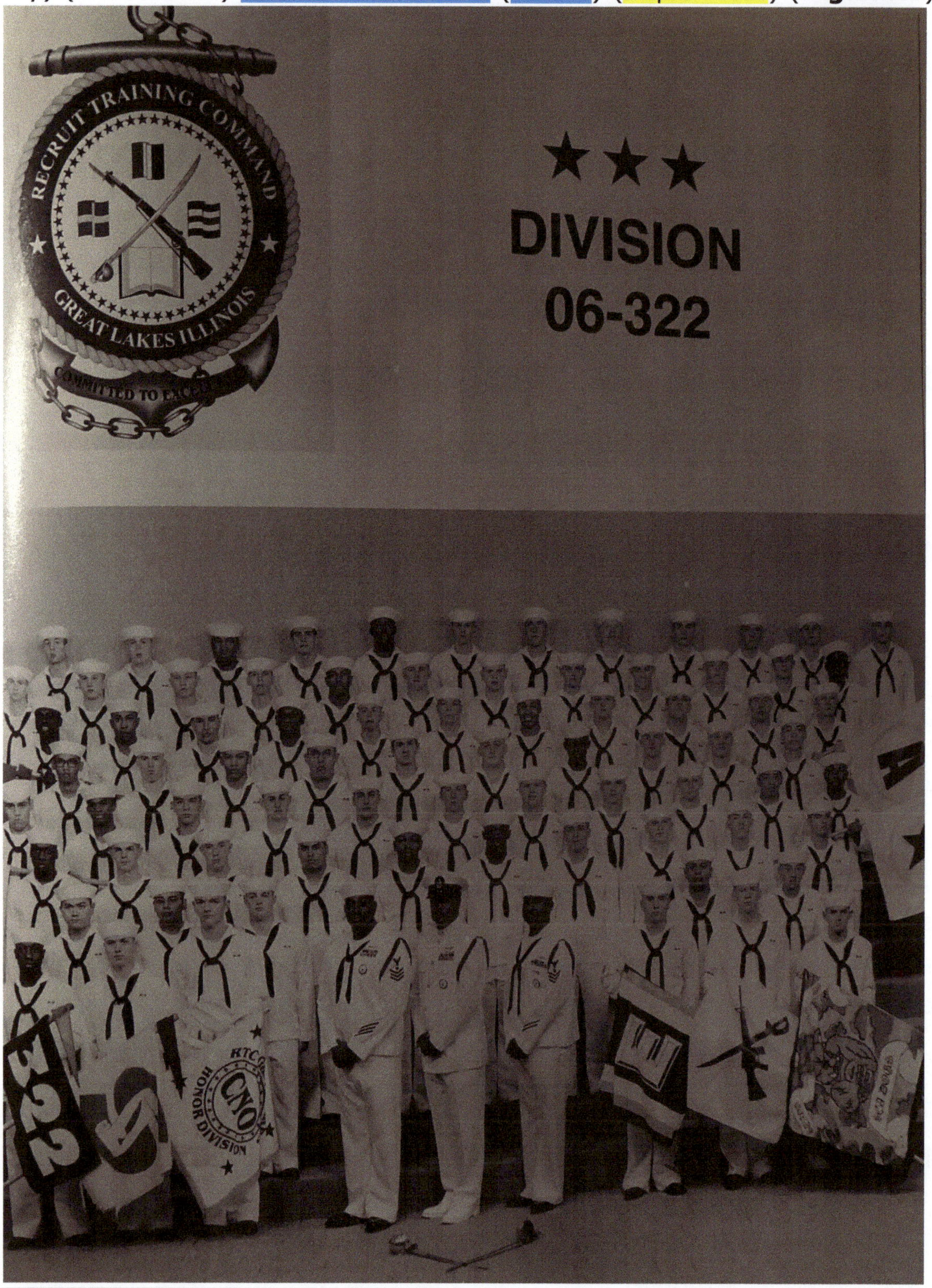

(Military) (Photo-Book) United States Sailor (U.S.N.) (Super-Hero) (**Vigilante**)

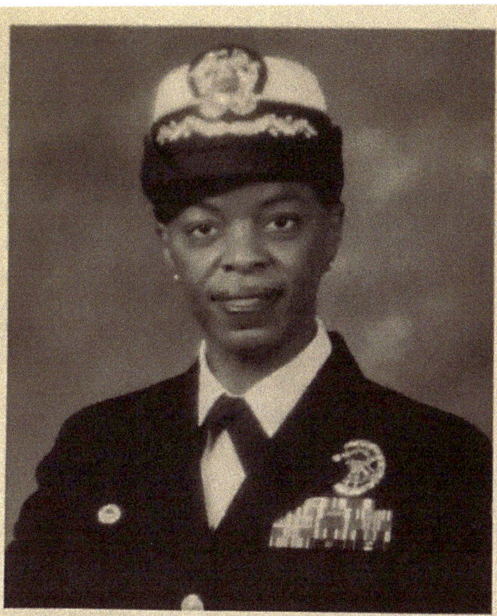

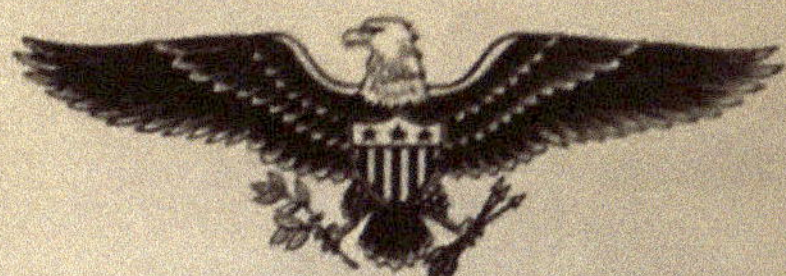

Captain Annie B. Andrews
Commanding Officer
Recruit Training Command

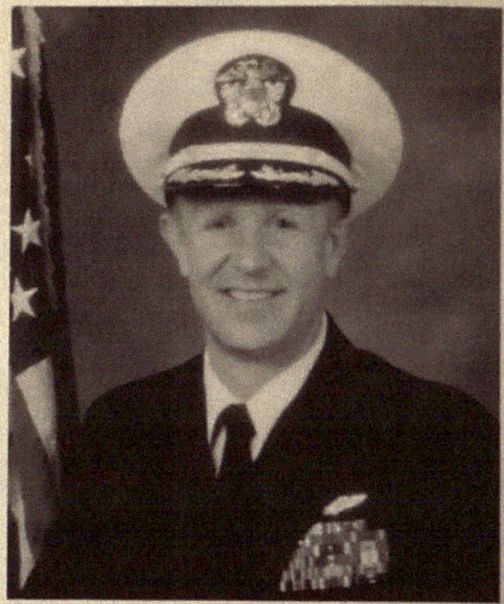

Commander Donald E. Czarapata
Executive Officer
Recruit Training Command

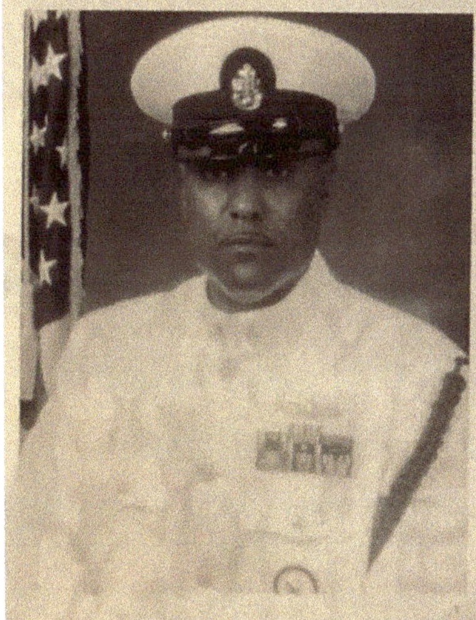

DIVISION COMMANDER
SKC(SW) SIMMONS

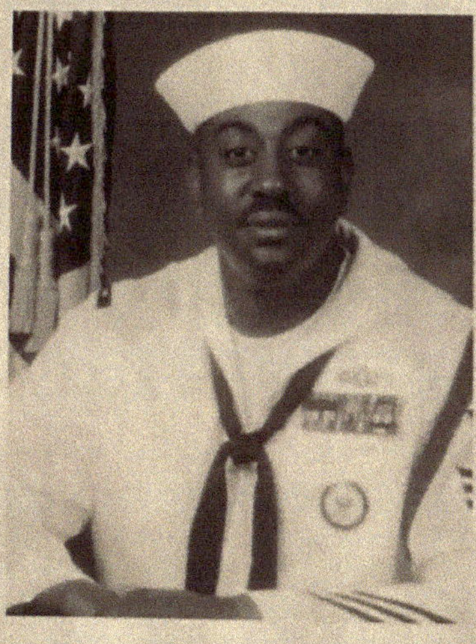

DIVISION COMMANDER
PS1(SW) MOULTRIE

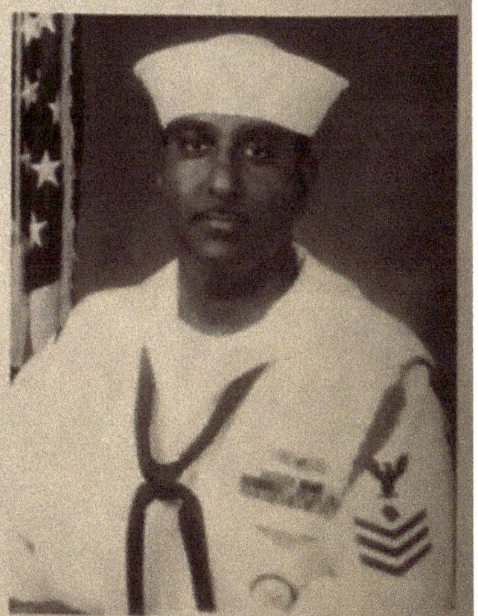

DIVISION COMMANDER
PS1(SW/AW) SOOKRAM

(Military) (Photo-Book) United States Sailor (U.S.N.) (Super-Hero) (**Vigilante**)

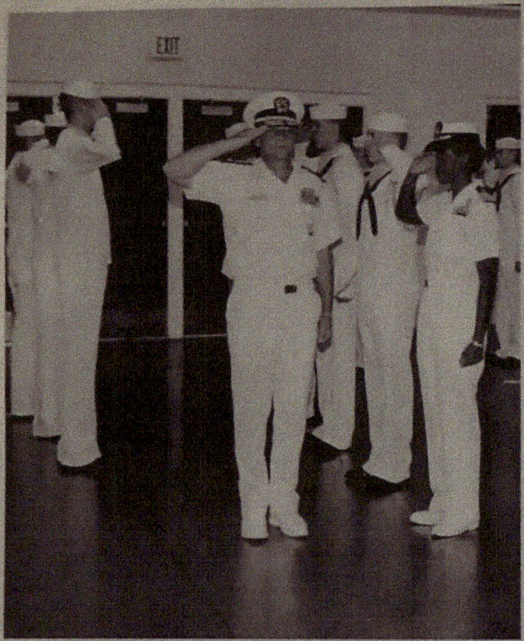

Piping Aboard Reviewing Officials

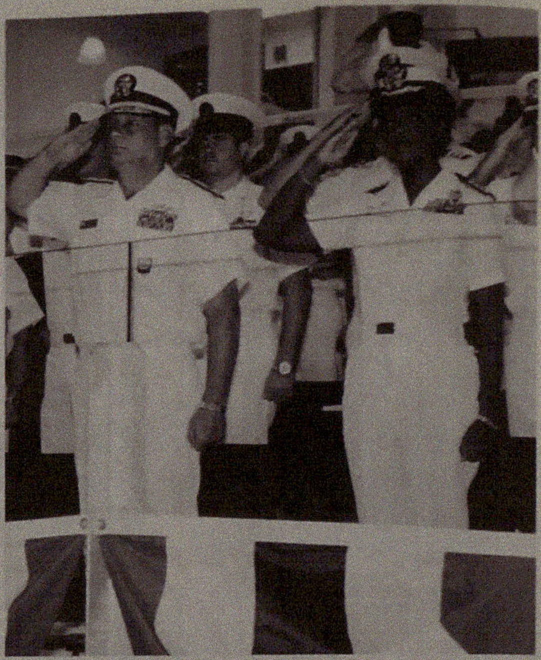

Reviewing Officials

Graduating Division

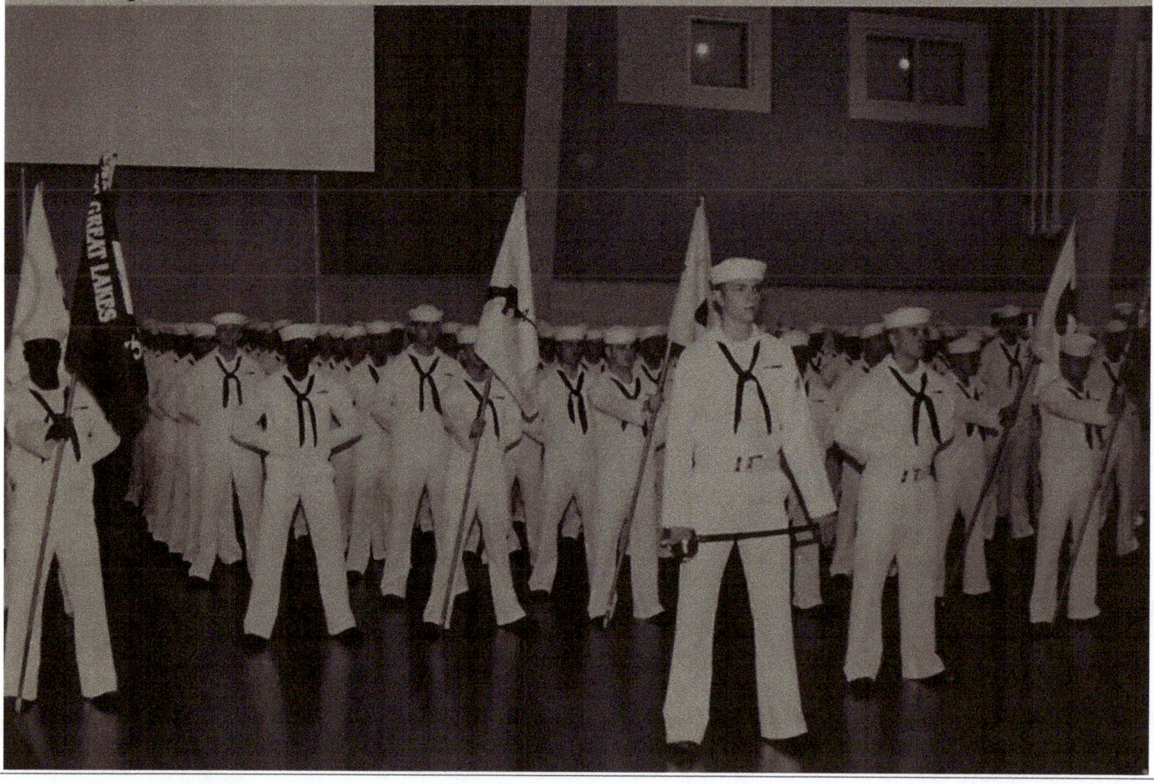

(Military) (Photo-Book) United States Sailor (U.S.N.) (Super-Hero) (**Vigilante**)

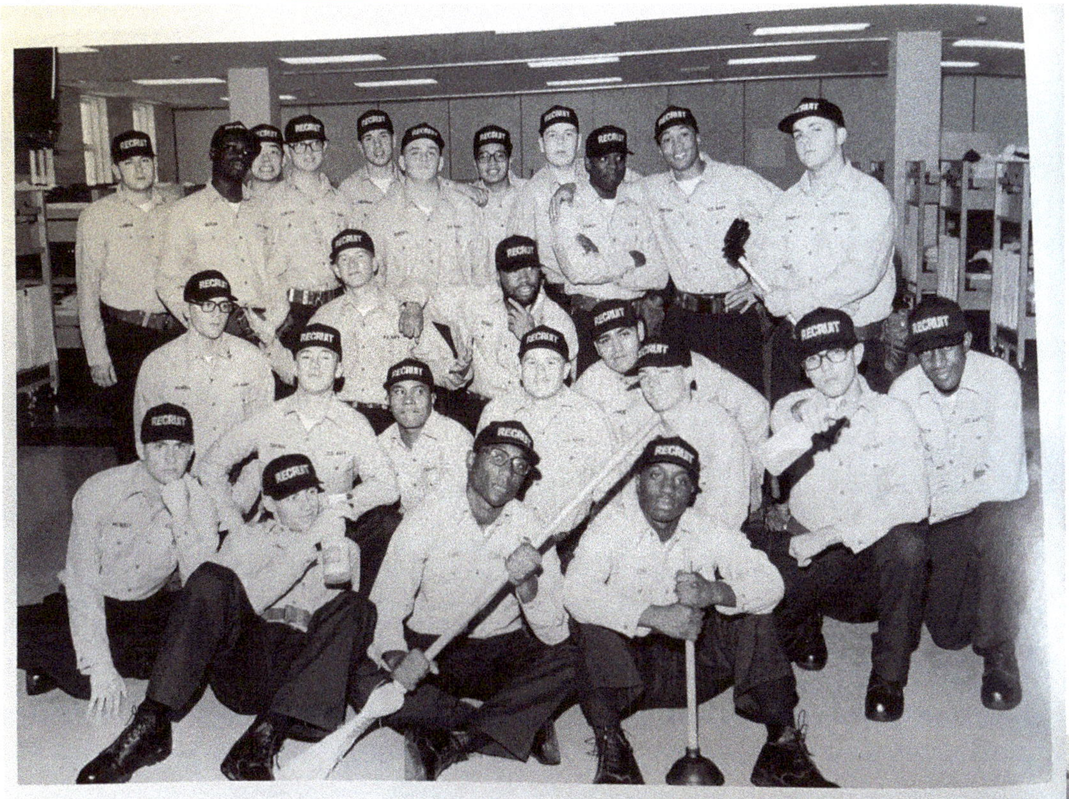

Shipmates

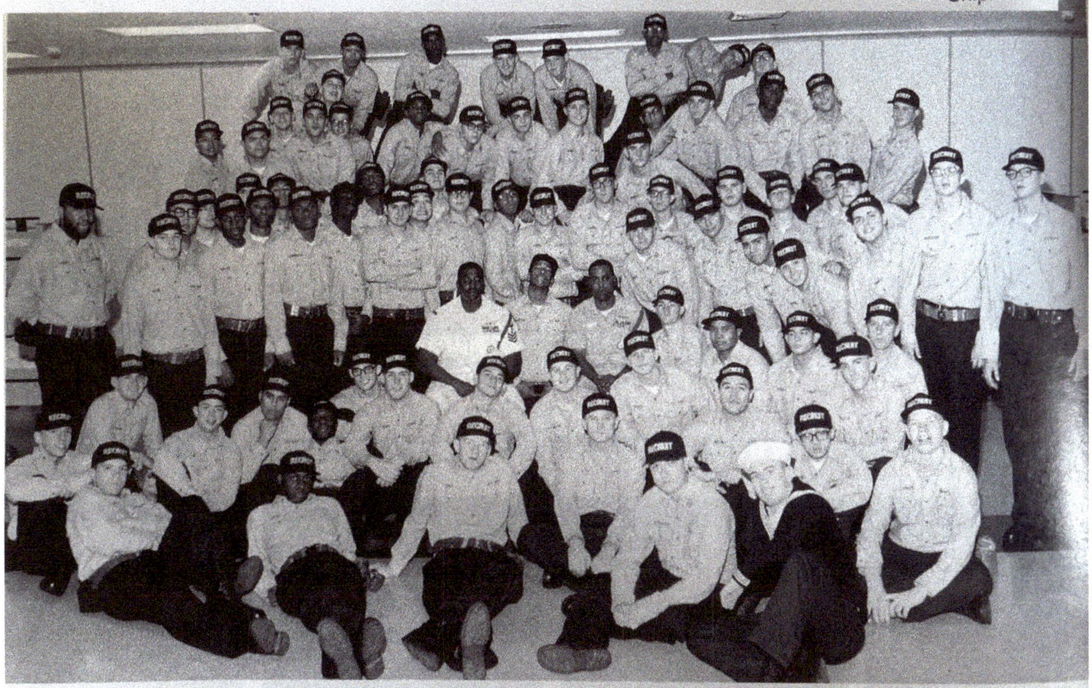

Shipmates

(Military) (Photo-Book) United States Sailor (U.S.N.) (Super-Hero) (**Vigilante**)

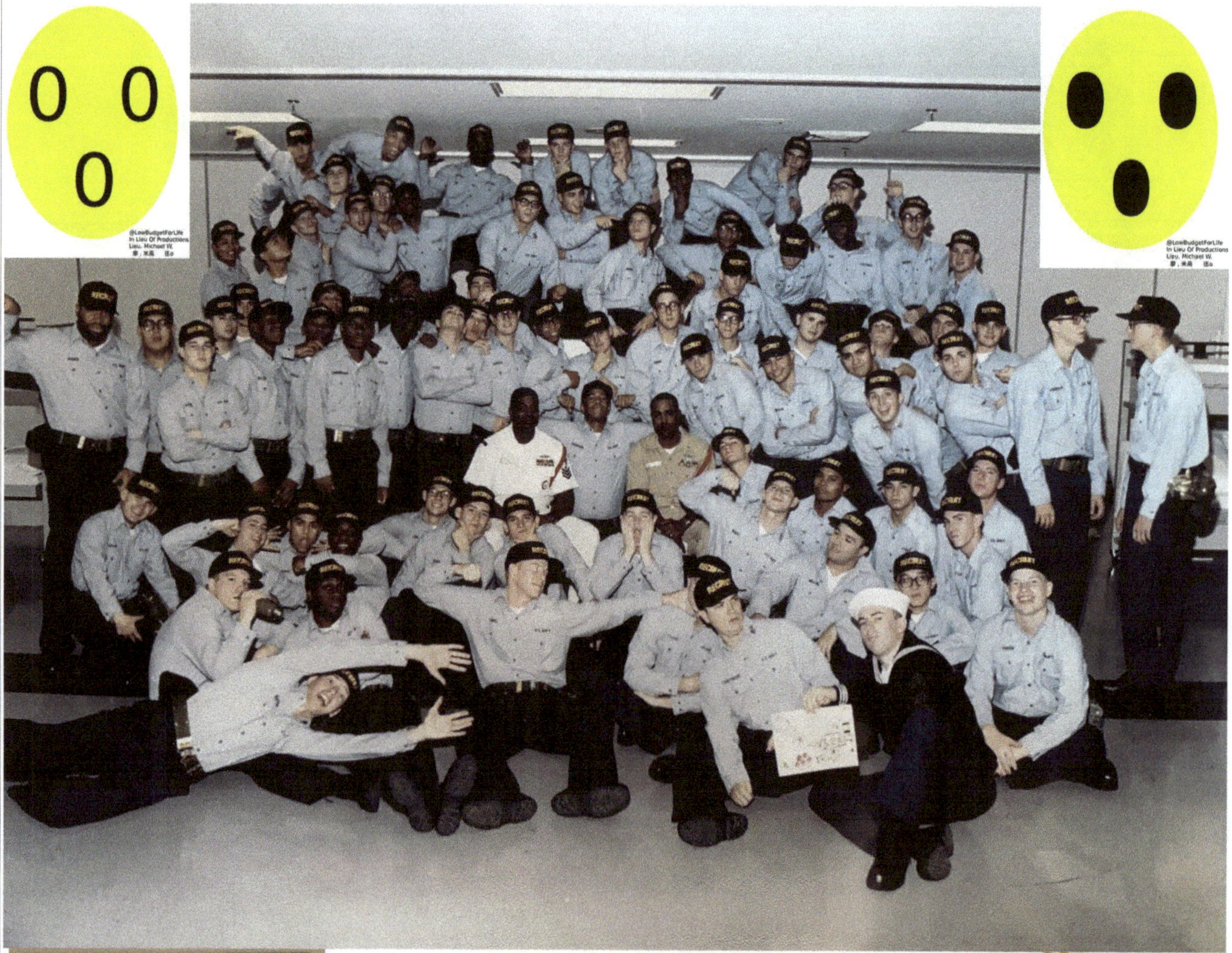

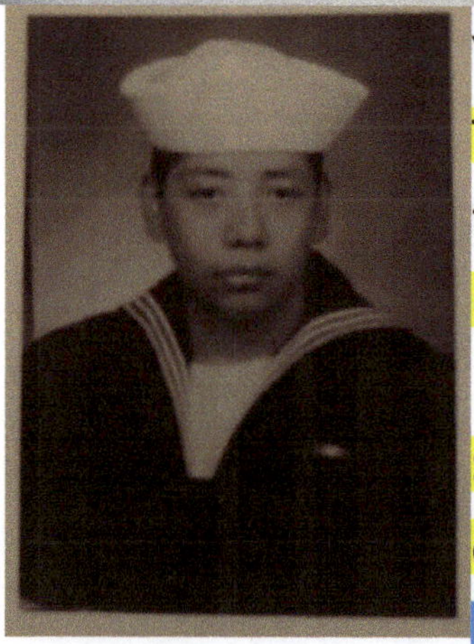

"Meeting Sisco, the future Electronics Technician (E.T.) MOS rate and co founder of Sisco Cisco Surprise Face (°o°) Yellow Broadcasting Networks.

Surprise face! (°o°) Left parenthesis degrees o degrees right parenthesis. U.S. Military Personal Manual (MilPerMan), Natural Asian male nerds for life!"

Sisco, Steven S.
Hagatna, Guam

(Contemporary) Earth's Global Government (EGG) Space Force
Joint Task Force (J.T.F.) (Military) United States Sailor (U.S.N.)
(Spy) (Super-Hero) (**Vigilante**) (War)

U.S.N. Sailor Class A School Training, Naval Station Great Lakes,
Military (M) Base, Basic Engineering Common Core (BECC),
North Chicago, Great Lakes, Illinois (ILL),
United **States** of North America (U.**S.** of N.A.), Planet Earth, Solar System.

U.S.N. Sailor Military (M) Base, Basic Engineering Common Core (BECC) Status;
"Mission Failed"

New Reassignment Reporting Under U.S. DOW, Navy, Military Official Orders;
"U.S.N. Sailor Military (M) Base In Norfolk, Virginia (VIR) "The Fleet"
Onboard "EGG United States Ship (U.S.S.) Harry S. Truman Augment
(C.V.N. 75+)" United **States** of North America (U.**S.** of N.A.), Planet Earth."
In Support Of Earth's Global Government (EGG) Space Force
Joint Task Force (J.T.F.) Current War On Mars Campaign(s) Operation(s).

"EGG Space Force J.T.F. defines our Earth **that stands for** one value...
You are either **with** us or **against** us on the War On Mars,
War is hell, either they destroy us or we destroy them."
-Current EGG Space Force J.T.F. Fleet Admiral Keith D. Williams

Approved Travel For 2 (Two) Weeks Leave Of Absence Geolocation;
United States Army (U.S.A.) Fort Angel
Military (M) Bases, Surviving County,
Los Angeles (R) (D) County, Military (M) Base(s), South California (SCA),
United **States** of North America (U.**S.** of N.A.), Planet Earth, Solar System."

(Military) (Photo-Book) U.S.N. Sailor (Spy) (Super-Hero) (**Vigilante**) (War)

U.S.N. Sailor Onboard "The Fleet"
Military (M) Base, Naval Station Norfolk, Virginia (VIR),
"EGG United States Ship (U.S.S.) Harry S. Truman Augment
(C.V.N. 75+) Joint Task Force (J.T.F.) Aircraft Carrier"
United **States** of North America (U.**S.** of N.A.), Planet Earth,
Solar System, Milky Way Galaxy."

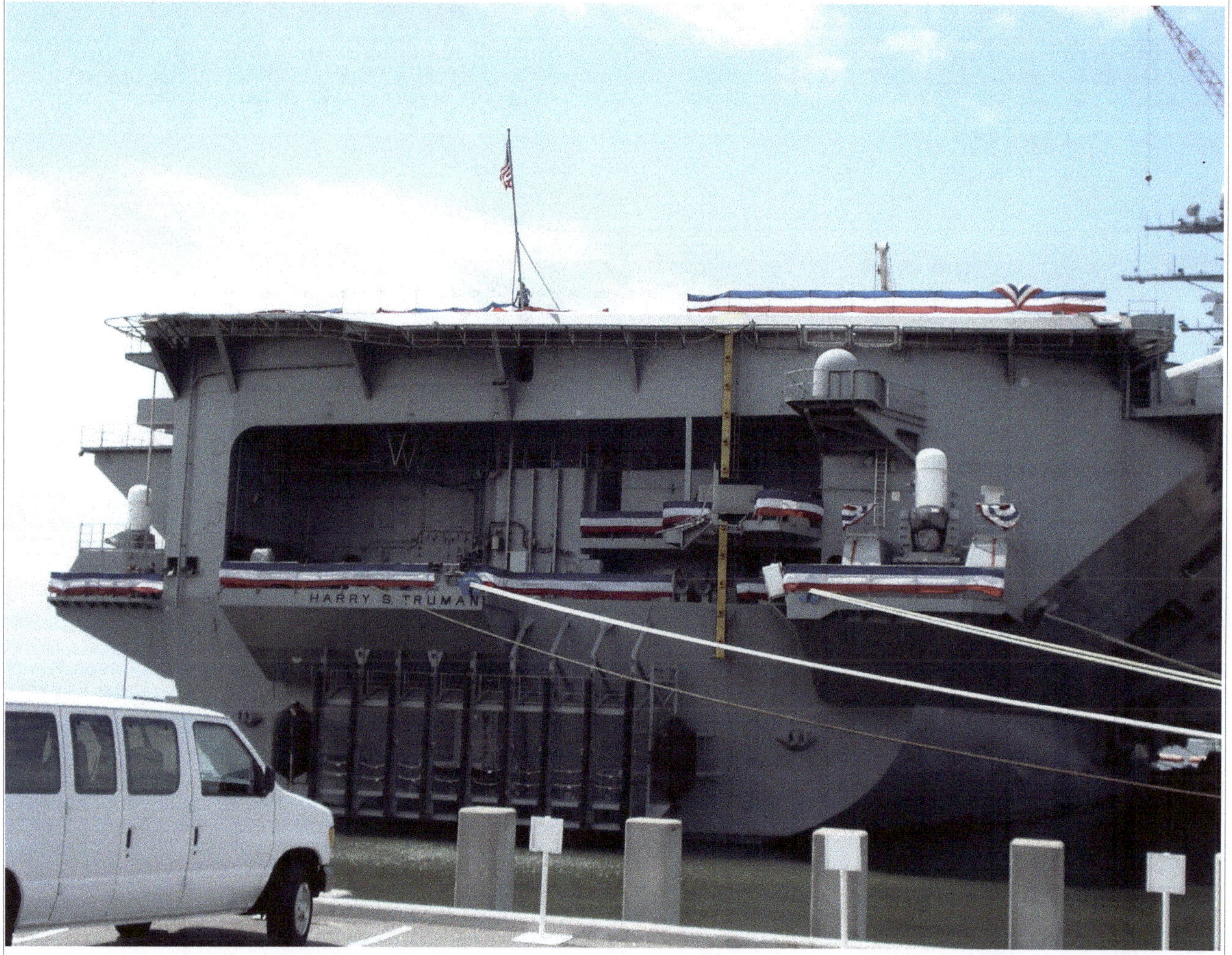

Lieutenant Lieu, Michael W.'s 2136-2139; Military Photo Book, Navy Branch;
"My first day on the fleet year one, twenty one thirty seven, February, on base Norfolk Virginia, AKA NoFuck Vagina as some servicemembers love to call it."

(Military) (Photo-Book) U.S.N. Sailor (Spy) (Super-Hero) (**Vigilante**) (War)

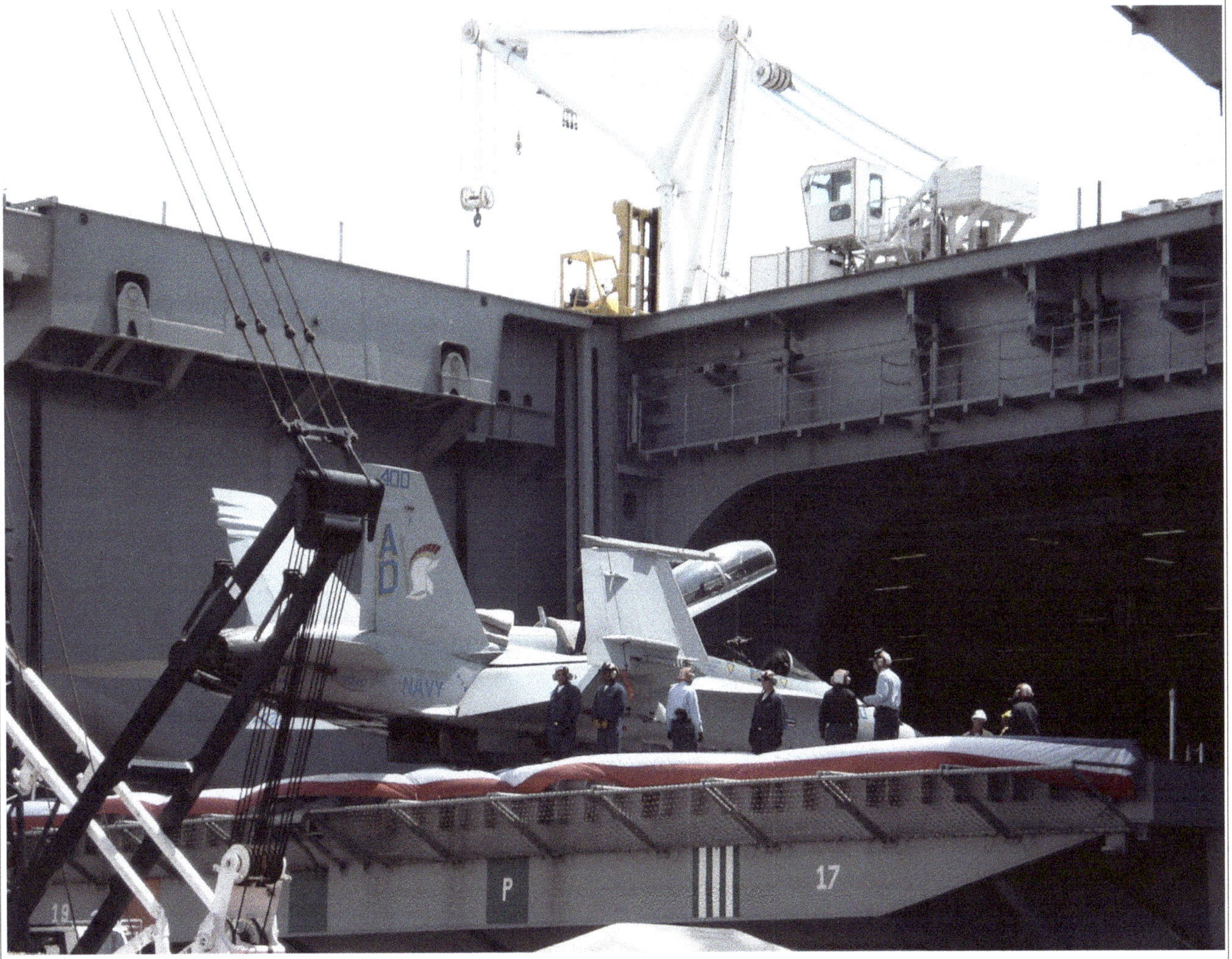

Lieutenant Lieu, Michael W.'s 2136-2139; Military Photo Book, Navy Branch;

Department; United States Department Of War (U.S. DOW)
Branch; Navy. **Year; Month (XXX)** = 2137; 02 (Feb),
Component; "Military (M) Active Duty, Servicemember Check In"

" I recall walking by and checking in on my first fleet day with Uncle Sam Military, Navy branch... What a sight... Check out that aircraft... and that elevator loading the aircraft up and down... How that works?

Just how many aircraft can this war ship even hold?

Wonder if Sisco made it to the fleet since he did not take any leave after reassignment orders in Illinois base, failing technician school..."

(Military) (Photo-Book) U.S.N. Sailor (Spy) (Super-Hero) (**Vigilante**) (War)

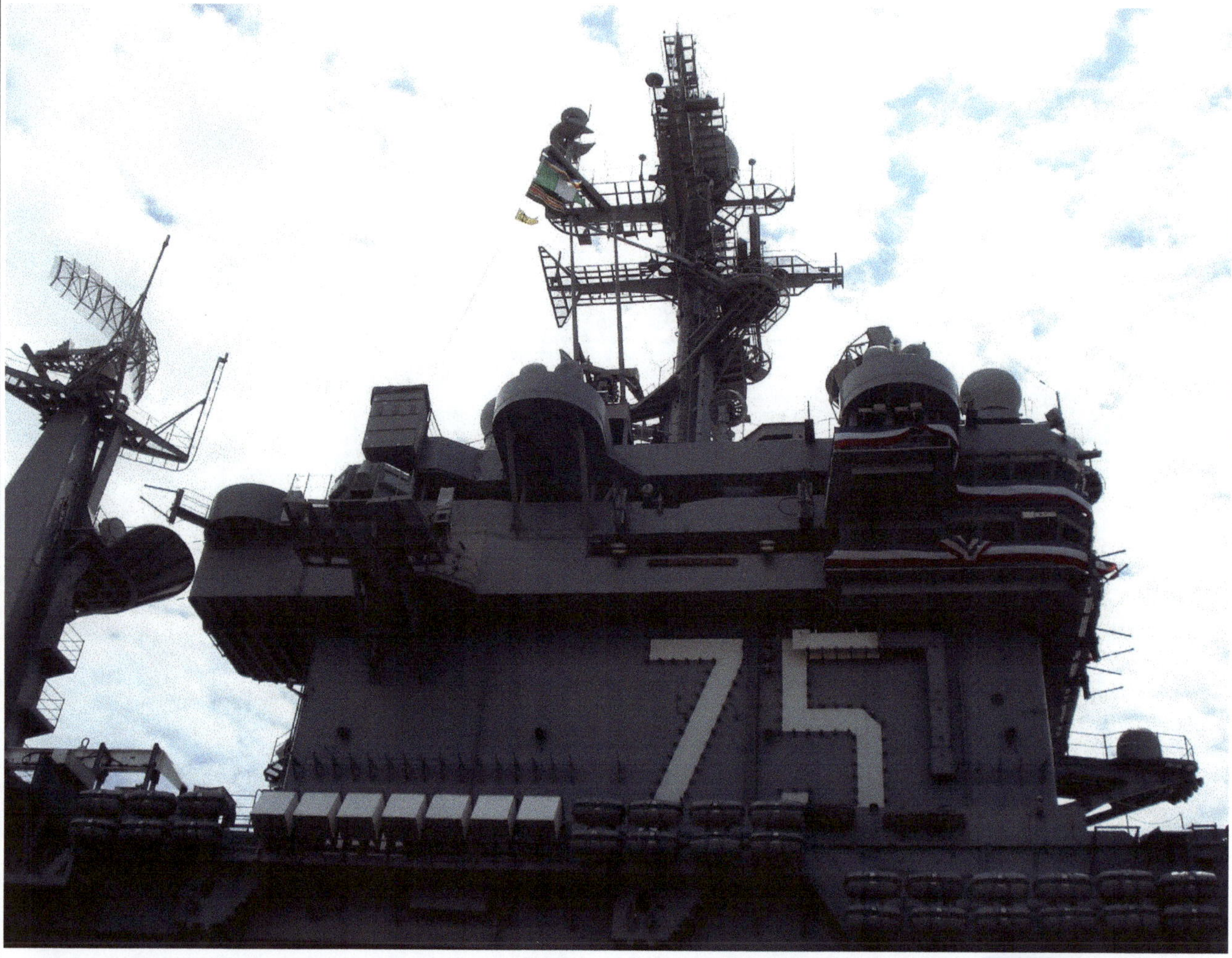

Lieutenant Lieu, Michael W.'s 2136-2139; Military Photo Book, Navy Branch;

Department; United States Department Of War (U.S. DOW)
Branch; Navy. **Year; Month (XXX)** = 2137; 02 (Feb),
Component; "Military (M) Active Duty, Servicemember Check In"

" Those broadcasting antennas and receivers!

I can most likely stare all day, gotta report in to the Officer Of Deck, O.O.D...

If Sisco, made Electronics Technician specialty he would be troubleshooting those broadcasting as a technician, his training takes about one year and a half long...

What if alternative history, in the fleet as engineers and technicians..."

(Military) (Photo-Book) U.S.N. Sailor (Spy) (Super-Hero) (**Vigilante**) (War)

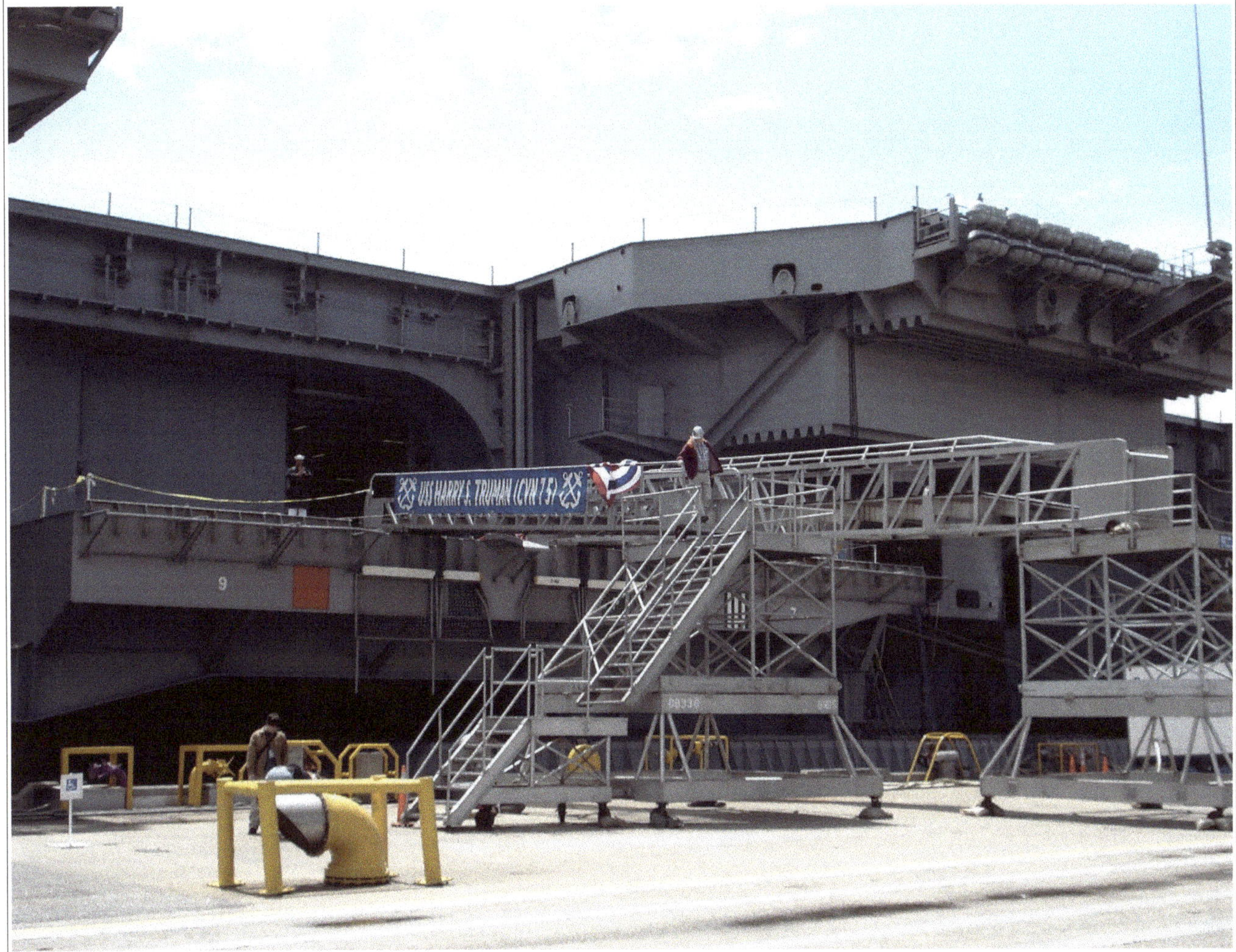

Lieutenant Lieu, Michael W.'s 2136-2139; Military Photo Book, Navy Branch;

Department; United States Department Of War (U.S. DOW)
Branch; Navy. **Year; Month (XXX)** = 2137; 02 (Feb),
Component; "Military (M) Active Duty, Servicemember Check In"

" C.V.N. seventy five... C.V. means Conventional. N. means Nuclear...

U.S.S. Harry S. Truman, war ship named after that President who took over President, F.D.R., Franklin D. Roosevelt who flatlined in office while serving.

On top of that, President Truman deter foreign foes from attacking United **States** of North America during World War Two that lead to a long Kold War and Mutual Assured Destruction, MAD, and space force of nineteen fifties...

(Military) (Photo-Book) U.S.N. Sailor (Spy) (Super-Hero) (**Vigilante**) (War) (**Political**) **British** His / Her Majesty's Ship (H.M.S.) Royal Navy War Ship.

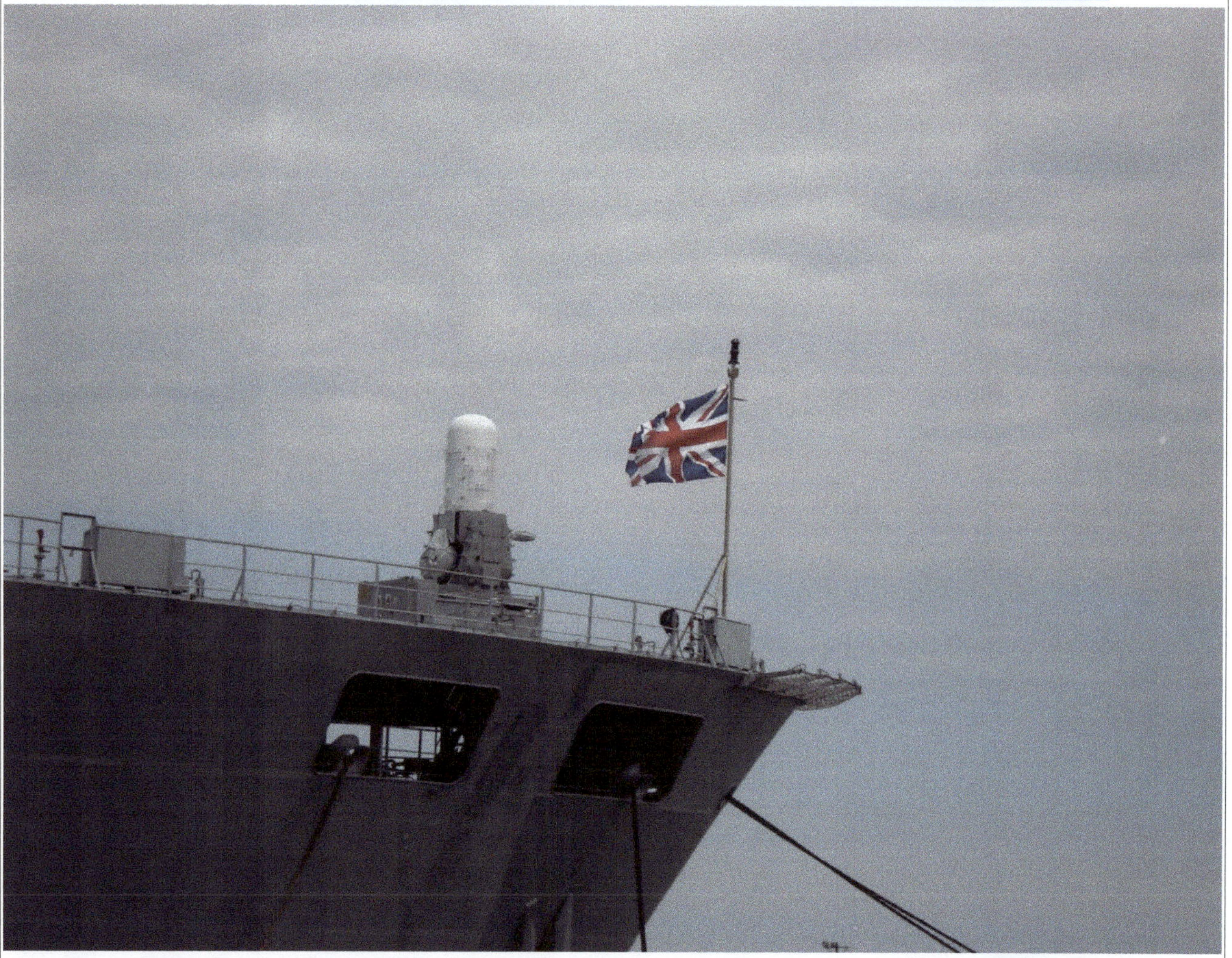

Lieutenant Lieu, Michael W.'s 2136-2139; Military Photo Book, Navy Branch;

Department; United States Department Of War (U.S. DOW)
Branch; Navy. **Year; Month (XXX)** = 2137; 02 (Feb),
Component; "Military (M) Active Duty, Servicemember Check In"

" The **British** are coming! The **British** are coming!! No, The **British** are here! The **British** are here! Station next to our war ship Truman...

As part of Earth's Global Government, EGG, it is interesting to see other surviving countries' war ships and flags being waived around...

Joint Task Force does has an interesting way to unite the survivors as one."

(Military) (Photo-Book) U.S.N. Sailor (Spy) (Super-Hero) (**Vigilante**) (War)

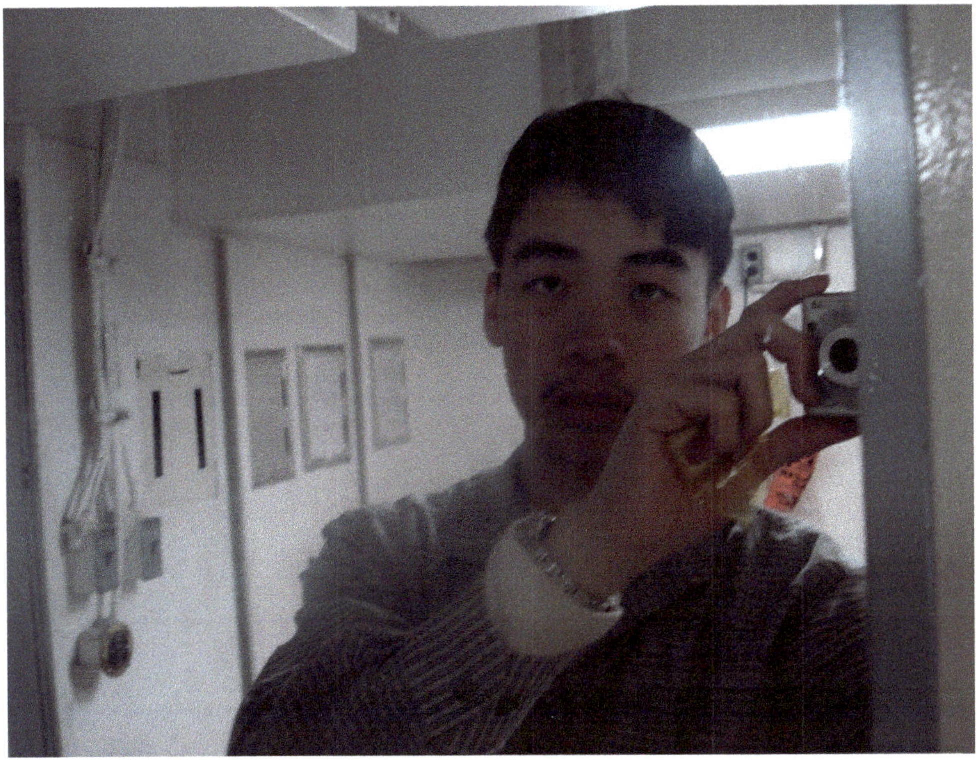

Lieutenant Lieu, Michael W.'s 2136-2139; Military Photo Book, Navy Branch;

Department; United States Department Of War (U.S. DOW)
Branch; Navy. **Year; Month (XXX)** = 2137; 02 (Feb),
Component; "Military (M) Active Duty, Servicemember Check In"

" Truman's O.O.D. says I am to report, and check in to Deck Department... since my military specialty rate is now undesignated, undes, going to be working along side sailors of Boatswain's Mate, B.M..

From Engineering Department Machinist's Mates to Deck Department Undesignated, best way to explain what undesignated is, no specialty... any and all duties as assigned per military managers orders, a janitor in the fleet...

Here to do what I can in public sec military, what an enlistment this is.

Gonna take a photo of myself before I head on out and explore this war ship.

Sisco is already here and been settled as since he did not take any leave from Illinois Base transfer, we got much to catch up and since I was two weeks off."

(Military) (Photo-Book) U.S.N. Sailor (Spy) (Super-Hero) (**Vigilante**) (War)

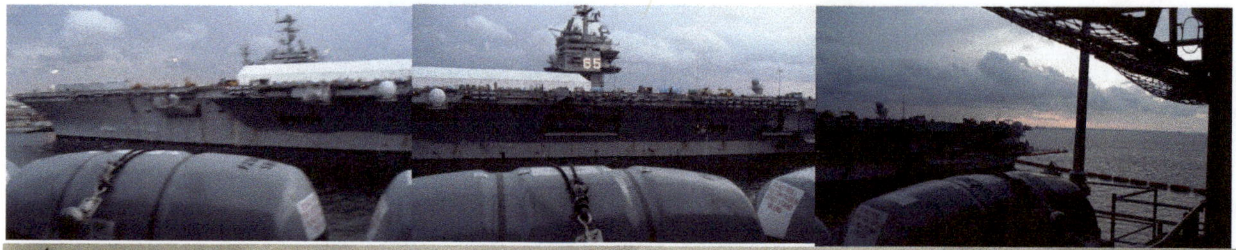

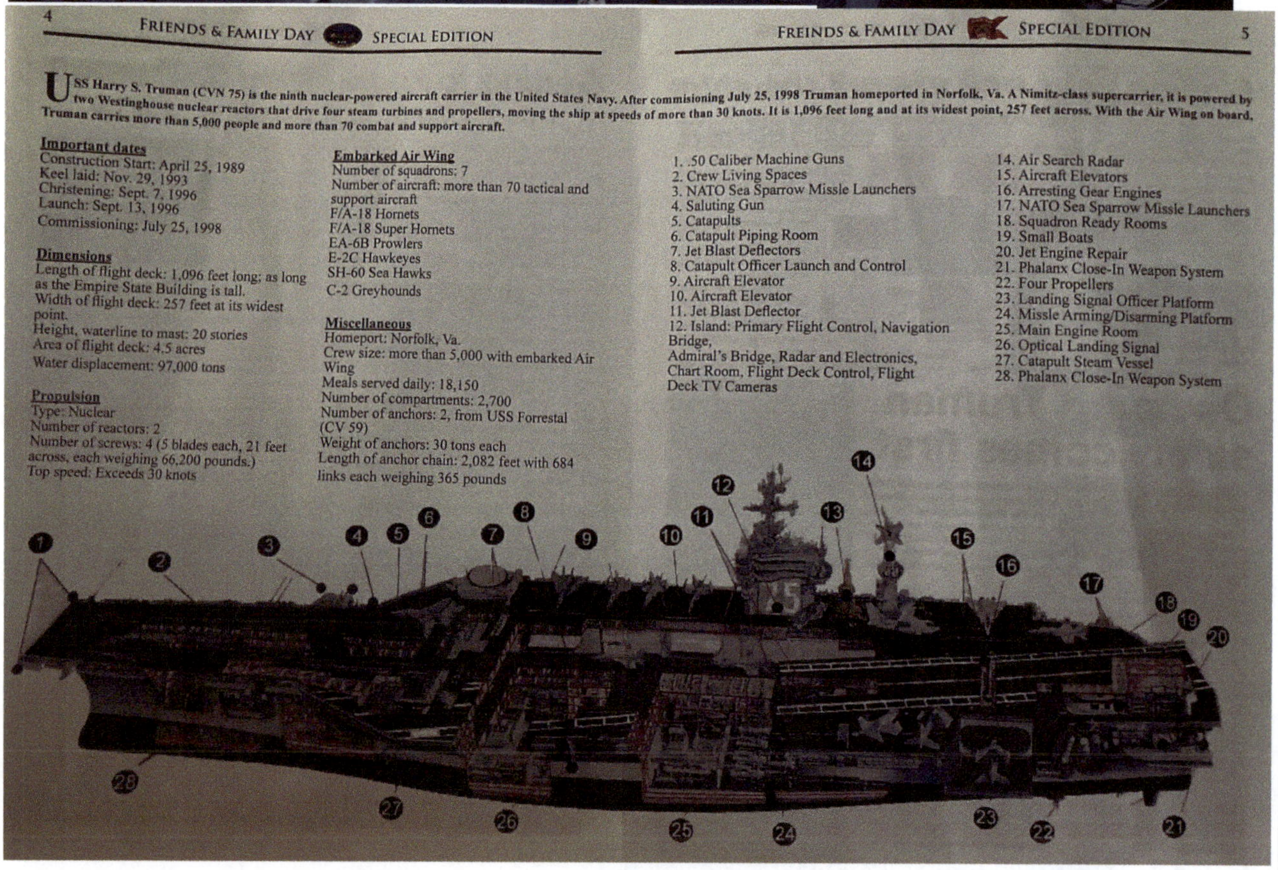

4 FRIENDS & FAMILY DAY SPECIAL EDITION

FREINDS & FAMILY DAY SPECIAL EDITION 5

USS Harry S. Truman (CVN 75) is the ninth nuclear-powered aircraft carrier in the United States Navy. After commisioning July 25, 1998 Truman homeported in Norfolk, Va. A Nimitz-class supercarrier, it is powered by two Westinghouse nuclear reactors that drive four steam turbines and propellers, moving the ship at speeds of more than 30 knots. It is 1,096 feet long and at its widest point, 257 feet across. With the Air Wing on board, Truman carries more than 5,000 people and more than 70 combat and support aircraft.

Important dates
Construction Start: April 25, 1989
Keel laid: Nov. 29, 1993
Christening: Sept. 7, 1996
Launch: Sept. 13, 1996
Commissioning: July 25, 1998

Dimensions
Length of flight deck: 1,096 feet long; as long as the Empire State Building is tall.
Width of flight deck: 257 feet at its widest point.
Height, waterline to mast: 20 stories
Area of flight deck: 4.5 acres
Water displacement: 97,000 tons

Propulsion
Type: Nuclear
Number of reactors: 2
Number of screws: 4 (5 blades each, 21 feet across, each weighing 66,200 pounds.)
Top speed: Exceeds 30 knots

Embarked Air Wing
Number of squadrons: 7
Number of aircraft: more than 70 tactical and support aircraft
F/A-18 Hornets
F/A-18 Super Hornets
EA-6B Prowlers
E-2C Hawkeyes
SH-60 Sea Hawks
C-2 Greyhounds

Miscellaneous
Homeport: Norfolk, Va.
Crew size: more than 5,000 with embarked Air Wing
Meals served daily: 18,150
Number of compartments: 2,700
Number of anchors: 2, from USS Forrestal (CV 59)
Weight of anchors: 30 tons each
Length of anchor chain: 2,082 feet with 684 links each weighing 365 pounds

1. .50 Caliber Machine Guns
2. Crew Living Spaces
3. NATO Sea Sparrow Missle Launchers
4. Saluting Gun
5. Catapults
6. Catapult Piping Room
7. Jet Blast Deflectors
8. Catapult Officer Launch and Control
9. Aircraft Elevator
10. Aircraft Elevator
11. Jet Blast Deflector
12. Island: Primary Flight Control, Navigation Bridge, Admiral's Bridge, Radar and Electronics, Chart Room, Flight Deck Control, Flight Deck TV Cameras
14. Air Search Radar
15. Aircraft Elevators
16. Arresting Gear Engines
17. NATO Sea Sparrow Missle Launchers
18. Squadron Ready Rooms
19. Small Boats
20. Jet Engine Repair
21. Phalanx Close-In Weapon System
22. Four Propellers
23. Landing Signal Officer Platform
24. Missle Arming/Disarming Platform
25. Main Engine Room
26. Optical Landing Signal
27. Catapult Steam Vessel
28. Phalanx Close-In Weapon System

Lieutenant Lieu, Michael W.'s 2136-2139; Military Photo Book, Navy Branch;

Department; United States Department Of War (U.S. DOW)
Branch; Navy. **Year; Month (XXX)** = 2137; 02 (Feb),
Component; "Military (M) Active Duty, Servicemember Check In"

" Sisco already completed the fleet's indoc for new sailors checking in onboard and provided me some of the war ship's basic layouts…

Got to explore and capture some kool military photos of C. V. N. + sixty five Enterprise war ship …That aircraft carrier size is Surprise face O ; (°o°)."

(Military) (Photo-Book) U.S.N. Sailor (Spy) (Super-Hero) (**Vigilante**) (War)

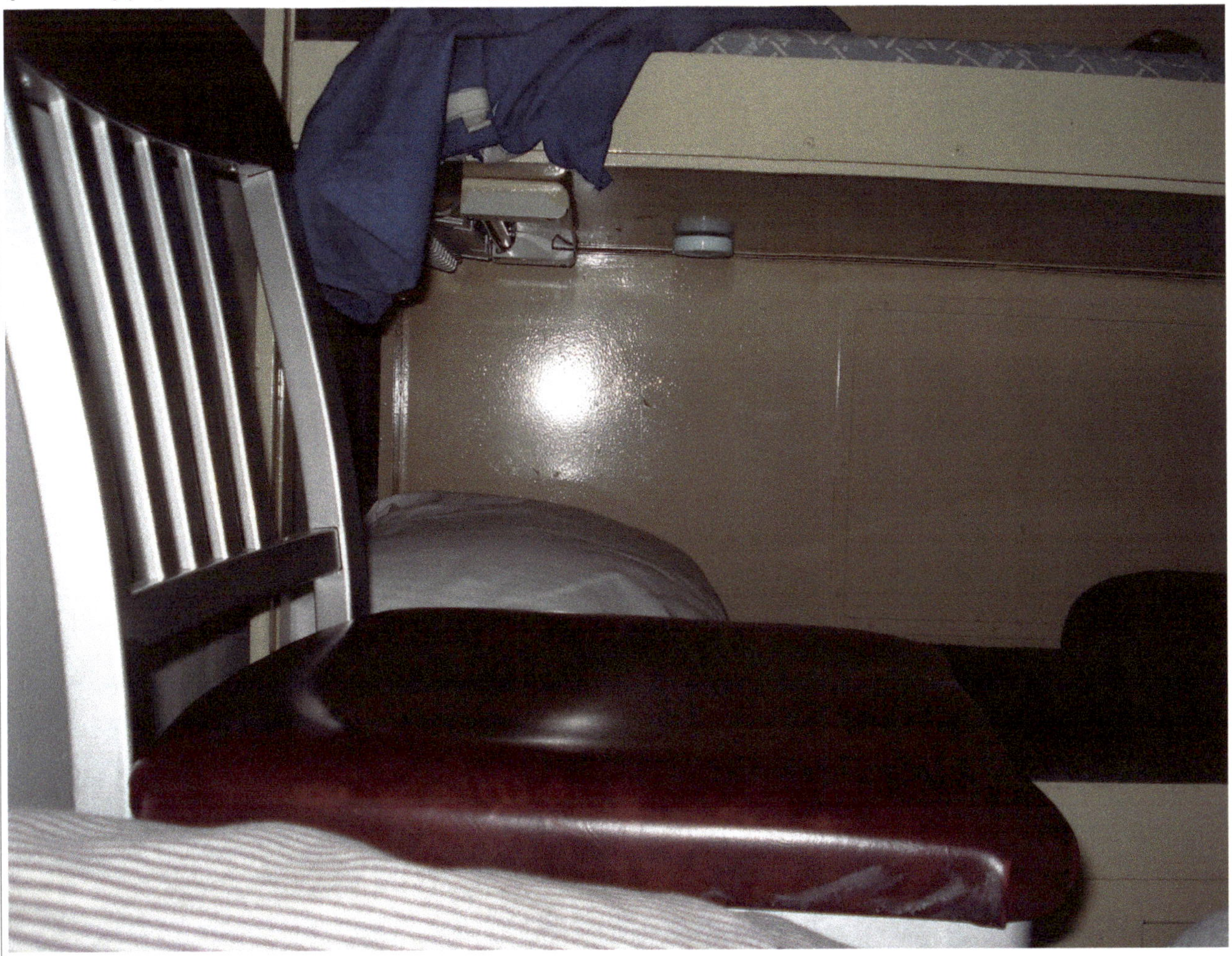

Lieutenant Lieu, Michael W.'s 2136-2139; Military Photo Book, Navy Branch;

Department; United States Department Of War (U.S. DOW)
Branch; Navy. **Year; Month (XXX)** = 2137; 02 (Feb),
Component; "Military (M) Active Duty, Servicemember Check In"

" This is my resting quarters, military Navy calls it "berthing" and this is my rack, I am going to be stationed at for the time I am in the fleet… that chair in between gives us a good perspective on how much space I have when resting.

Fresh **white new pillow case** and black warm wool blanket…

Got to prep up for my indoc dates soon… Sisco says this Deck Department is kind of a Dick Department, consequence for failing engineering, tech school."

(Military) (Photo-Book) U.S.N. Sailor (Spy) (Super-Hero) (**Vigilante**) (War)

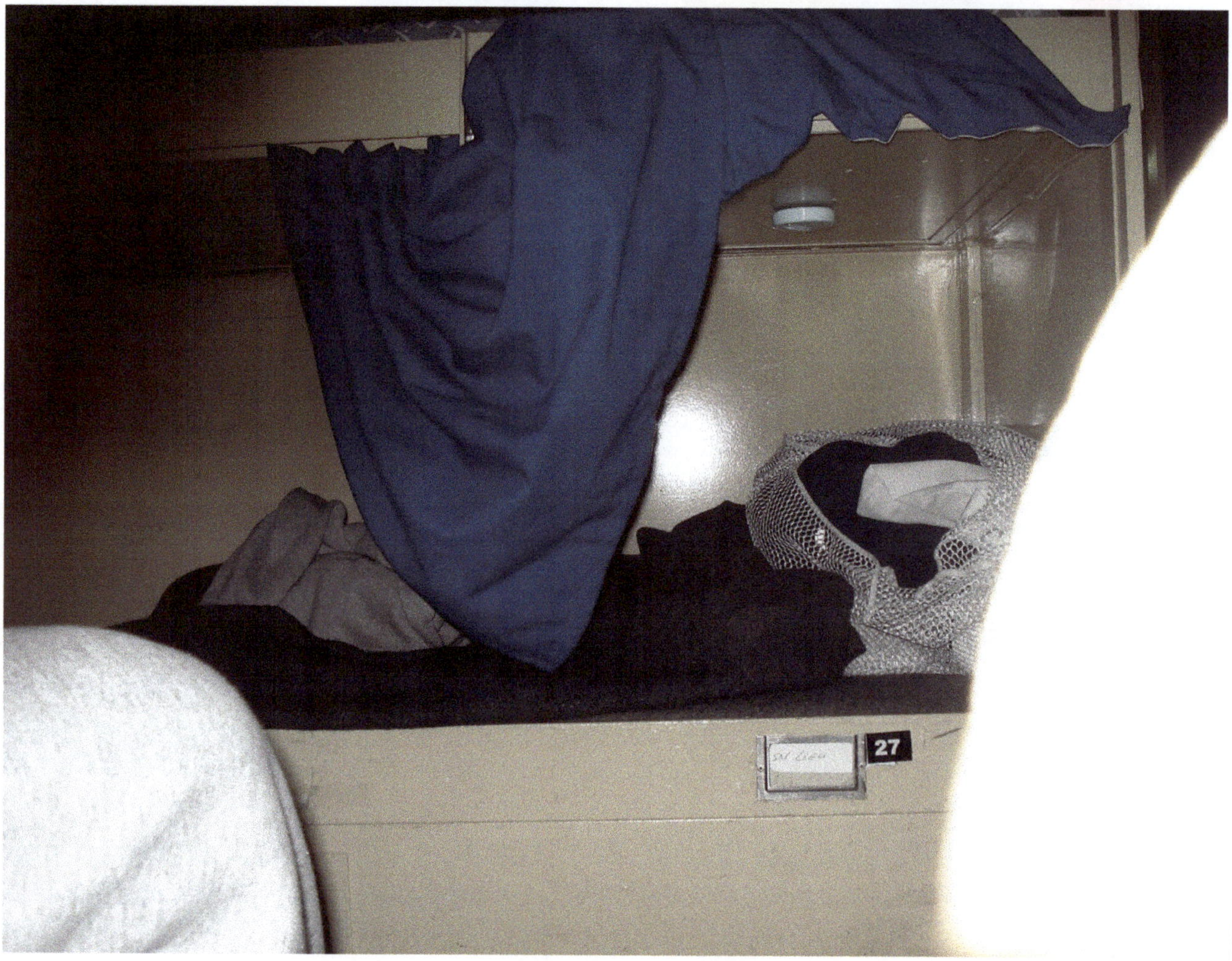

Lieutenant Lieu, Michael W.'s 2136-2139; Military Photo Book, Navy Branch;

Department; United States Department Of War (U.S. DOW)
Branch; Navy. **Year; Month (XXX)** = 2137; 02 (Feb),
Component; "Military (M) Active Duty, Servicemember Check In"

" Since failing out of Engineering Department as M.M., my military specialty rate is now Seaman, S.N., most specifically, Seaman Recruit, S.R., since I started as Enlisted One, E-1, sailor rank.

Got my things settled in, got to catch up with Sisco, time to see what the fleet is going to offer me in terms of public sec military Navy...

Let us see how being a sailor undesignated life style is going to be..."

(Military) (Photo-Book) U.S.N. Sailor (Spy) (Super-Hero) (**Vigilante**) (War) (Crime Drama)

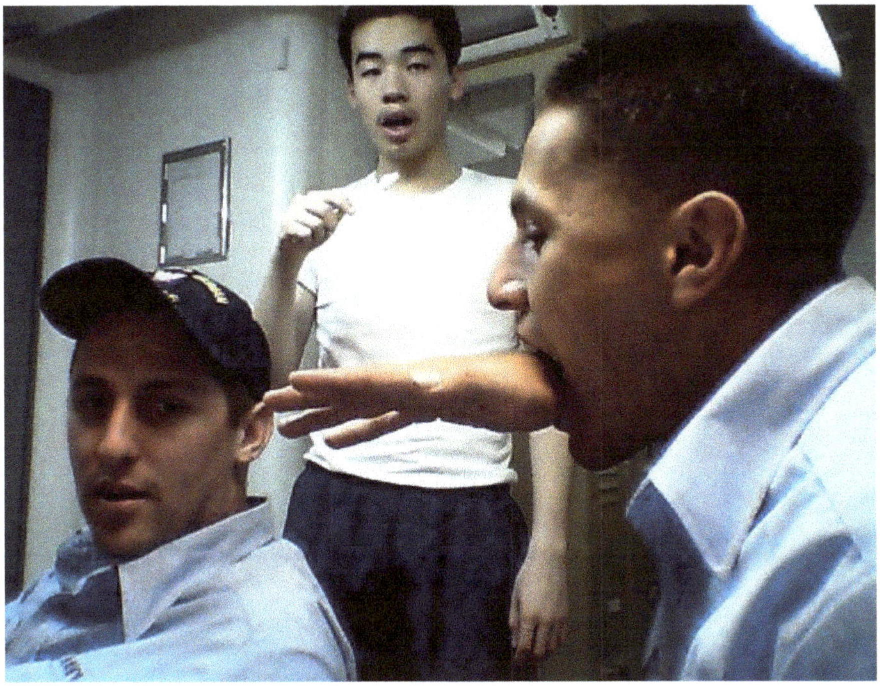

Lieutenant Lieu, Michael W.'s 2136-2139; Military Photo Book, Navy Branch;

Department; United States Department Of War (U.S. DOW)
Branch; Navy. **Year; Month (XXX)** = 2137; 02 (Feb),
Component; "Military (M) Active Duty, Servicemember Check In"

" Meeting interesting people in public sec military Navy in **photo book forms**...

Good thing I going to have these military photos to remind me of the... different types of people that arrive from all across North America and surviving county bases to form the fifty five United States military Navy fleet forces plus Earth's Space Force and other **military branches** as well.

Well, better get well rested and settled in since first thing for me to get is my indoc dates so I get a better understanding of the fleet forces, as well as this aircraft carrier war ship... The Truman... "The Buck Stops Here".

NoFuck Vagina, which ever Joe Navy came up with that name for this base?

ZzZz... (-_-)"

(Military) (Photo-Book) U.S.N. Sailor (Spy) (Super-Hero) (**Vigilante**) (War) (Crime Drama)

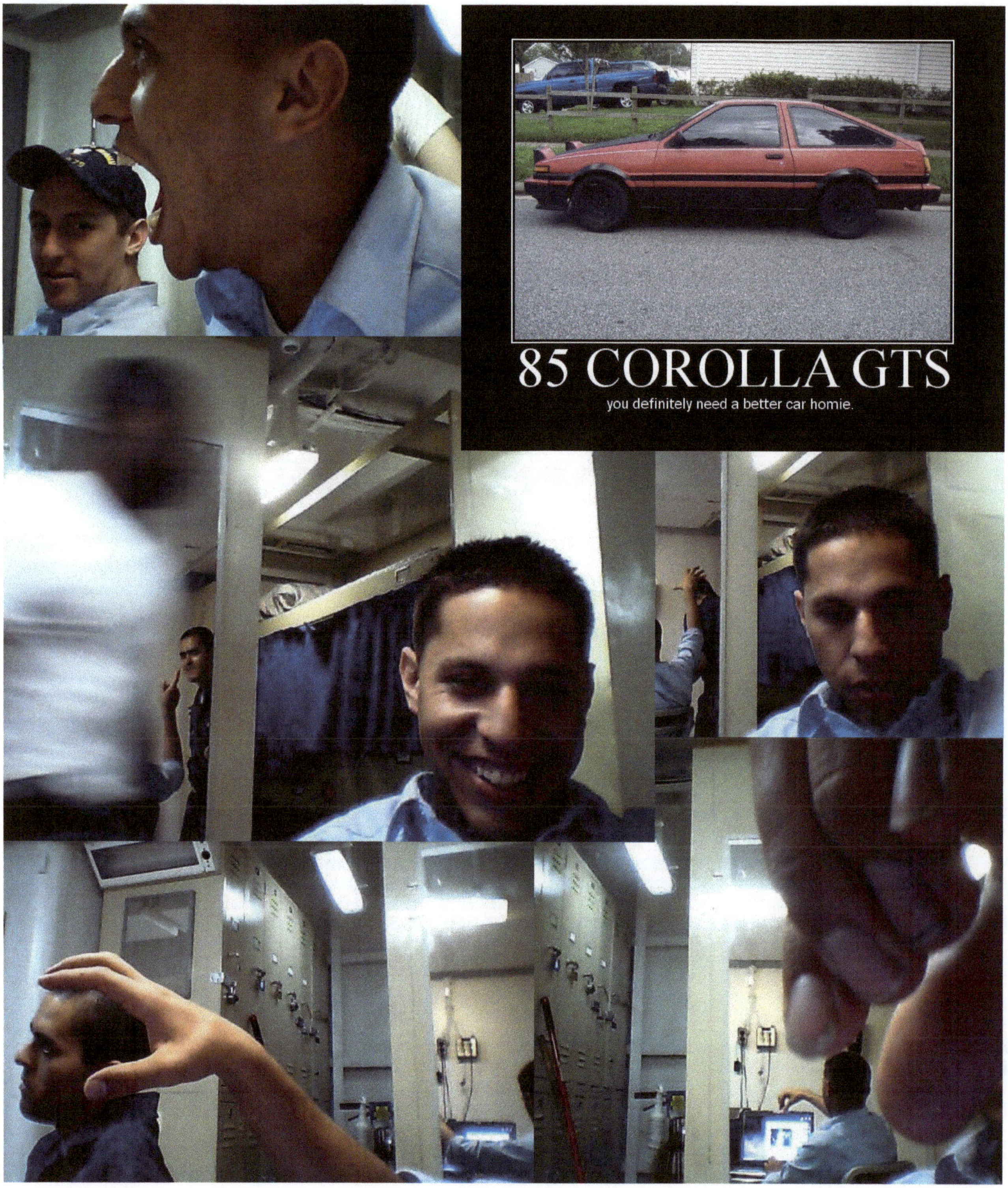

56 \ **2154;** Lieu, 廖, 'WhiteWolf 白狼' Left to Right (L→R)

(Military) (Photo-Book) U.S.N. Sailor (Spy) (Super-Hero) (**Vigilante**) (War)

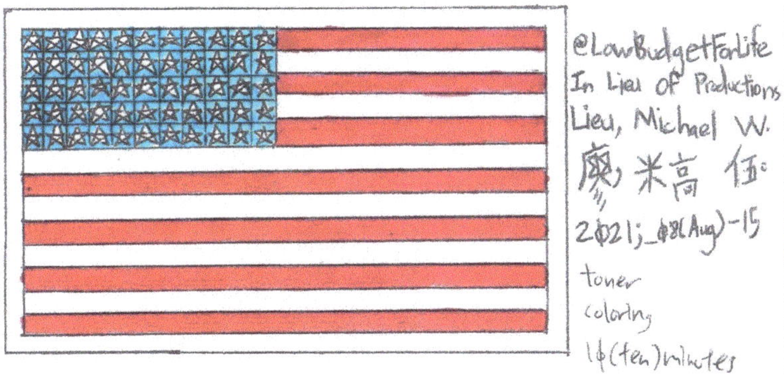

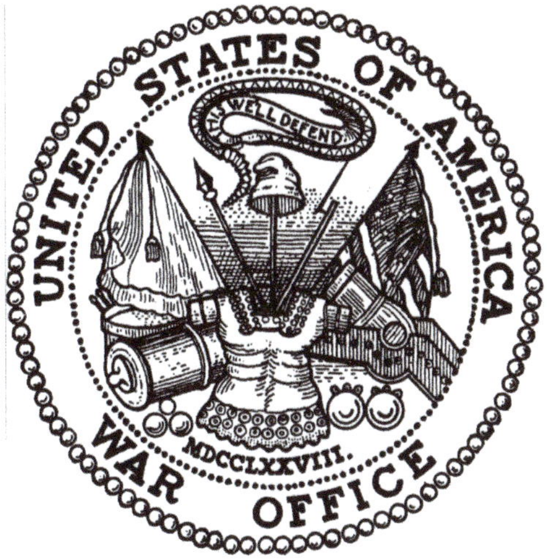

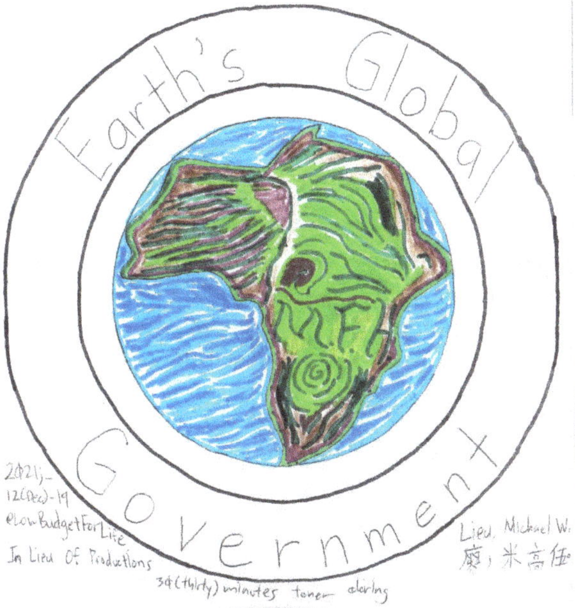

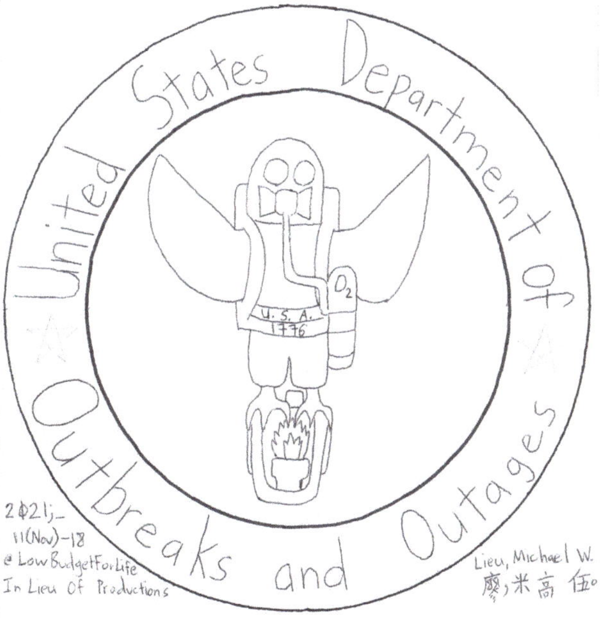

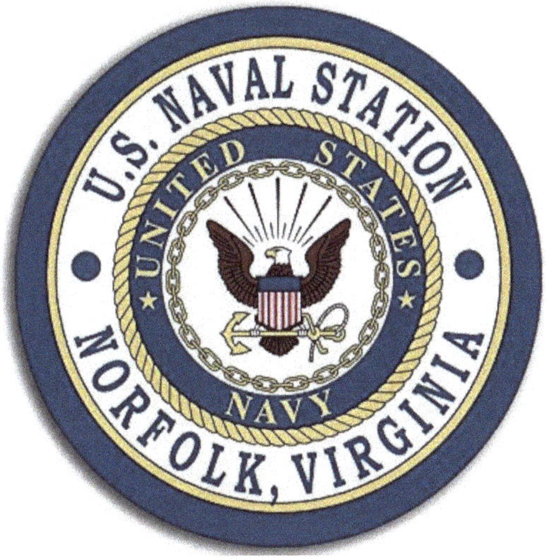

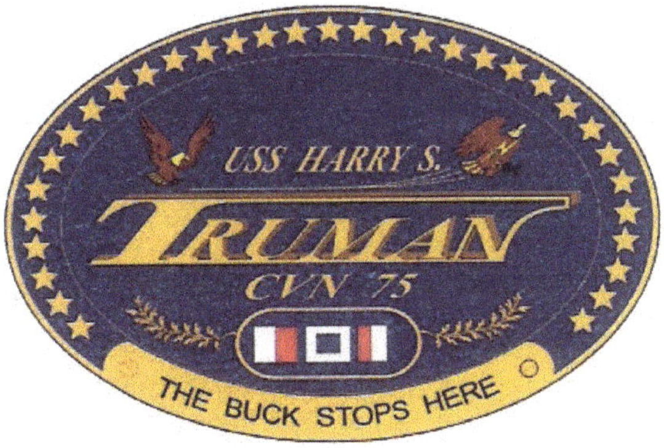

(Military) (Photo-Book) U.S.N. Sailor (Spy) (Super-Hero) (**Vigilante**) (War)

United States **Department** Of War (U.S. DOW), **Branch;** Navy AKA United States Navy (U.S.N.), **Component**; Military (M) Active Duty 2136-2139,

"EGG United States Ship (U.S.S.) Harry S. Truman Augment (C.V.N. 75+) J.T.F. Aircraft Carrier" Indoctrination "Indoc", Norfolk, Virginia (VIR), N.A.

Welcome From The Captain

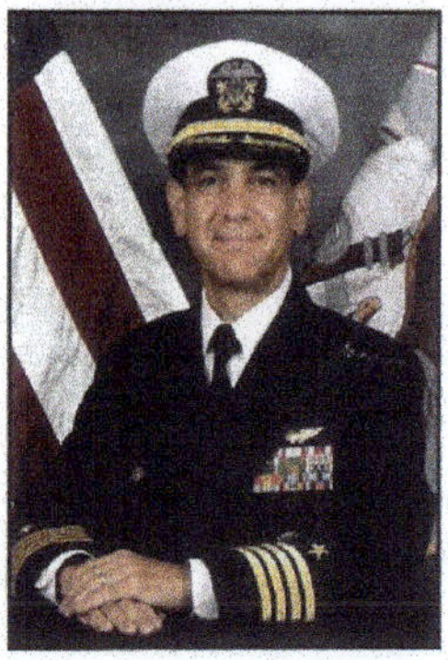

Greetings friends and families!

The long awaited "USS Harry S. Truman (CVN 75) Friends and Family Day Cruise" is finally here. This is perhaps the single greatest day at sea for our Sailors because it gives us the opportunity to show you, our friends and families, what we proudly do each day. In addition to having the unique opportunity to spend a day at sea onboard America's finest aircraft carrier, we will be offering plenty to keep you busy. Today's events include an impressive Air Power Demonstration by our friends in Carrier Air Wing Three, tours of the ship, a DJ and a host of other activities for the entire family.

HST spent nearly all of 2006 in Norfolk Naval Shipyards undergoing renovations and upgrades to ensure she is combat ready and prepared for deployment. With the ship and air wing spending the coming months continuing to sharpen our skills, we will be spending many more weeks at sea. This is our opportunity to show off this magnificent ship and superb crew to our families and friends, and even more importantly, show you firsthand how we are standing-by to answer the call of the Commander-in-Chief on short notice to defend freedom and U.S. interests around the world.

I can assure you that this crew has met or exceeded every challenge our operational commanders have placed on us. I've never served with men and women who show so much pride in what they do. It's an inspiration to me personally and a source of motivation to me professionally. Your support is paramount to that attitude. Taking care of business on the home front allows your Sailor to focus on the important work at sea. For that support, I thank you.

Please take advantage of the "HST Friends and Family Day Cruise" to learn a little more about the ship and the men and women who serve her. This is your opportunity to ask a lot of questions, take a lot of pictures and explore this great vessel.

Once again, thank you for your support and welcome aboard!

Captain Herman Shelanski
Commanding Officer
USS Harry S. Truman (CVN 75)

Rainbow Warriors

The flight deck of an aircraft carrier is a Technicolor ballet involving hundreds of people performing intricate moves in a dangerous, ever-changing environment. To avoid any possible confusion, Sailors working on the flight deck wear jerseys color coded to their job.

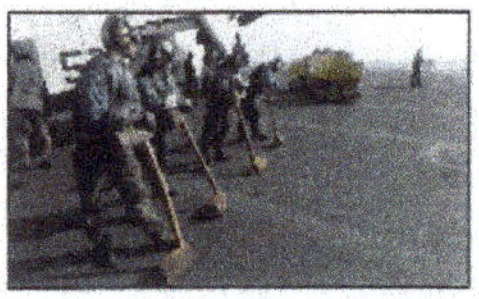

Blue Shirts
-Plane Handlers
-Tractor Drivers
-Aircraft Elevator Operators
-Messengers
-Phone Talkers

Brown Shirts
-Air Wing Plane Captains
-Air Wing Leading Petty Officers

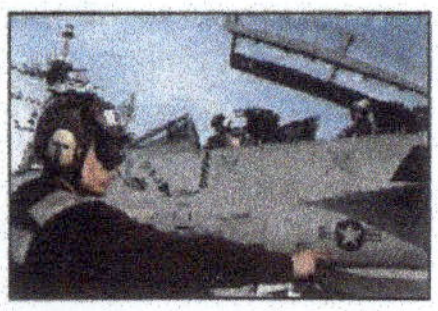

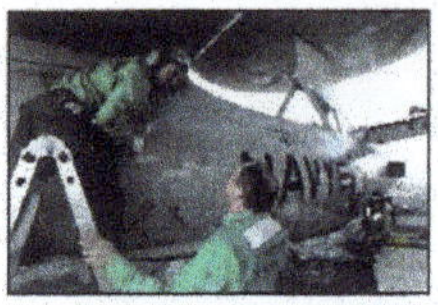

Green Shirts
-Catapult and Arresting Gear
-Air Wing Maintenance
-Photographers
-Air Wing Quality Control
-Cargo Handlers
-Ground Support Equipment
-Hook Runners
-Helicopter Landing Signal Enlisted

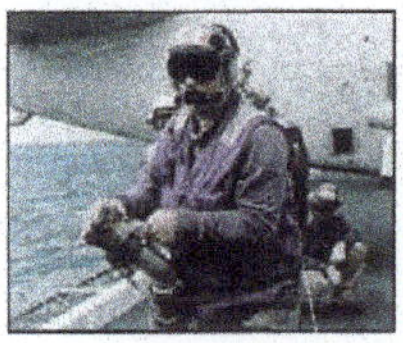

Purple Shirts
-Aviation Fuels

Red Shirts
-Ordnance
-Crash and Salvage
-Explosive Ordnance
-Disposal

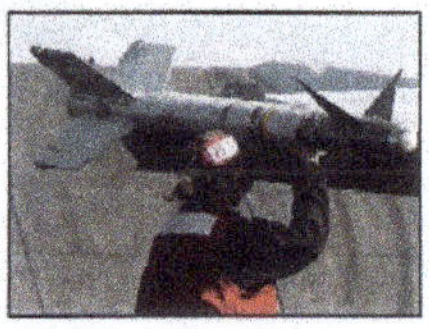

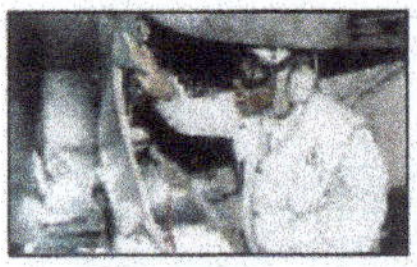

White Shirts
-Landing Signal Officer
-Safety Observers
-Medical
-Air Transfer Officers
-Liquid Oxygen Crews

Yellow Shirts
-Plane Directors
-Catapult and Arresting Gear
-Officers
-Aircraft Handling Officers

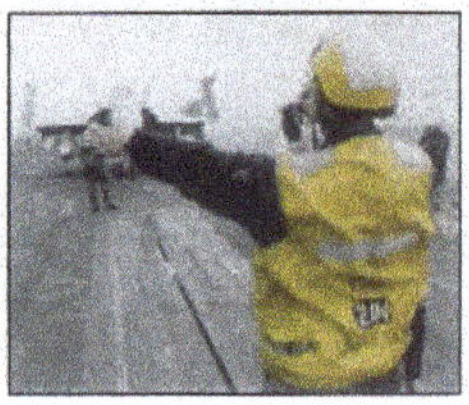

(Military) (Photo-Book) U.S.N. Sailor (Spy) (Super-Hero) (**Vigilante**) (War)

Help Keep Truman FOD Free

FOD is a Navy acronym for Foreign Object Damage or Debris. FOD is anything that could cause damage by getting sucked into a piece of equipment. It can be as simple as a paper cup clogging a commode or a hairpin that's sucked into a jet intake and damages the engine.

FOD can cause problems that can range in cost from a few cents to millions of dollars. It not only can prevent us from performing our mission efficiently, but it can affect you, the taxpayer, as well.

Therefore, we ask you to please avoid having food or drinks around electronic equipment and be careful not to leave anything unsecured, especially where it might get away from you. We take pride in our ship. Please make sure all FOD stays off the flight deck and hangar bay.

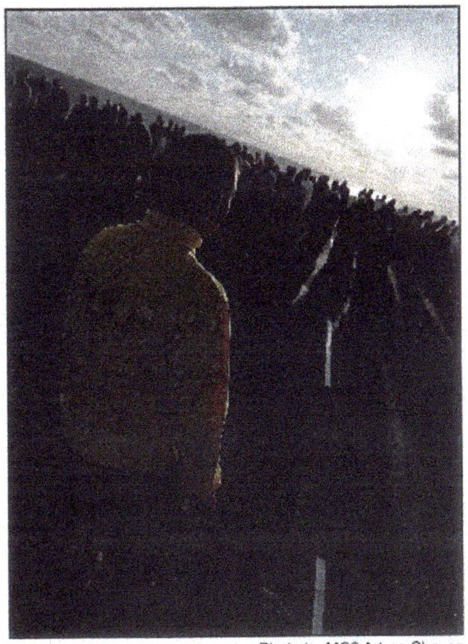

Photo by MC2 Arturo Chavez

Sailors onboard HST participate in a FOD walk-down, an evolution where they collect FOD and properly dispose of it.

GIVE 'EM HELL HERALD

COMMANDING OFFICER	Capt. Herm Shelanski
EXECUTIVE OFFICER	Cmdr. Ronald Reis
COMMAND MASTER CHIEF	CMDCM(AW/SW) Clarence Frye
PUBLIC AFFAIRS OFFICER	Lt. Matthew Conerly
PHOTOGRAPHIC OFFICER	Ens. Dave Lloyd
LEADING CHIEF PETTY OFFICER	MCC(AW/SW) Gil Dobison
PRODUCTION MANAGER	MCC(SW/AW) Michael Roberts
EDITOR	MC3 Michael Wilken

Views and opinions do not necessarily reflect those of the U.S. government or U.S. Navy. All articles received are subject to editing for content.

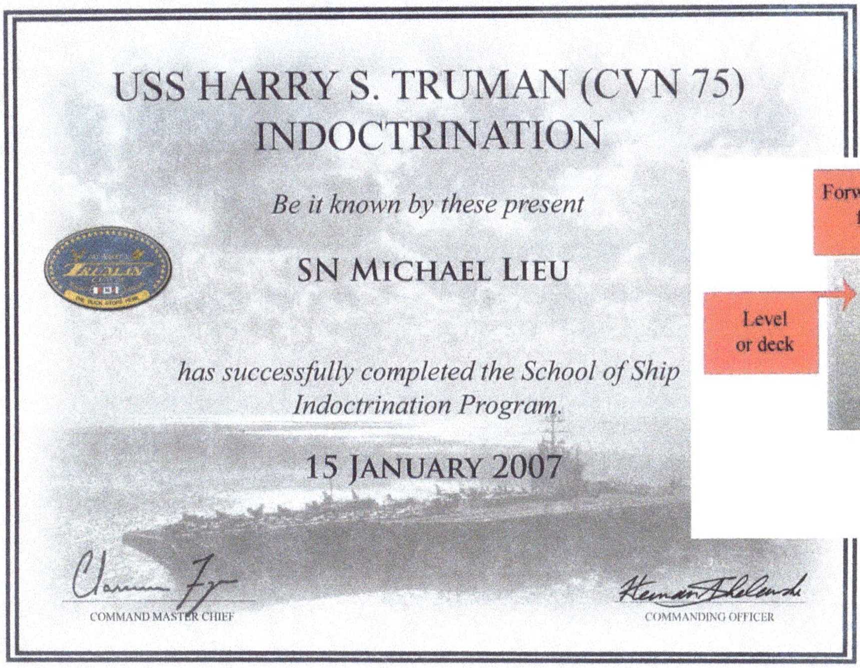

USS HARRY S. TRUMAN (CVN 75)
INDOCTRINATION

Be it known by these present

SN MICHAEL LIEU

has successfully completed the School of Ship Indoctrination Program.

15 JANUARY 2007

COMMAND MASTER CHIEF

COMMANDING OFFICER

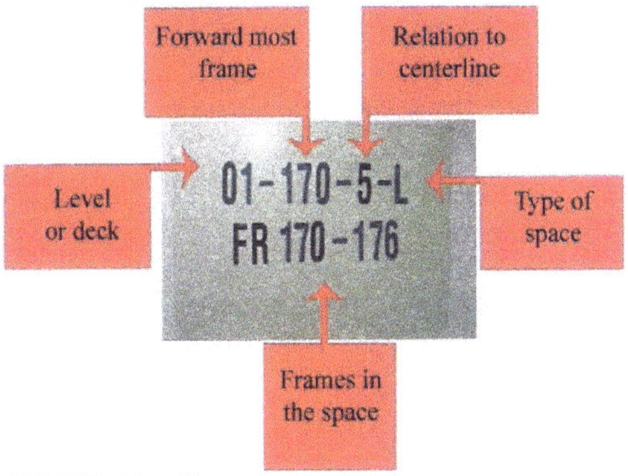

(Military) (Photo-Book) U.S.N. Sailor (Spy) (Super-Hero) (**Vigilante**) (War)

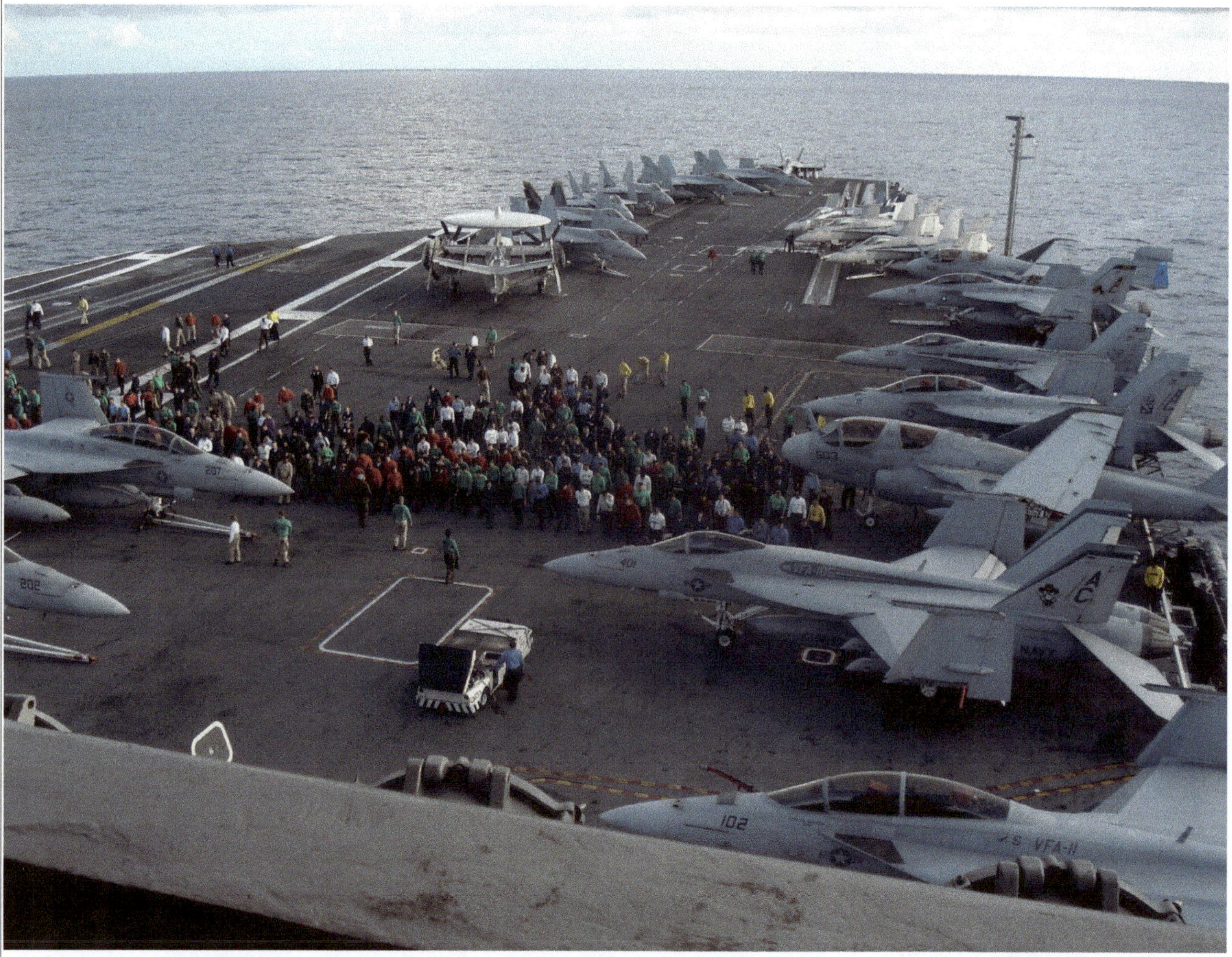

Lieutenant Lieu, Michael W.'s 2136-2139; Military Photo Book, Navy Branch;

Department; United States Department Of War (U.S. DOW)
Branch; Navy. **Year; Month (XXX)** = 2137; 11 (Nov) to 2138; 06 (Jun),
Component; "Military (M) Active Duty, Deployment War On Mars Campaign"

" Been a while since I updated my military photo book... This is now war deployment of twenty one thirty seven November to twenty one thirty eight June time frame... We were deployed for over half year plus on this war campaign.

Truman's Aviation Department and their rainbow warriors doing Foreign Object Debris, FOD, walk down, they look so down when looking down for FOD.

I **spy** them with my **photo** cam!"

(Military) (Photo-Book) U.S.N. Sailor (Spy) (Super-Hero) (**Vigilante**) (War)

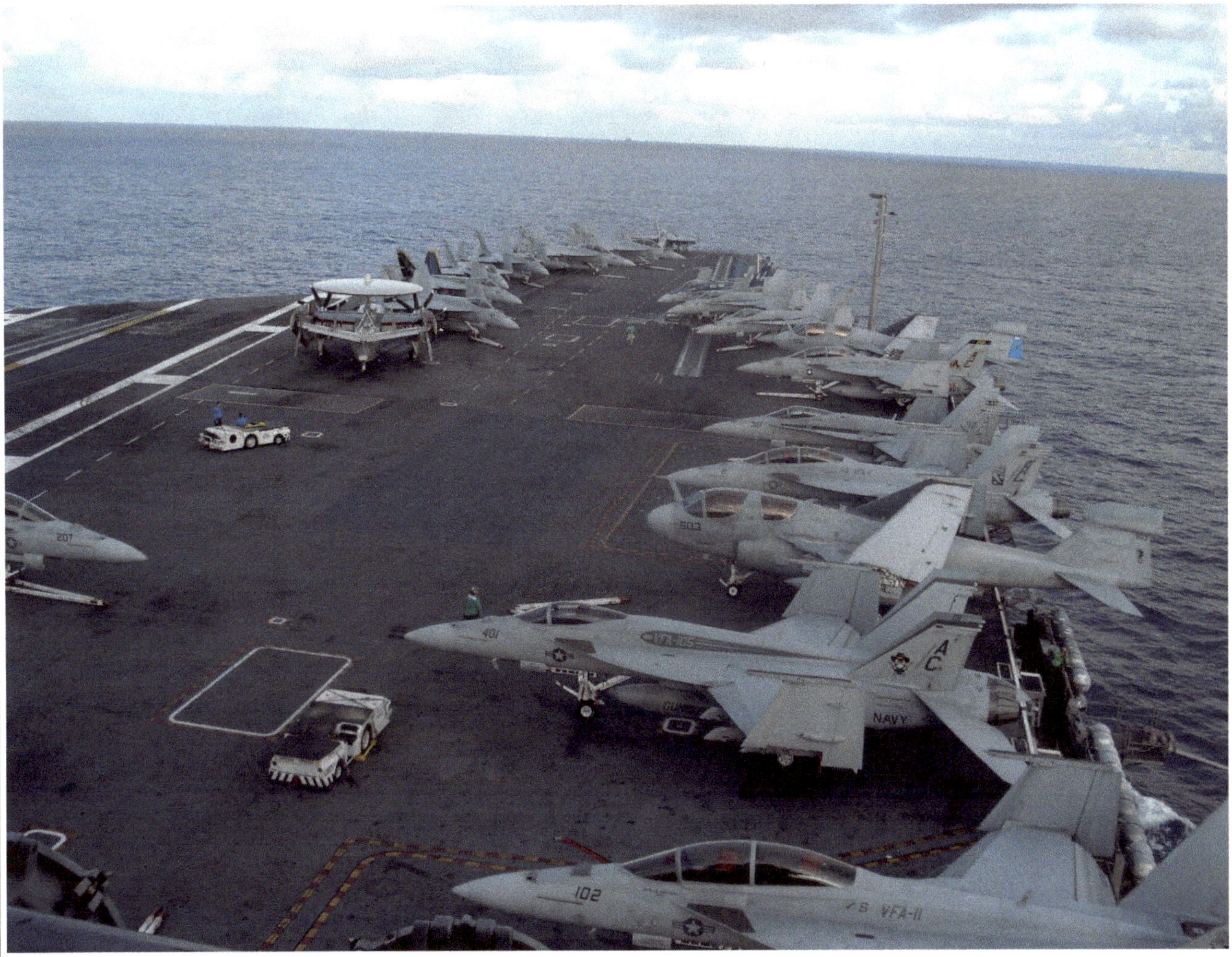

Lieutenant Lieu, Michael W.'s 2136-2139; Military Photo Book, Navy Branch;

Department; United States Department Of War (U.S. DOW)
Branch; Navy. **Year; Month (XXX)** = 2137; 11 (Nov) to 2138; 06 (Jun),
Component; "Military (M) Active Duty, Deployment War On Mars Campaign"

" Now the Truman's rainbow warriors are done and gone with FOD...

Flight deck is going to be nice and quiet, while I look out and stare at the sea, water, sky, clouds...

Wonder how are things back in Fort Angel against on coming space rocks... Meteorites bombardment, C.B.R., E.M.P. agents... Contamination is high.

It is now about one year since I been enlisted in public sec military Navy.

(Military) (Photo-Book) U.S.N. Sailor (Spy) (Super-Hero) (**Vigilante**) (War)

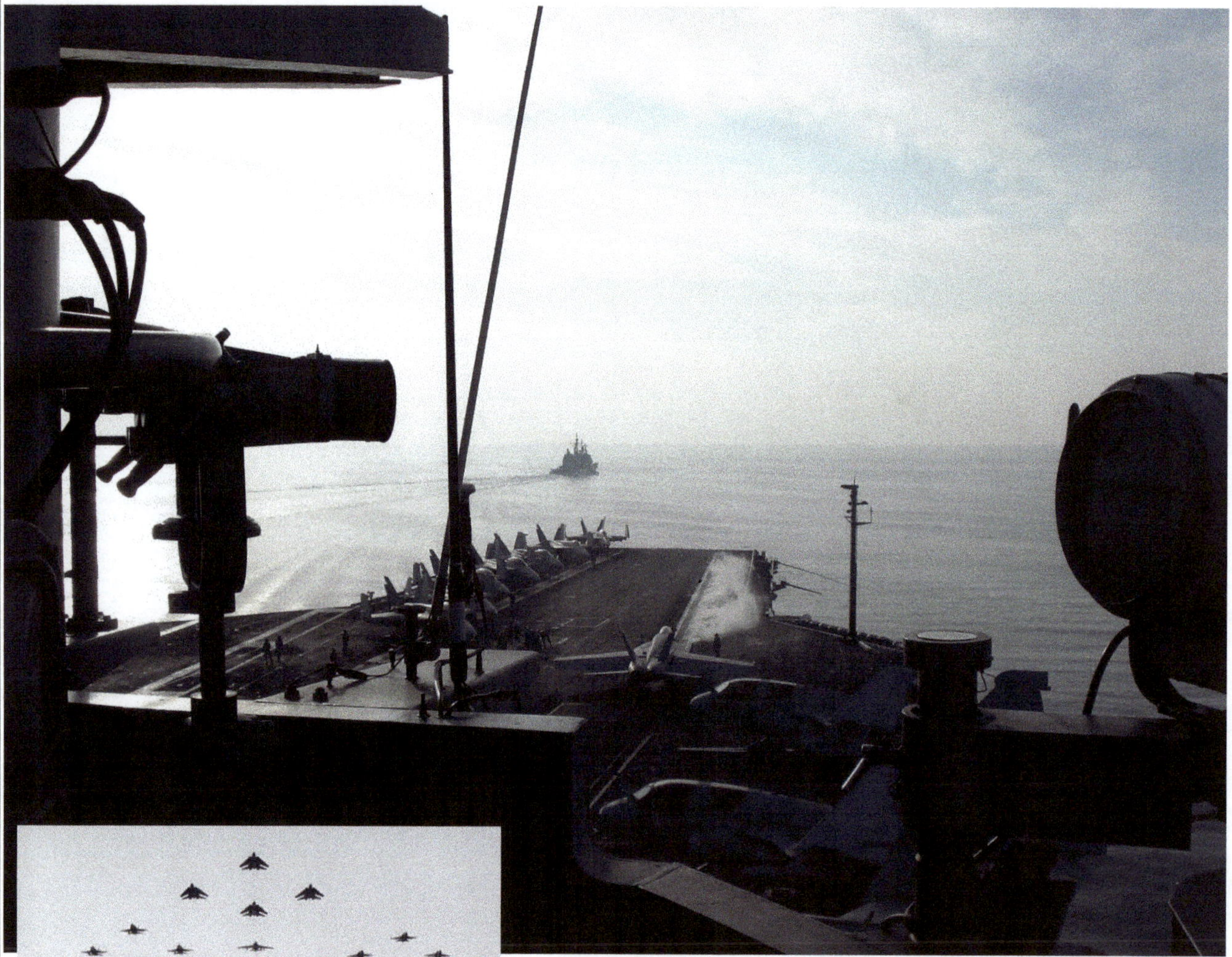

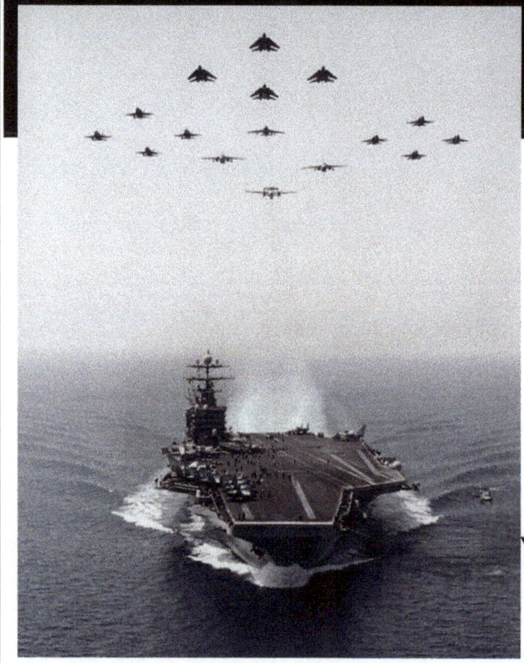

Lieutenant Lieu, Michael W.'s 2136-2139;

Military Photo Book, Navy Branch;

Department; United States Department Of War (U.S. DOW) **Branch**; Navy. **Year; Month (XXX)** = 2137; 11 (Nov) to 2138; 06 (Jun), **Component**; "Military (M) Active Duty,

Deployment War On Mars Campaign"

" A Uncle Sam war ship within our sights..."

← U.S. DOW, War Media Activity Official Photo.

(Military) (Photo-Book) U.S.N. Sailor (Spy) (Super-Hero) (**Vigilante**) (War)

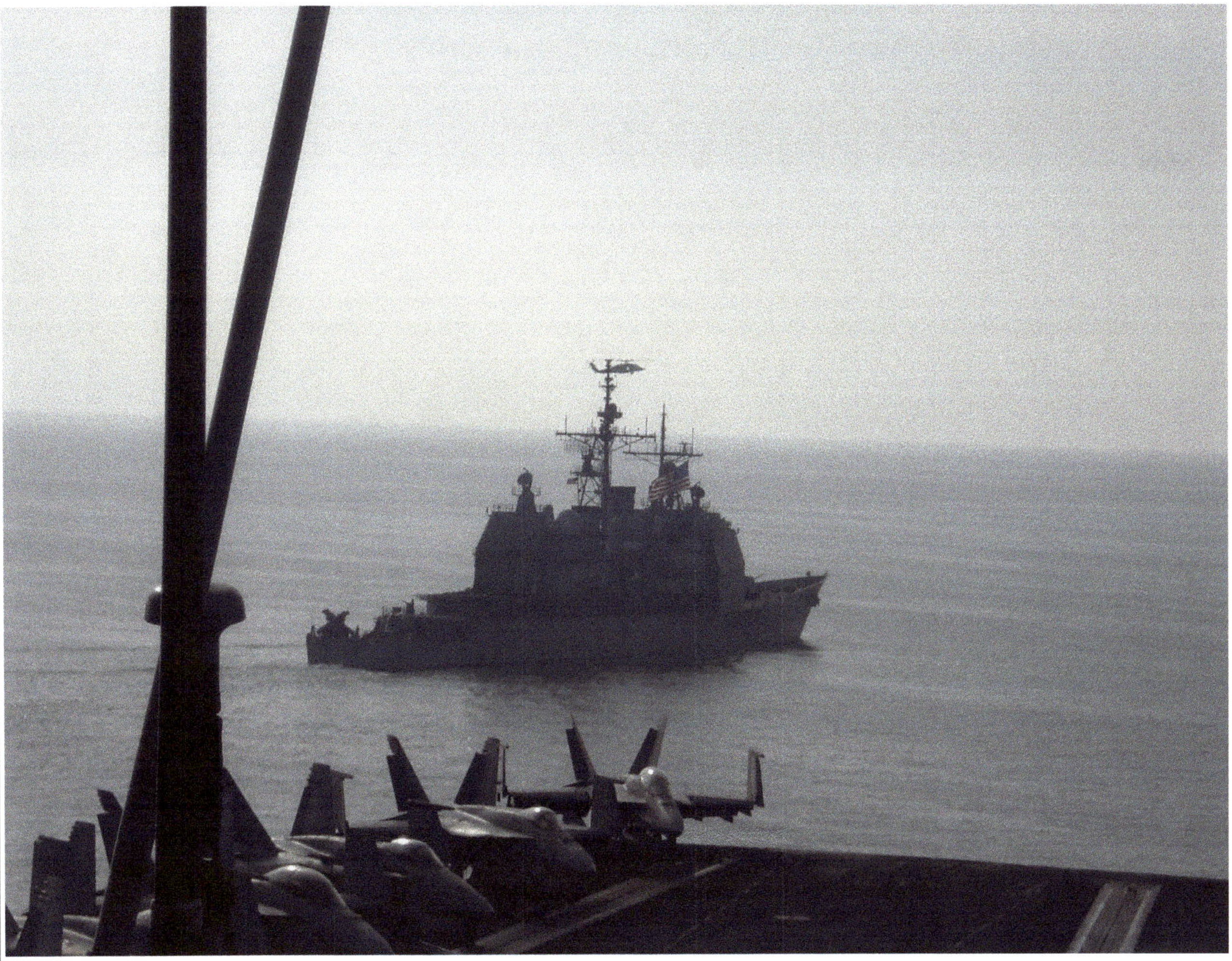

Lieutenant Lieu, Michael W.'s 2136-2139; Military Photo Book, Navy Branch;

Department; United States Department Of War (U.S. DOW)
Branch; Navy. **Year; Month (XXX)** = 2137; 11 (Nov) to 2138; 06 (Jun),
Component; "Military (M) Active Duty, Deployment War On Mars Campaign"

" Uncle Sam's flag on

Uncle Sam Ship(s)... U.S.S.,

AKA United States Ship(s),

Joe Navy Time."

(Military) (Photo-Book) U.S.N. Sailor (Spy) (Super-Hero) (**Vigilante**) (War)

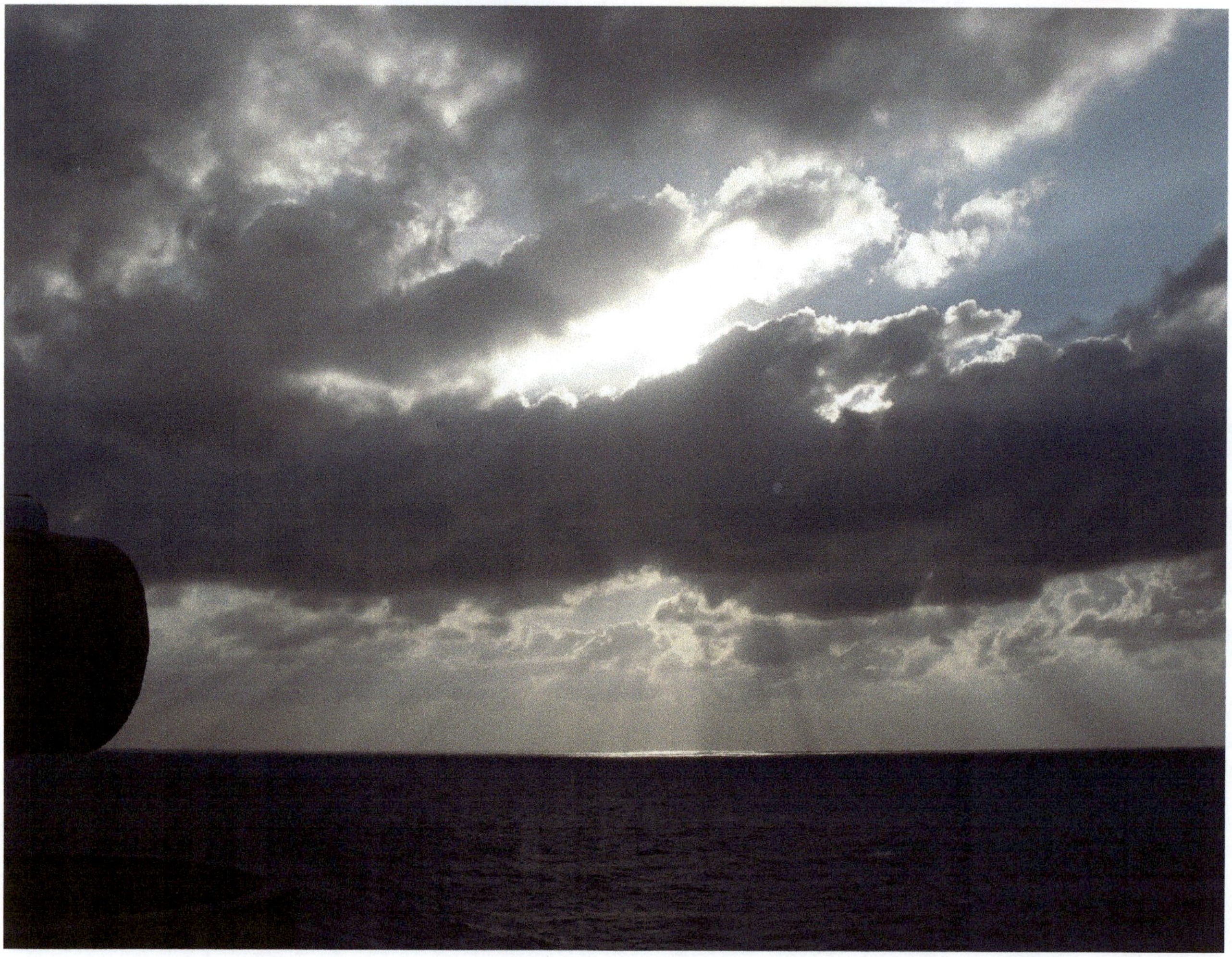

Lieutenant Lieu, Michael W.'s 2136-2139; Military Photo Book, Navy Branch;

Department; United States Department Of War (U.S. DOW)
Branch; Navy. **Year; Month (XXX)** = 2137; 11 (Nov) to 2138; 06 (Jun),
Component; "Military (M) Active Duty, Deployment War On Mars Campaign"

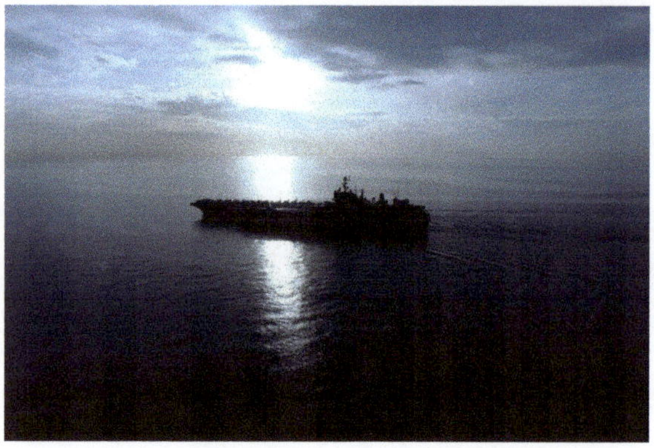

U.S. DOW, War Media Activity, Official Photo(s)← →

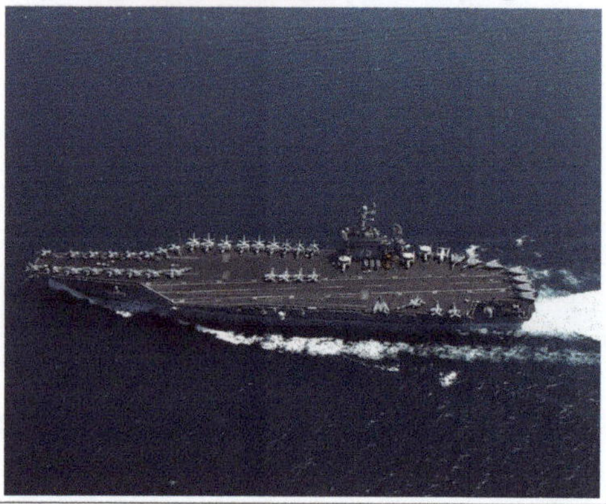

(Military) (Photo-Book) U.S.N. Sailor (Spy) (Super-Hero) (**Vigilante**) (War)

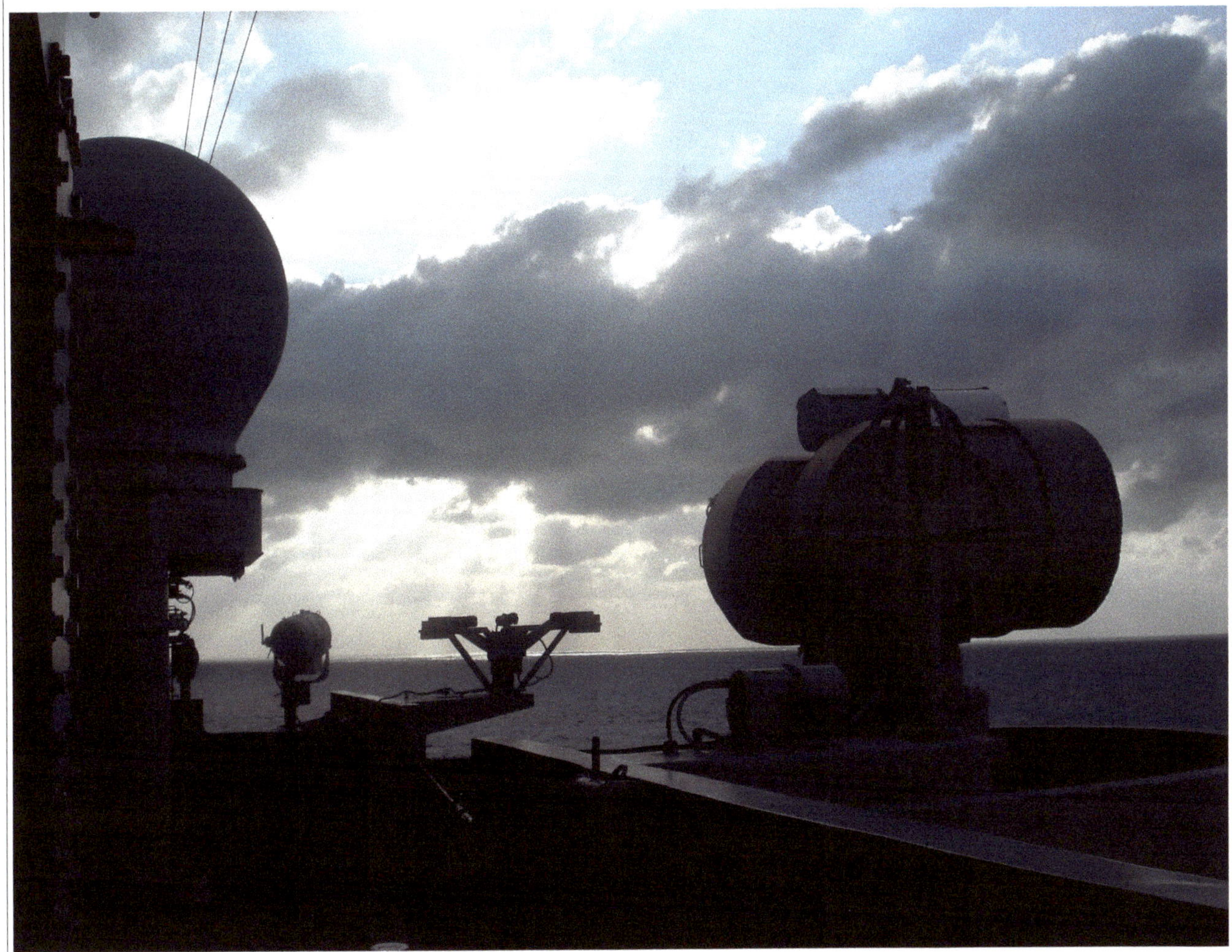

Lieutenant Lieu, Michael W.'s 2136-2139; Military Photo Book, Navy Branch;

Department; United States Department Of War (U.S. DOW)
Branch; Navy. **Year; Month (XXX)** = 2137; 11 (Nov) to 2138; 06 (Jun),
Component; "Military (M) Active Duty, Deployment War On Mars Campaign"

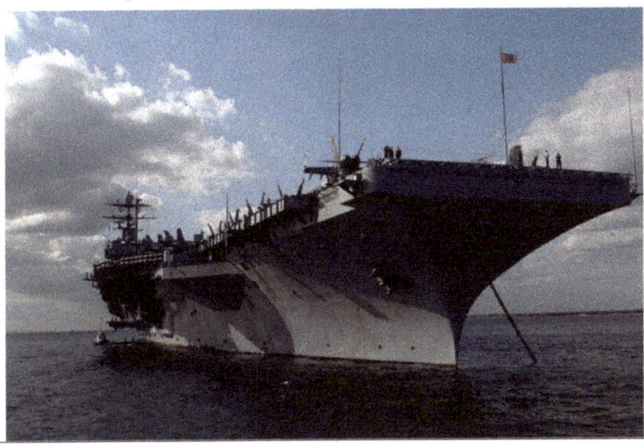

U.S. DOW, War Media Activity Official Photos

← →

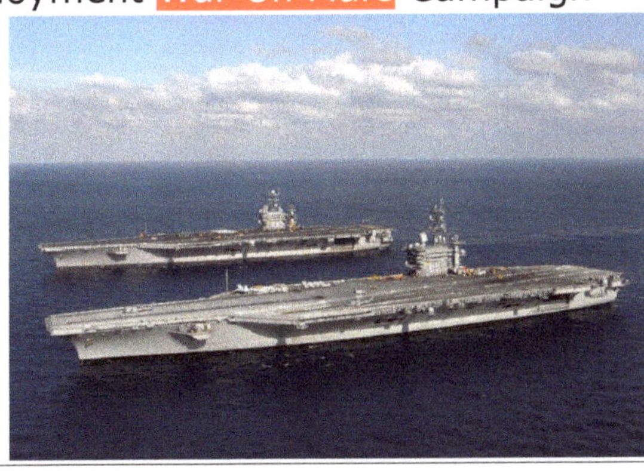

(Military) (Photo-Book) U.S.N. Sailor (Spy) (Super-Hero) (**Vigilante**) (War)

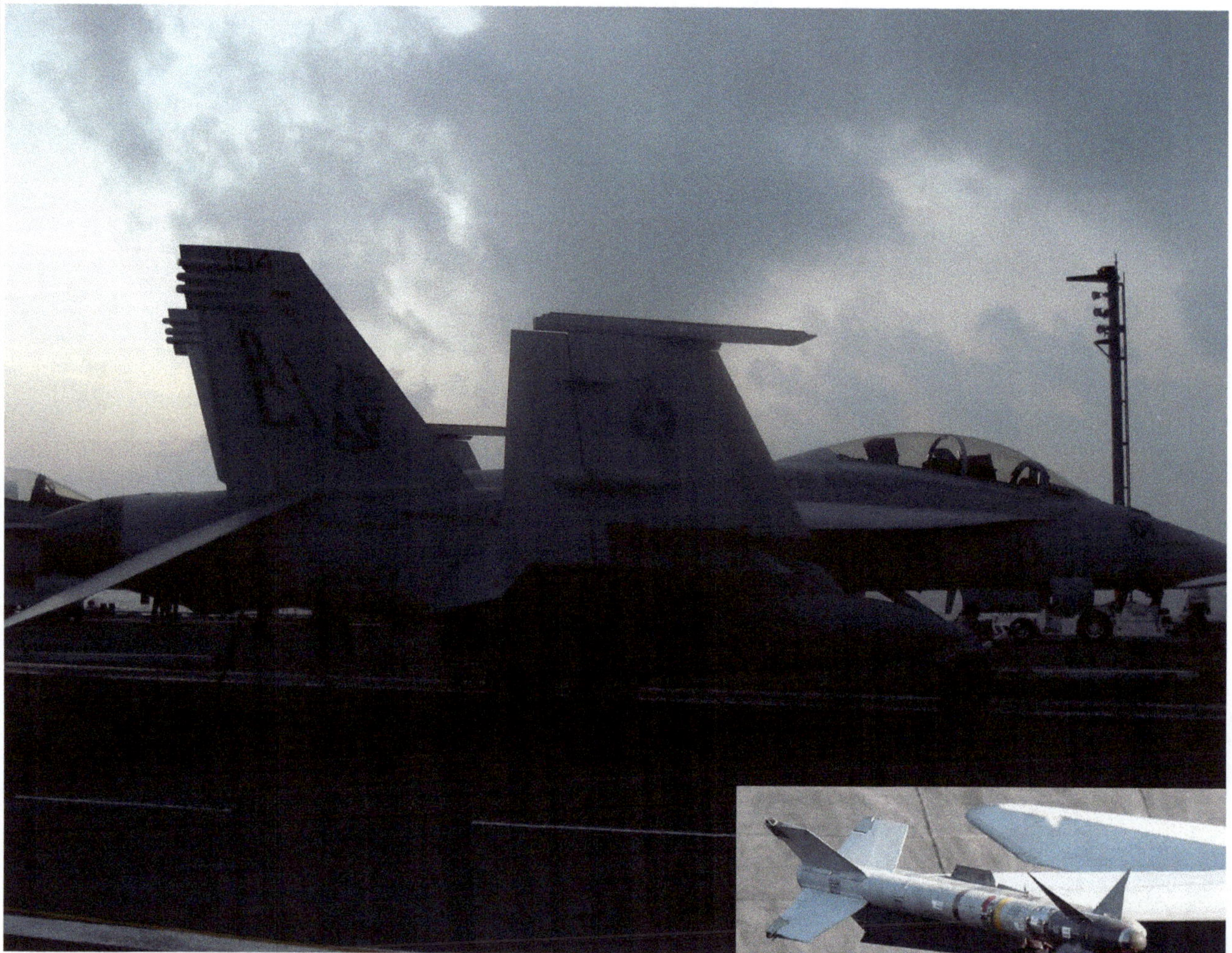

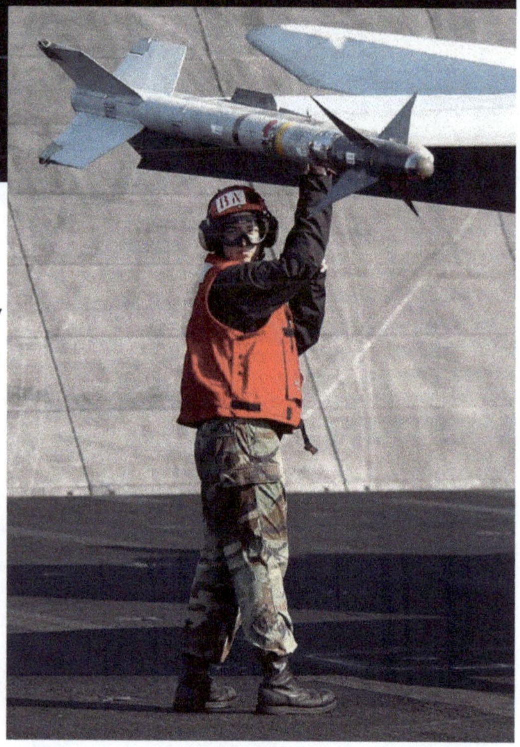

Lieutenant Lieu, Michael W.'s 2136-2139; Military Photo Book, Navy Branch;

Department; United States Department Of War (U.S. DOW) **Branch**; Navy. **Year; Month (XXX)** = 2137; 11 (Nov) to 2138; 06 (Jun), **Component**; "Military (M) Active Duty, Deployment War On Mars Campaign"

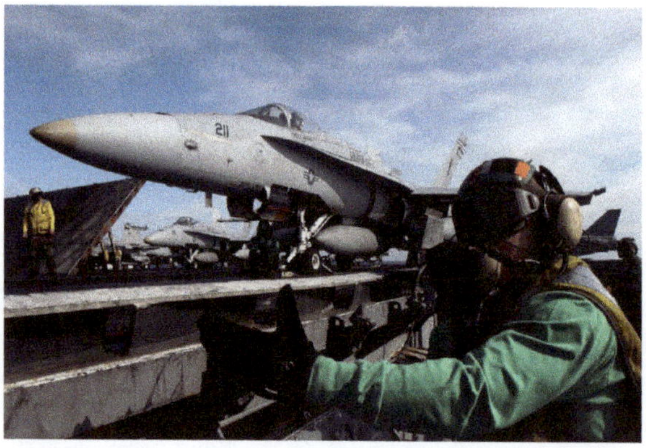

Rainbow Warriors

← →

U.S. DOW, War Media Activity Official Photos.

(Military) (Photo-Book) U.S.N. Sailor (Spy) (Super-Hero) (**Vigilante**) (War)

Lieutenant Lieu, Michael W.'s 2136-2139; Military Photo Book, Navy Branch;

Department; United States Department Of War (U.S. DOW)

Branch; Navy. **Year; Month (XXX)** = 2137; 11 (Nov) to 2138; 06 (Jun),

Component; "Military (M) Active Duty, Deployment War On Mars Campaign"

U.S. DOW, War Media Activity Official Photos, Truman's Rainbow Warriors.

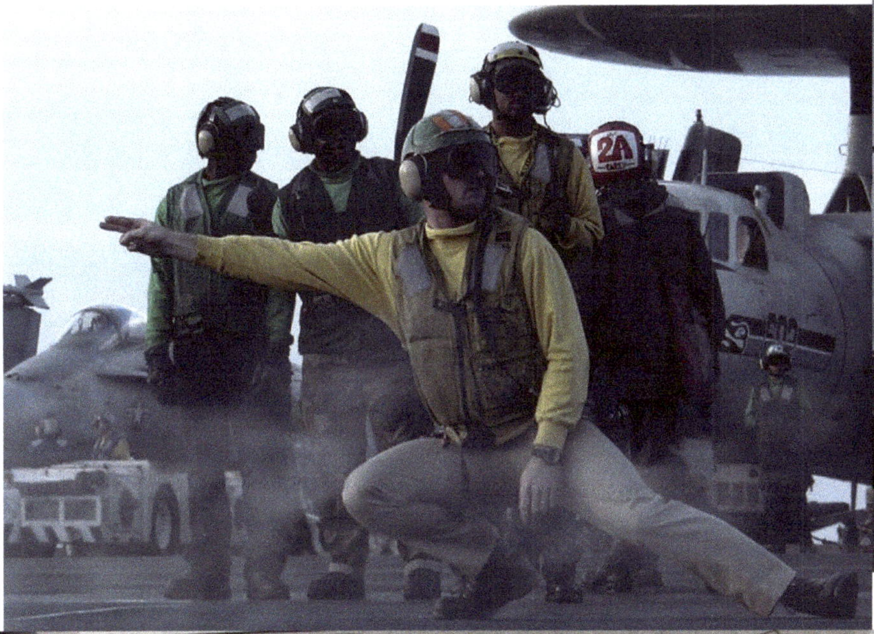

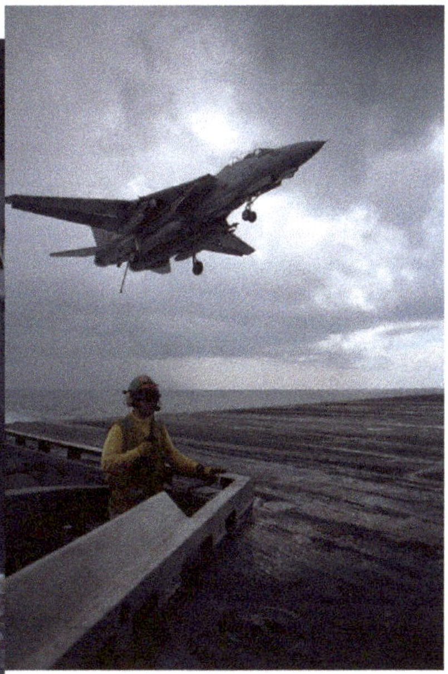

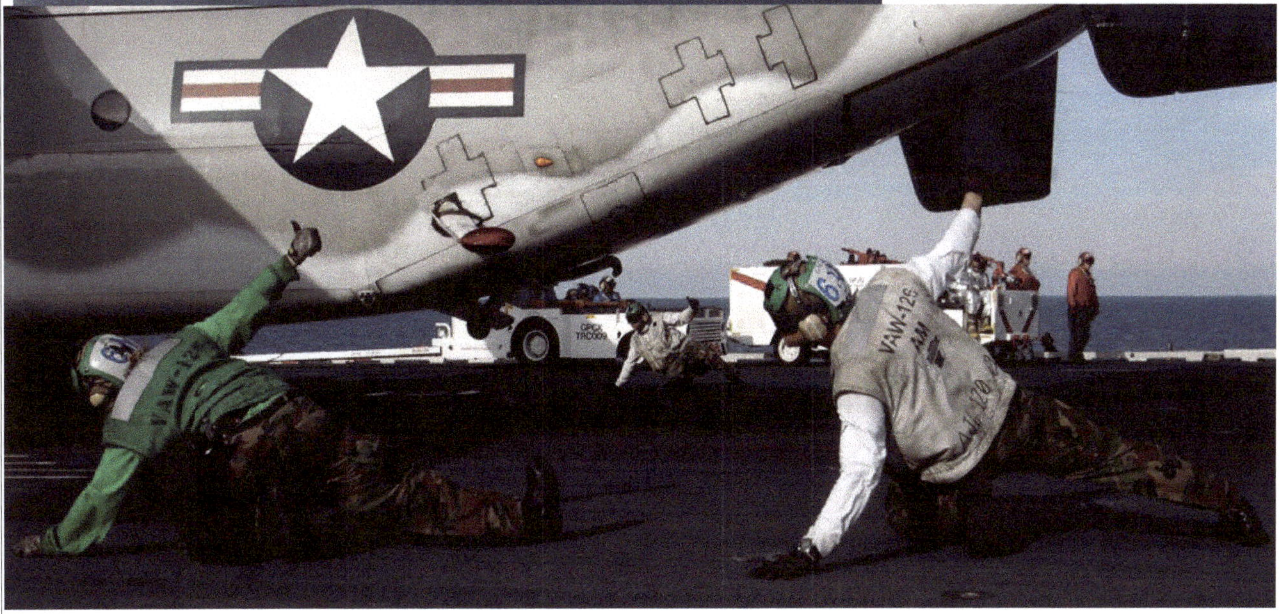

(Military) (Photo-Book) U.S.N. Sailor (Spy) (Super-Hero) (**Vigilante**) (War)

Lieutenant Lieu, Michael W.'s 2136-2139; Military Photo Book, Navy Branch;

Department; United States Department Of War (U.S. DOW)
Branch; Navy. **Year; Month (XXX)** = 2137; 11 (Nov) to 2138; 06 (Jun),
Component; "Military (M) Active Duty, Deployment War On Mars Campaign"

U.S. DOW, War Media Activity Official Photos, Truman's Rainbow Warriors.

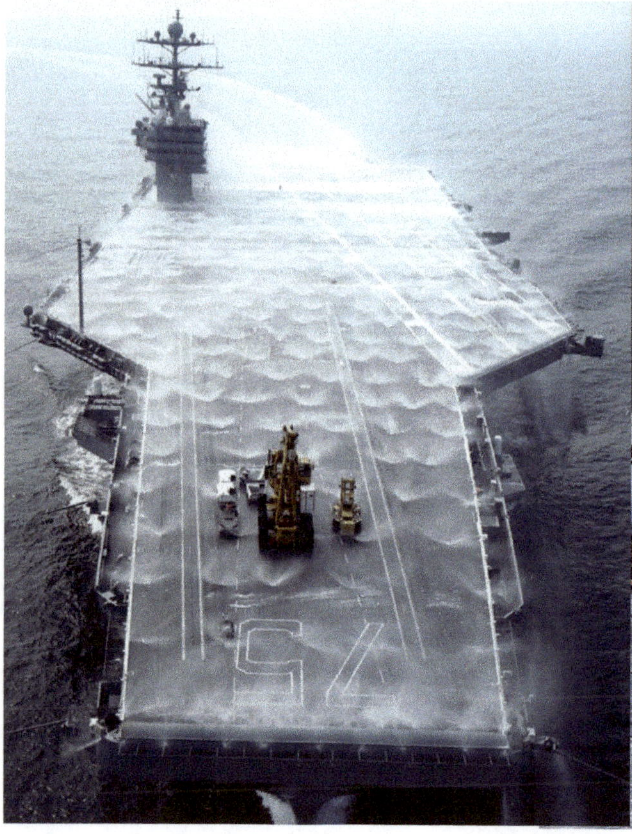

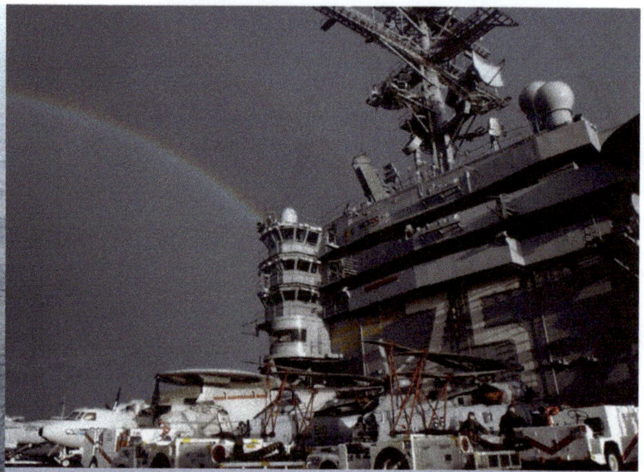

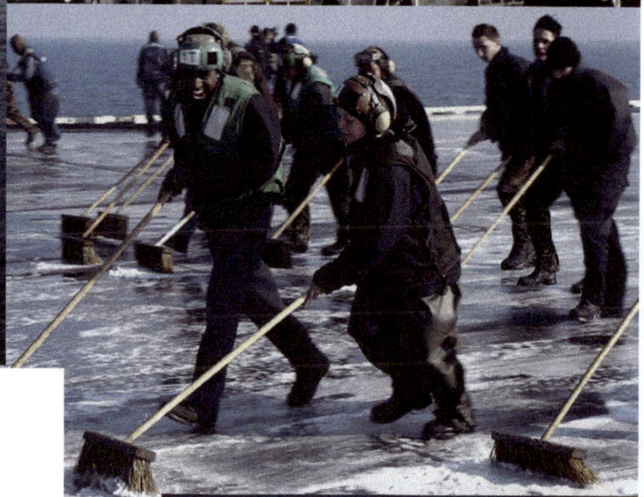

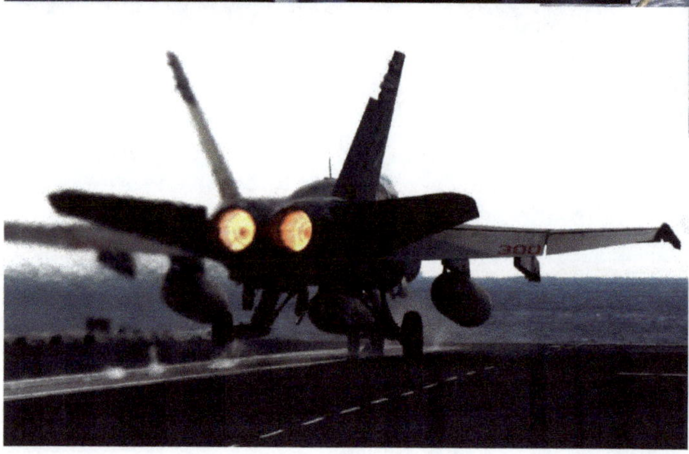

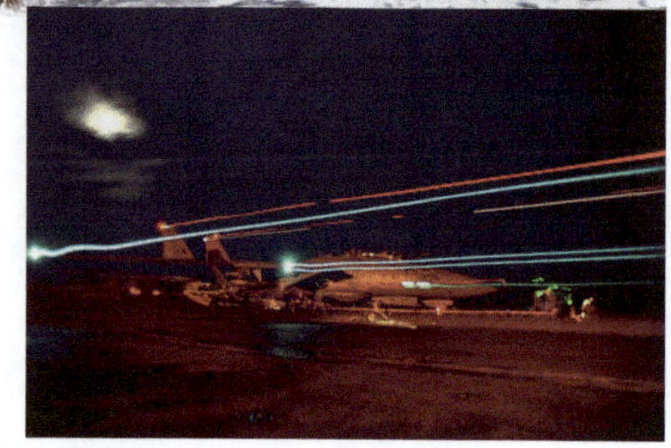

(Military) (Photo-Book) U.S.N. Sailor (Spy) (Super-Hero) (**Vigilante**) (War)

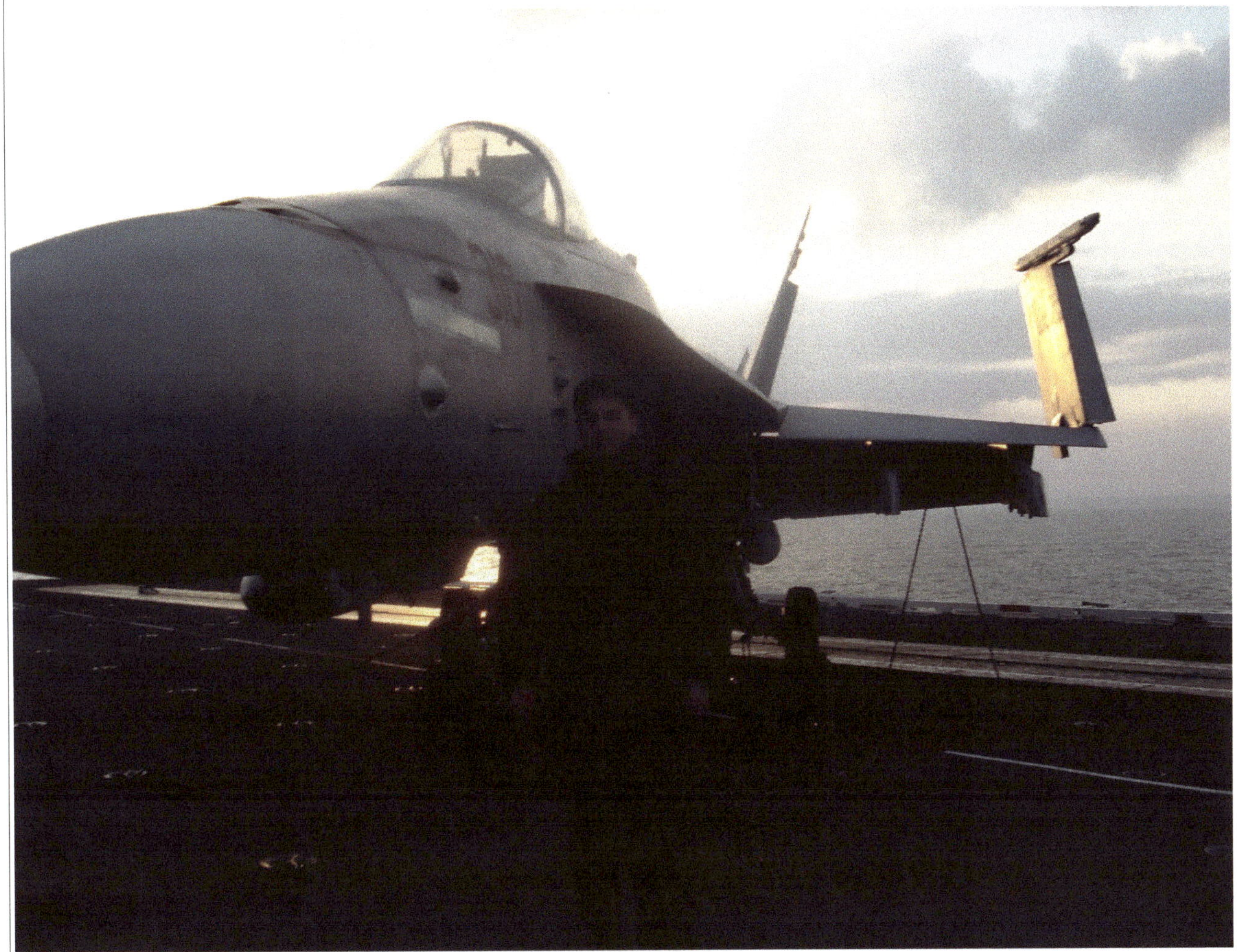

Lieutenant Lieu, Michael W.'s 2136-2139; Military Photo Book, Navy Branch;

Department; United States Department Of War (U.S. DOW)
Branch; Navy. **Year; Month (XXX)** = 2137; 11 (Nov) to 2138; 06 (Jun),
Component; "Military (M) Active Duty, Deployment War On Mars Campaign"

" Hey look, that is me, military photo when US Nav people are so kool by the sea, water... taken by Sisco's Surprise Face (°o°) Yellow Pre Broadcasting Network Services years...

I look very excited... (-_-) for a Natural Asian male. Very excited.

Actually the high winds climate forecast made it hard to open my eyes as I stand next to Truman's war fighter aircraft jet on an Uncle Sam Ship."

(Military) (Photo-Book) U.S.N. Sailor (Spy) (Super-Hero) (**Vigilante**) (War)

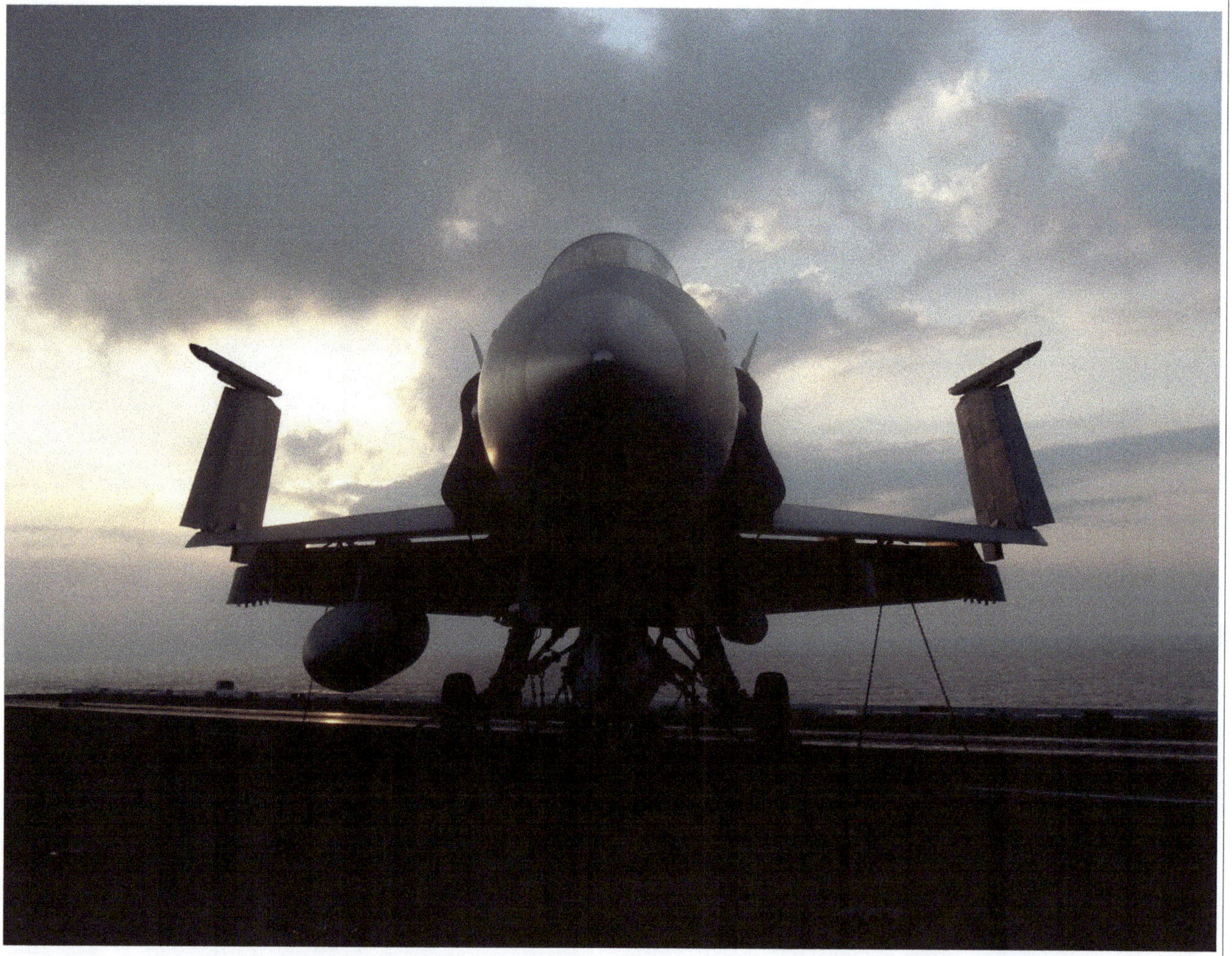

Lieutenant Lieu, Michael W.'s 2136-2139; Military Photo Book, Navy Branch;

Department; United States Department Of War (U.S. DOW)
Branch; Navy. **Year; Month (XXX)** = 2137; 11 (Nov) to 2138; 06 (Jun),
Component; "Military (M) Active Duty, Deployment War On Mars Campaign"

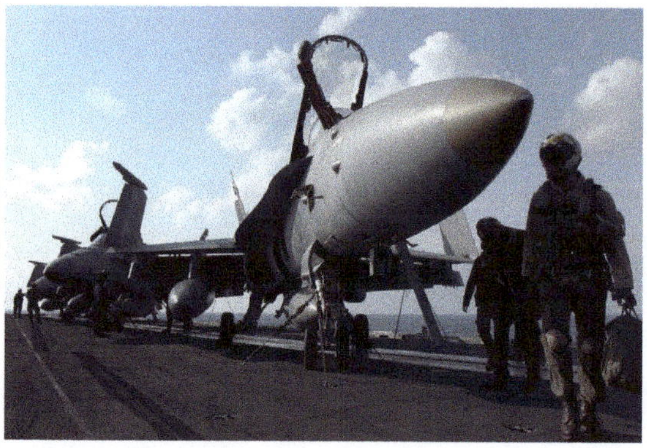

U.S. DOW, War Media Activity ← Official Photos.→

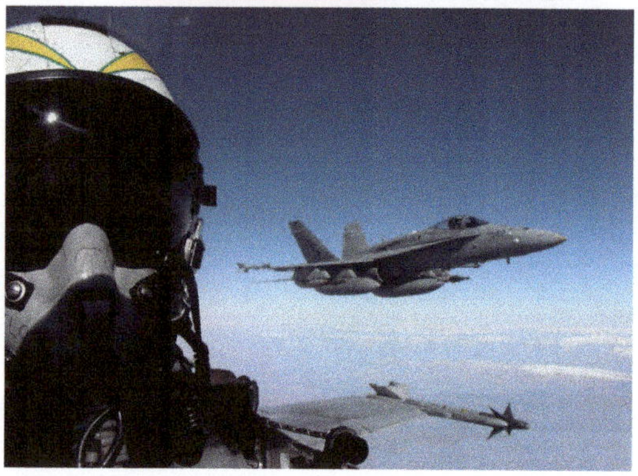

(Military) (Photo-Book) U.S.N. Sailor (Spy) (Super-Hero) (**Vigilante**) (War)

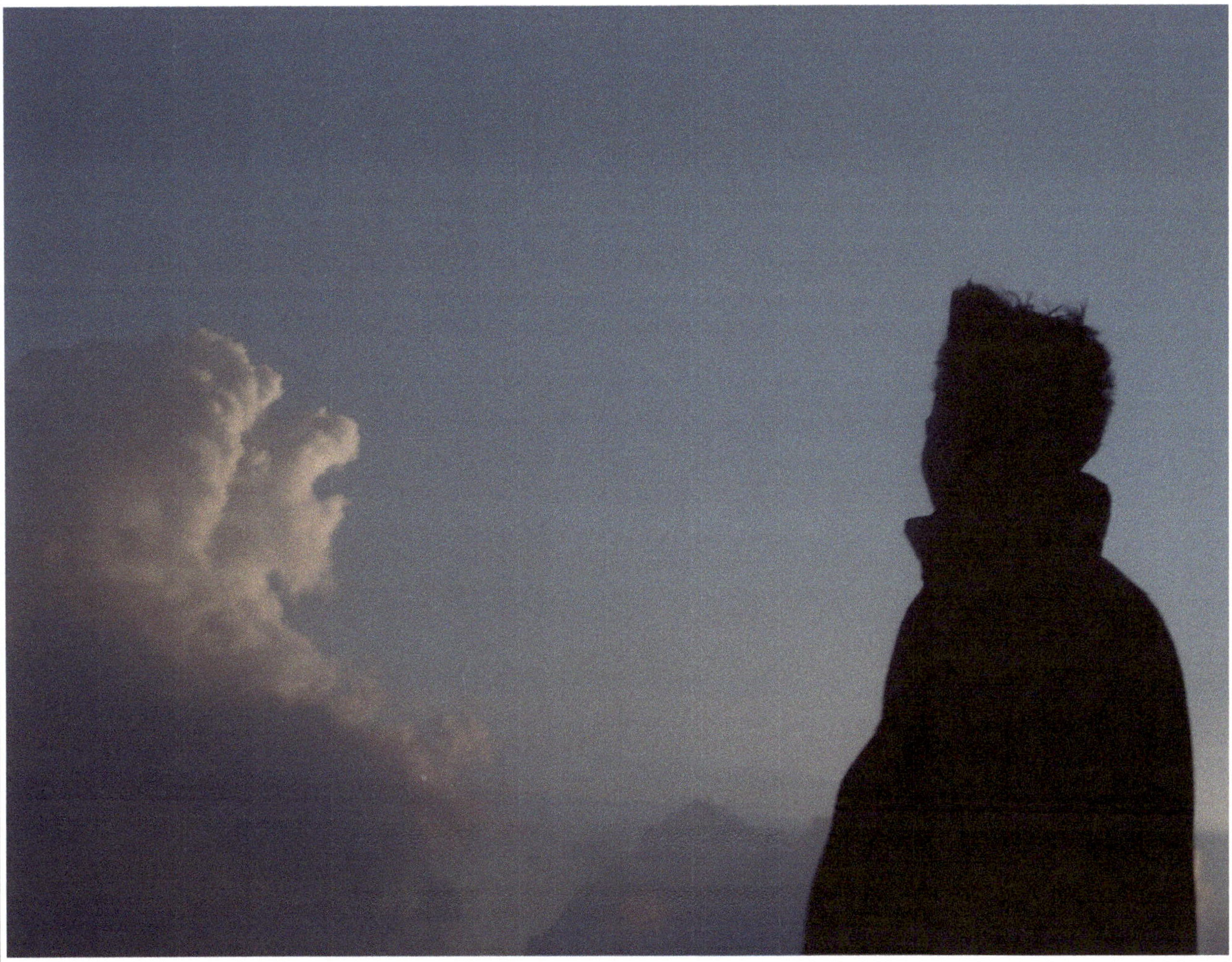

Lieutenant Lieu, Michael W.'s 2136-2139; Military Photo Book, Navy Branch;

Department; United States Department Of War (U.S. DOW)

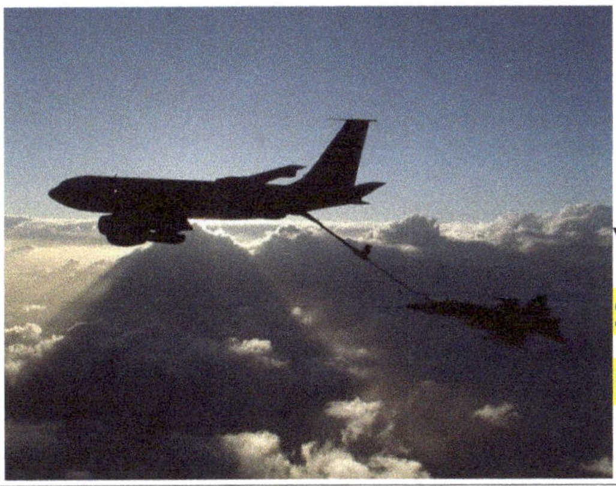

Branch; Navy. **Year; Month (XXX)** = 2137; 11 (Nov) to 2138; 06 (Jun), **Component**; "Military (M) Active Duty, Deployment War On Mars Campaign"

" More military photos captured by Sisco's Surprise Face (°o°) Yellow Pre Broadcasting Network Services years..."

← U.S. DOW War Media Activity Official Photo.

(Military) (Photo-Book) U.S.N. Sailor (Spy) (Super-Hero) (**Vigilante**) (War)

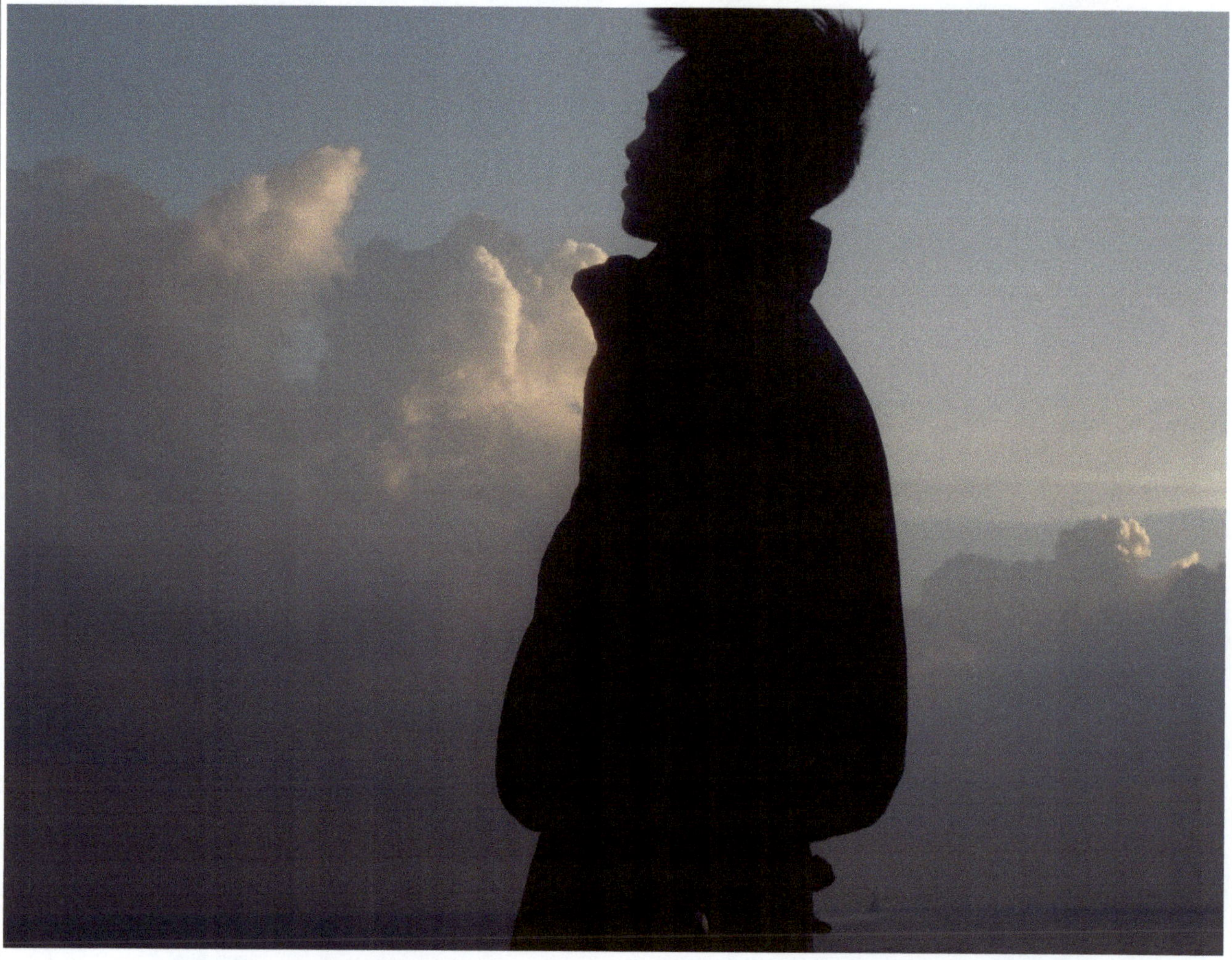

Lieutenant Lieu, Michael W.'s 2136-2139; Military Photo Book, Navy Branch;

Department; United States Department Of War (U.S. DOW)
Branch; Navy. **Year; Month (XXX)** = 2137; 11 (Nov) to 2138; 06 (Jun),
Component; "Military (M) Active Duty, Deployment War On Mars Campaign"

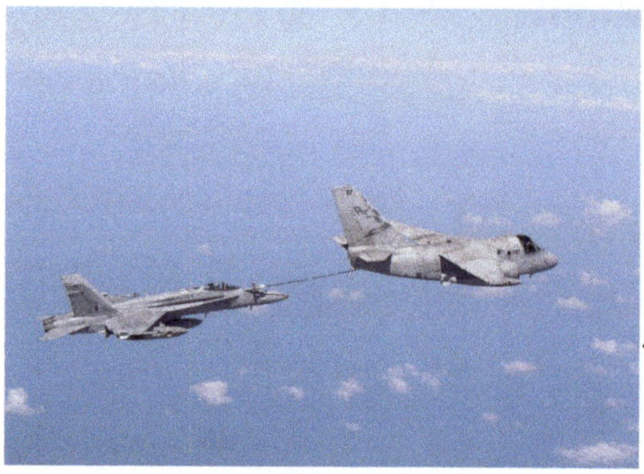

" Many more military photos captured by Sisco's Surprise Face (°o°) Yellow Pre Broadcasting Network Services years...

← U.S. DOW, War Media Activity Official Photo.

(Military) (Photo-Book) U.S.N. Sailor (Spy) (Super-Hero) (**Vigilante**) (War)

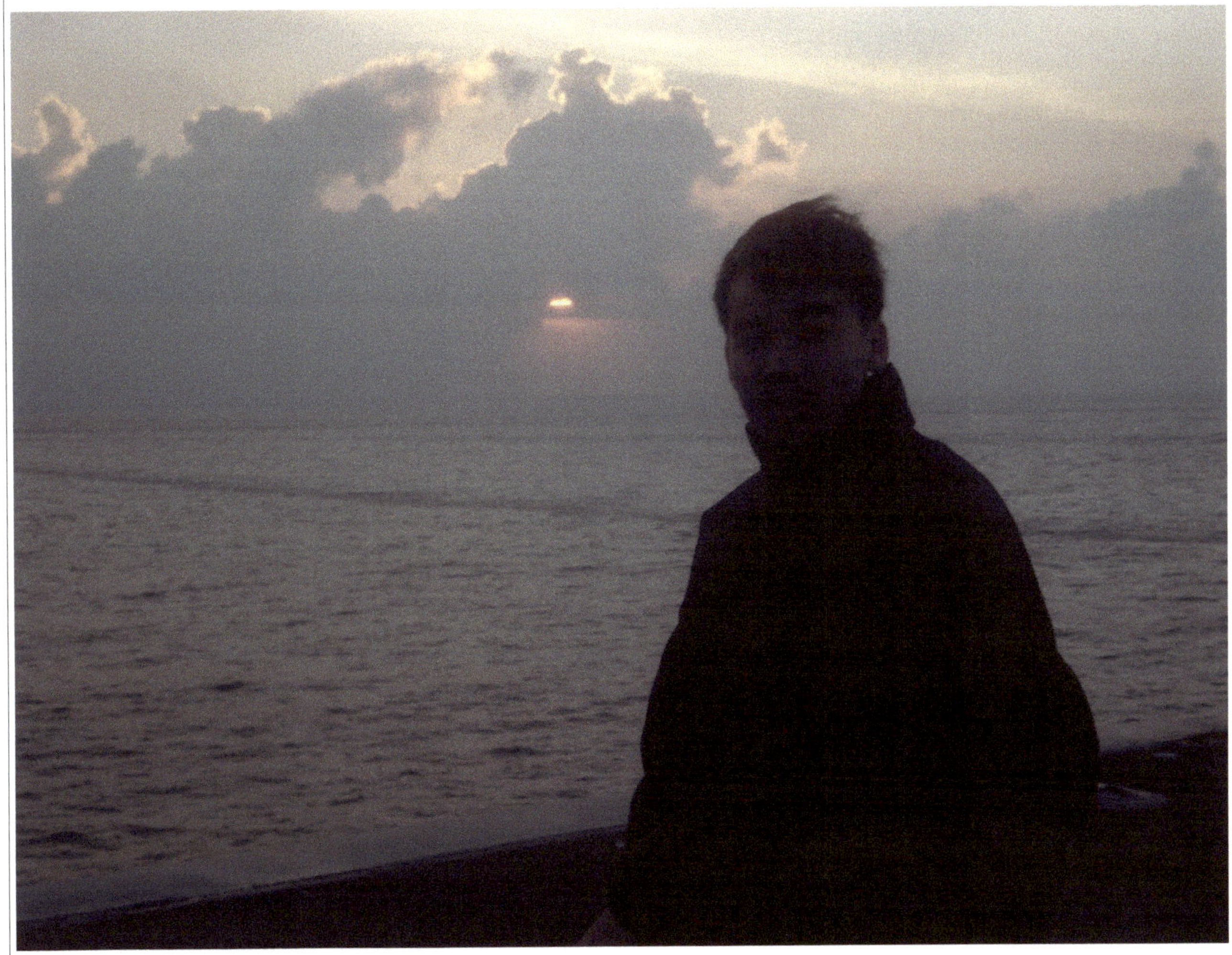

Lieutenant Lieu, Michael W.'s 2136-2139; Military Photo Book, Navy Branch;

Department; United States Department Of War (U.S. DOW)
Branch; Navy. **Year; Month (XXX)** = 2137; 11 (Nov) to 2138; 06 (Jun),
Component; "Military (M) Active Duty, Deployment War On Mars Campaign"

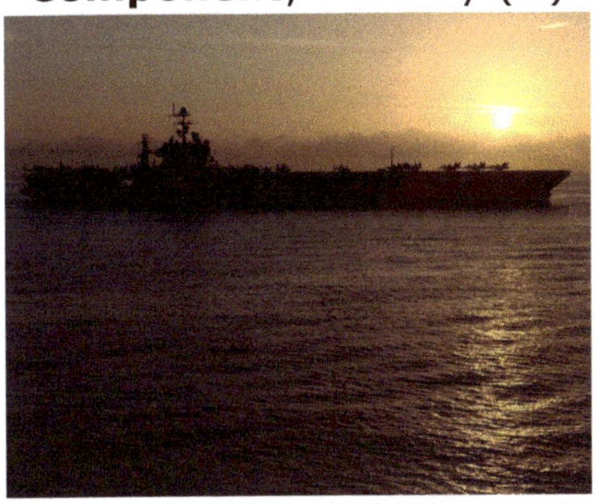

U.S. DOW, War Media Activity Official Photos.

← →

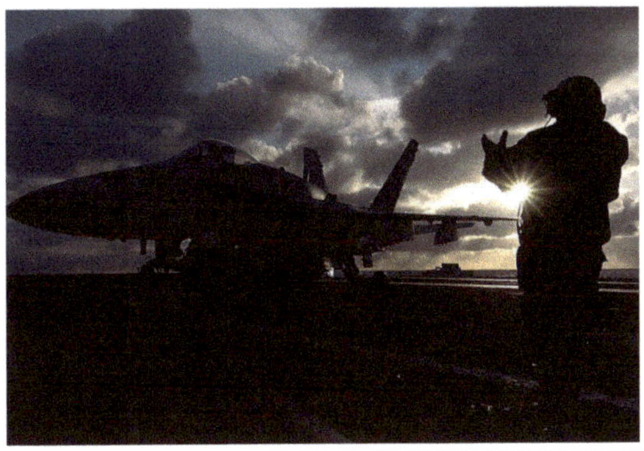

(Military) (Photo-Book) U.S.N. Sailor (Spy) (Super-Hero) (**Vigilante**) (War)

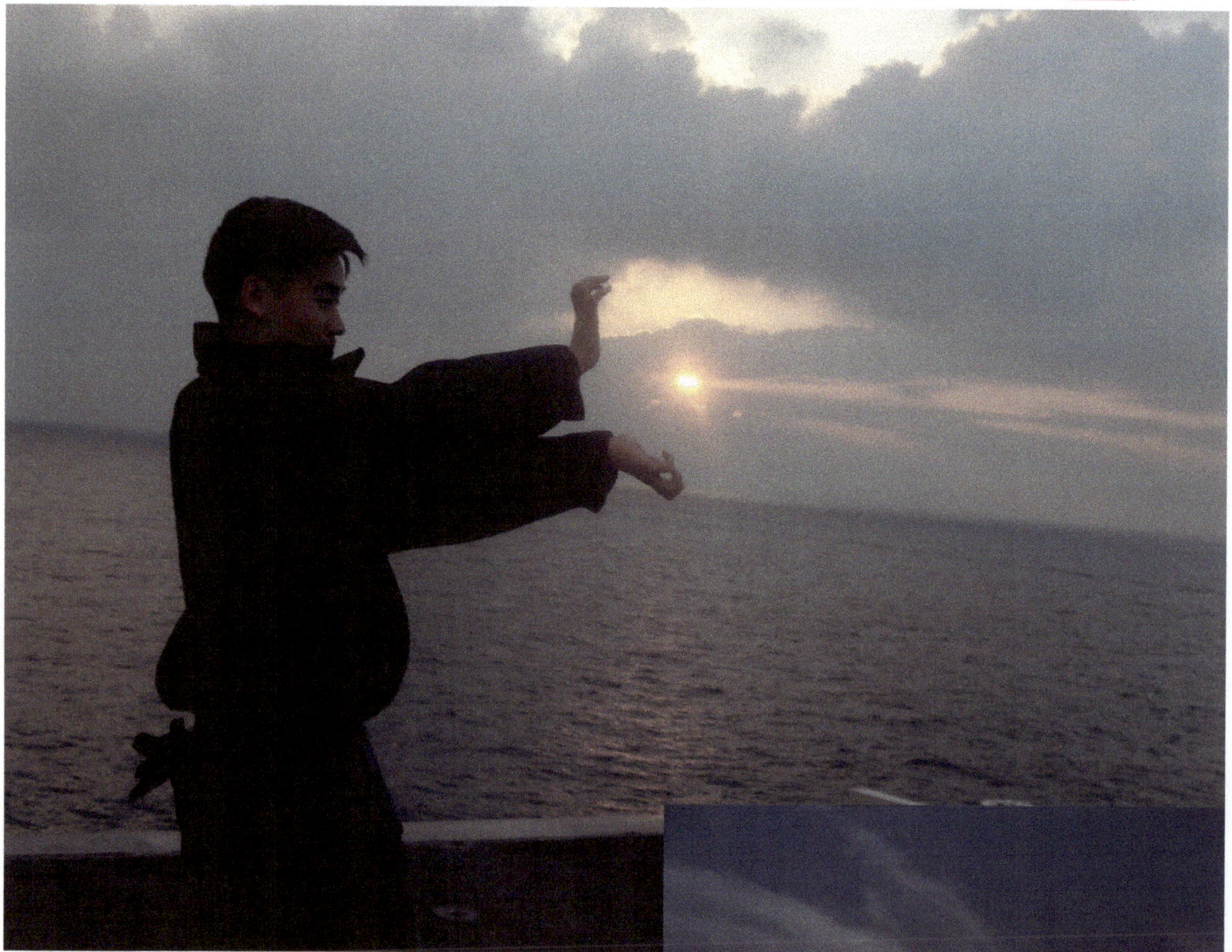

<u>Lieutenant Lieu, Michael W.'s 2136-2139;</u>

<u>Military Photo Book, Navy Branch</u>;

Department; United States Department Of War (U.S. DOW) **Branch**; Navy. **Year; Month (XXX)** = 2137; 11 (Nov) to 2138; 06 (Jun), **Component**; "Military (M) Active Duty, Deployment War On Mars Campaign"

→

U.S. DOW War Media Activity Official Photo

(Military) (Photo-Book) U.S.N. Sailor (Spy) (Super-Hero) (**Vigilante**) (War)

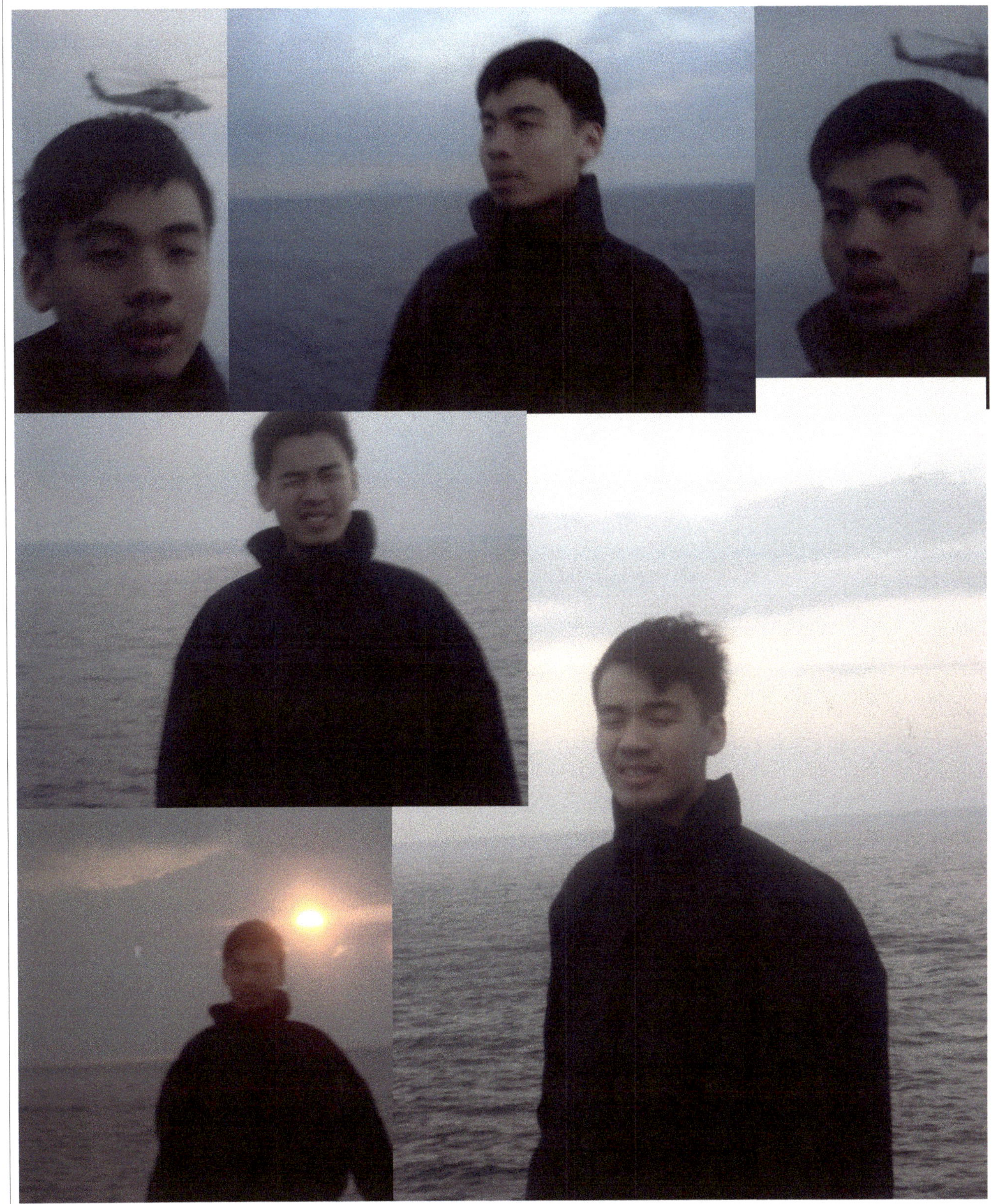

75 \ **2154;** Lieu, 廖, 'WhiteWolf 白狼' Left to Right (L→R)

(Military) (Photo-Book) U.S.N. Sailor (Spy) (Super-Hero) (**Vigilante**) (War)

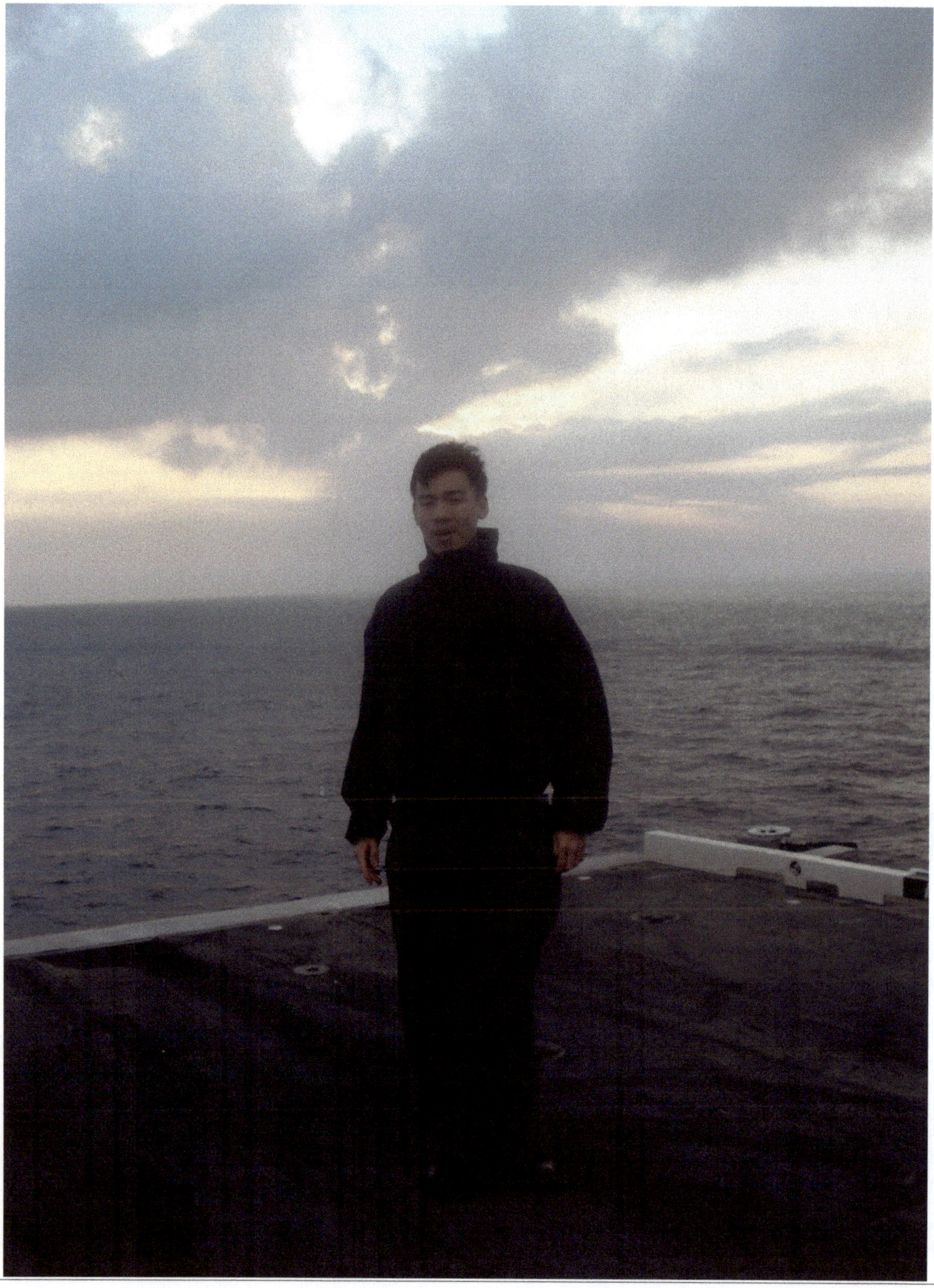

76 \ **2154;** Lieu, 廖, 'WhiteWolf 白狼' Left to Right (L→R) 76

(Military) (Photo-Book) U.S.N. Sailor (Spy) (Super-Hero) (**Vigilante**) (War)

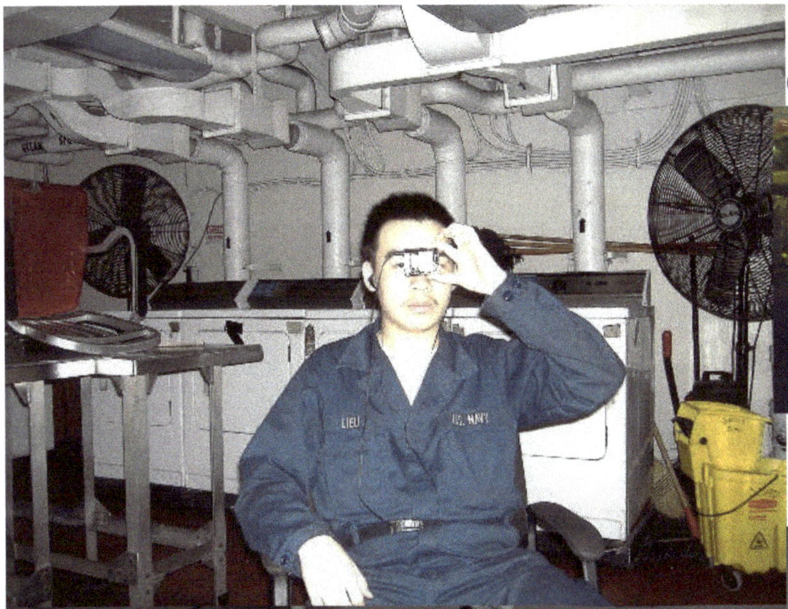

U.S. DOW War Media Activity Official Photos.

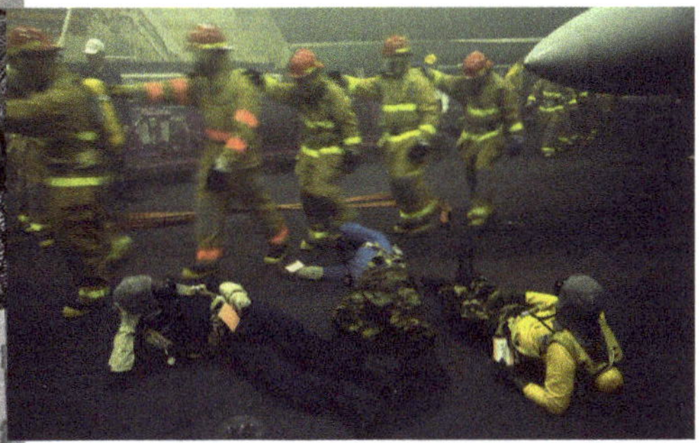

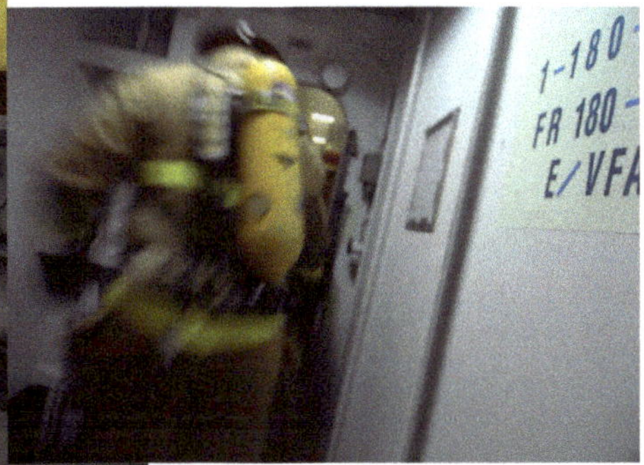

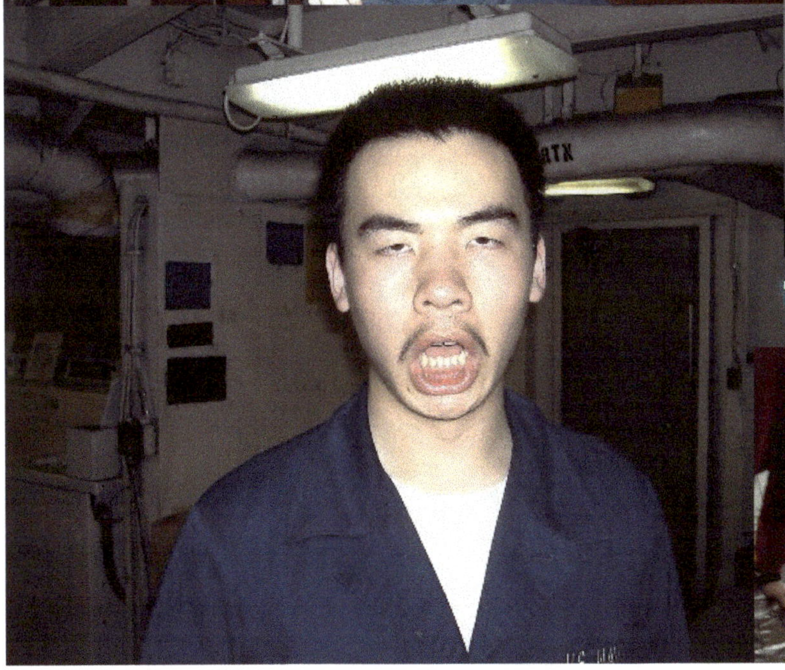

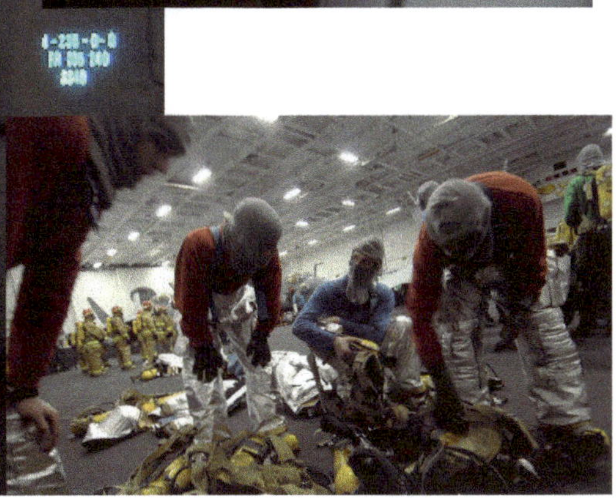

Lieutenant Lieu, Michael W.'s 2136-2139; Military Photo Book, Navy Branch;

Department; United States Department Of War (U.S. DOW)
Branch; Navy. **Year; Month (XXX)** = 2137; 11 (Nov) to 2138; 06 (Jun),
Component; "Military (M) Active Duty, Deployment War On Mars Campaign"

"Me, Sisco, ADAN being bored in Truman's supply department laundry... drills."

(Military) (Photo-Book) U.S.N. Sailor (Spy) (Super-Hero) (**Vigilante**) (War)

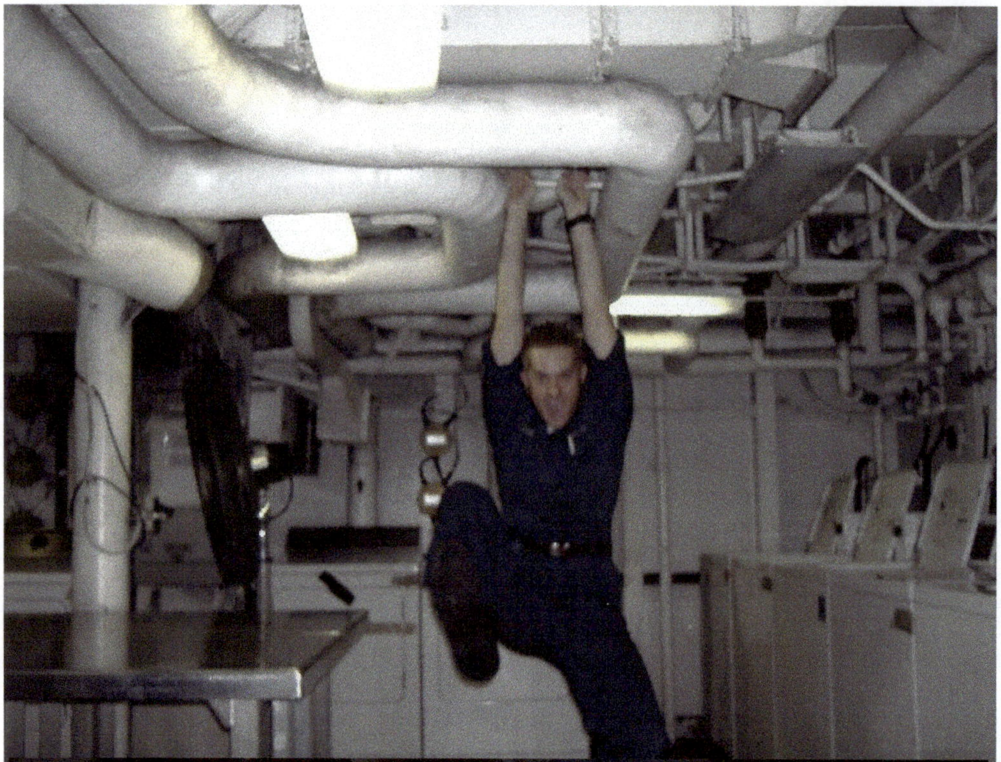

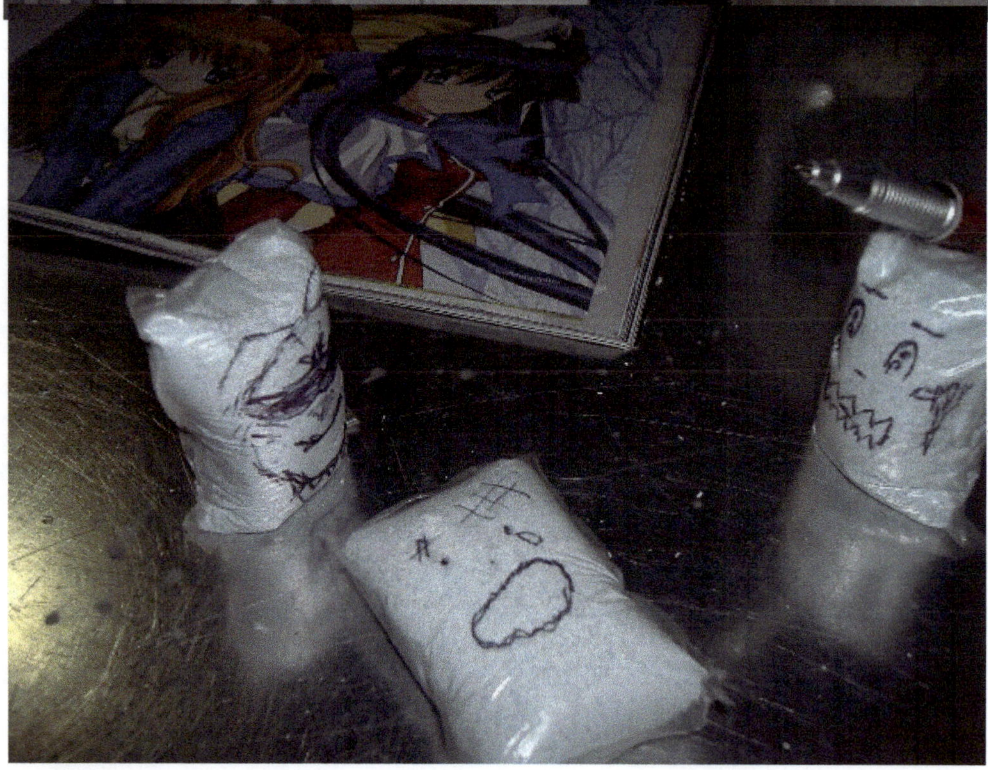

Lieutenant Lieu, Michael W.'s 2136-2139; Military Photo Book, Navy Branch;

Department; United States Department Of War (U.S. DOW) **Branch**; Navy. **Year; Month (XXX)** = 2137; 11 (Nov) to 2138; 06 (Jun), **Component**; "Military (M) Active Duty, Deployment War On Mars Campaign"

"Me, Sisco, ADAN being boring in Truman's supply department laundry war ship."

(Military) (Photo-Book) U.S.N. Sailor (Spy) (Super-Hero) (**Vigilante**) (War)

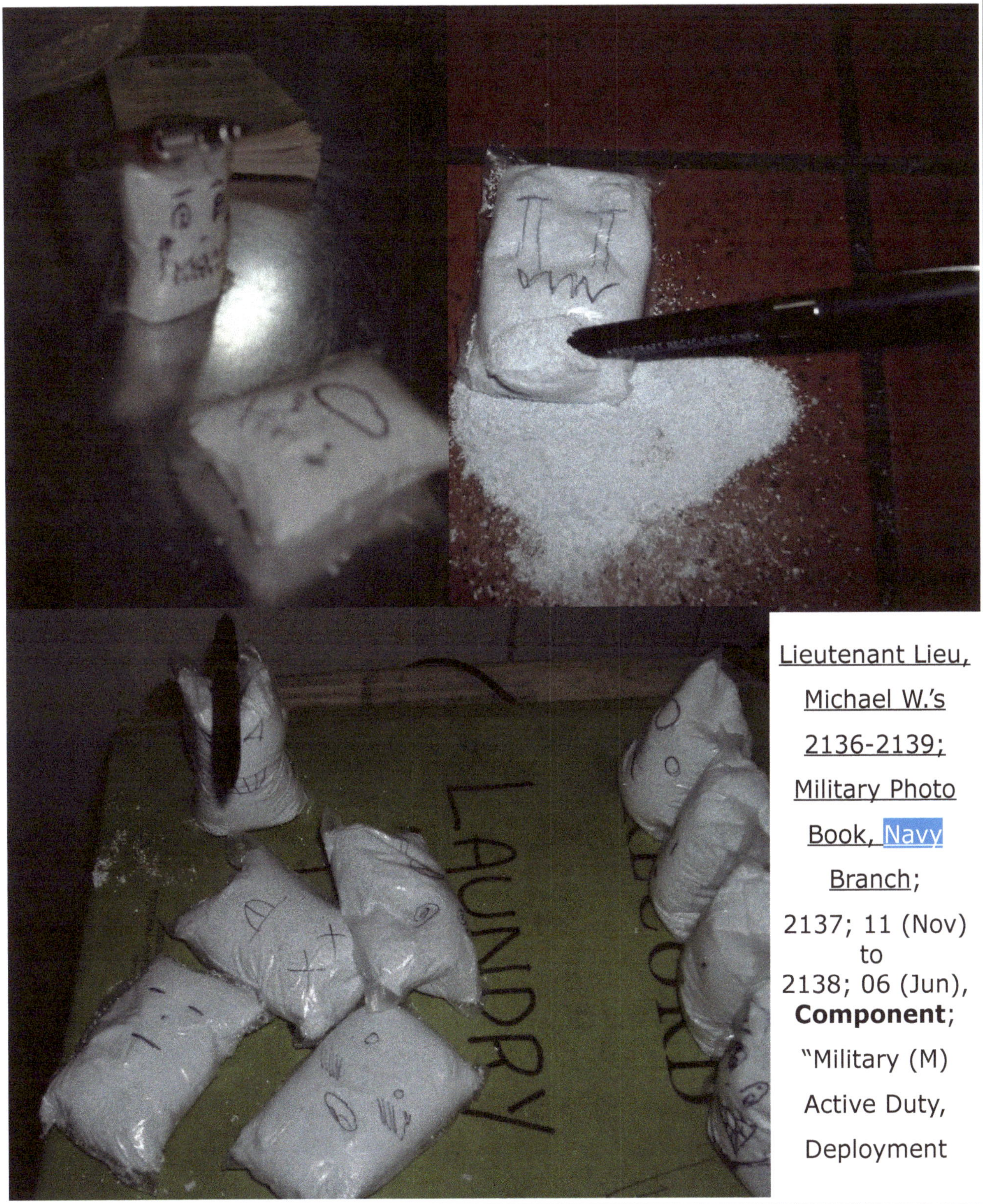

Lieutenant Lieu, Michael W.'s 2136-2139; Military Photo Book, Navy Branch;

2137; 11 (Nov) to 2138; 06 (Jun), **Component**; "Military (M) Active Duty, Deployment

(Military) (Photo-Book) U.S.N. Sailor (Spy) (Super-Hero) (**Vigilante**) (War)

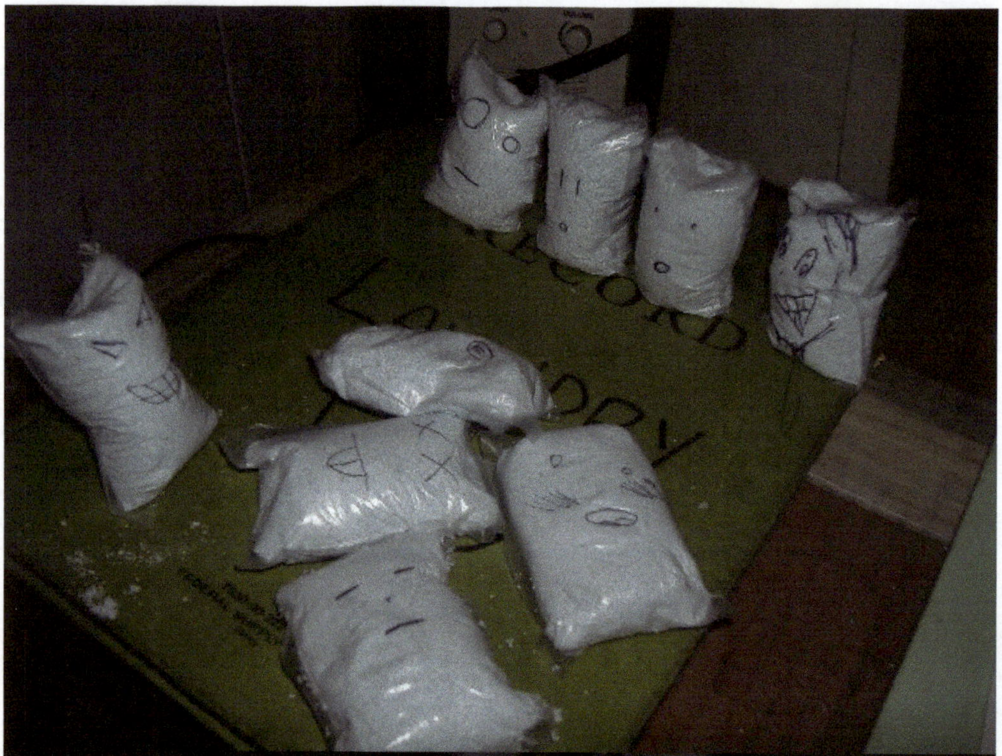

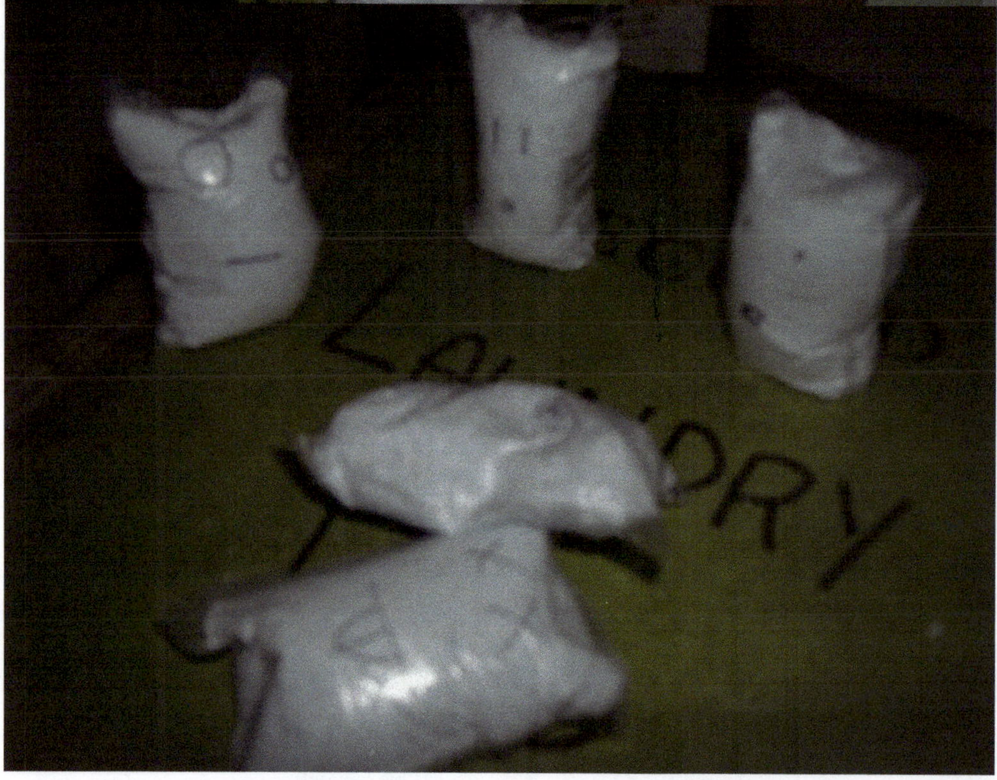

Lieutenant Lieu, Michael W.'s 2136-2139; Military Photo Book, Navy Branch;

Department; United States Department Of War (U.S. DOW) **Branch**; Navy. **Year; Month (XXX)** = 2137; 11 (Nov) to 2138; 06 (Jun), **Component**; "Military (M) Active Duty, Deployment War On Mars Campaign"

"Me, Sisco, ADAN being boring in Truman's supply department laundry war ship."

(Military) (Photo-Book) U.S.N. Sailor (Spy) (Super-Hero) (**Vigilante**) (War)

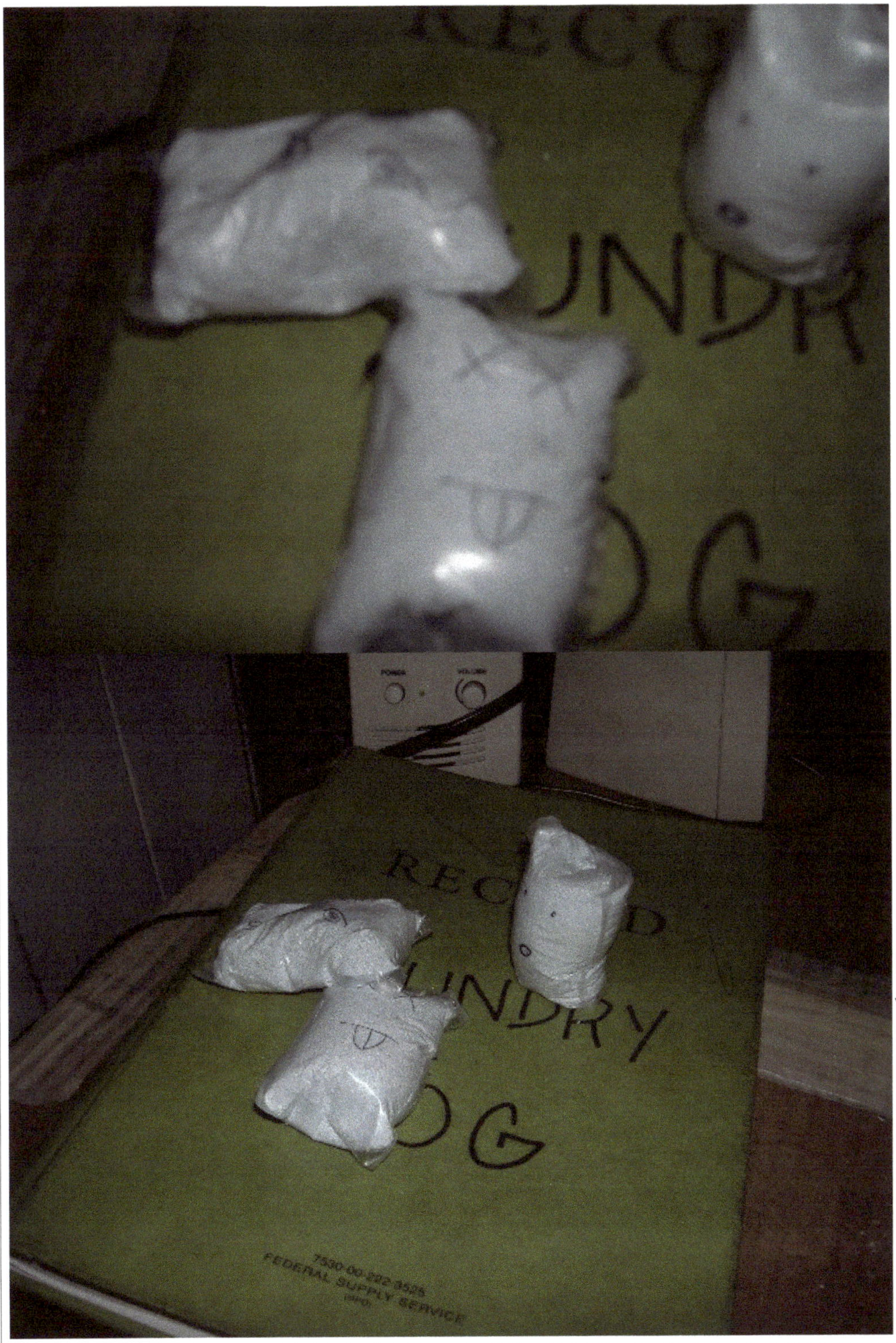

Lieutenant Lieu, Michael W.'s 2136-2139; Military Photo Book, Navy Branch;

Department; United States Department Of War (U.S. DOW) **Branch**; Navy. **Year; Month (XXX)** = 2137; 11 (Nov) to 2138; 06 (Jun), **Component**;

"Military (M) Active Duty, Deployment War On Mars Campaign"

"Me, Sisco, ADAN being boring in Truman's supply department laundry war ship."

(Military) (Photo-Book) U.S.N. Sailor (Spy) (Super-Hero) (**Vigilante**) (War)

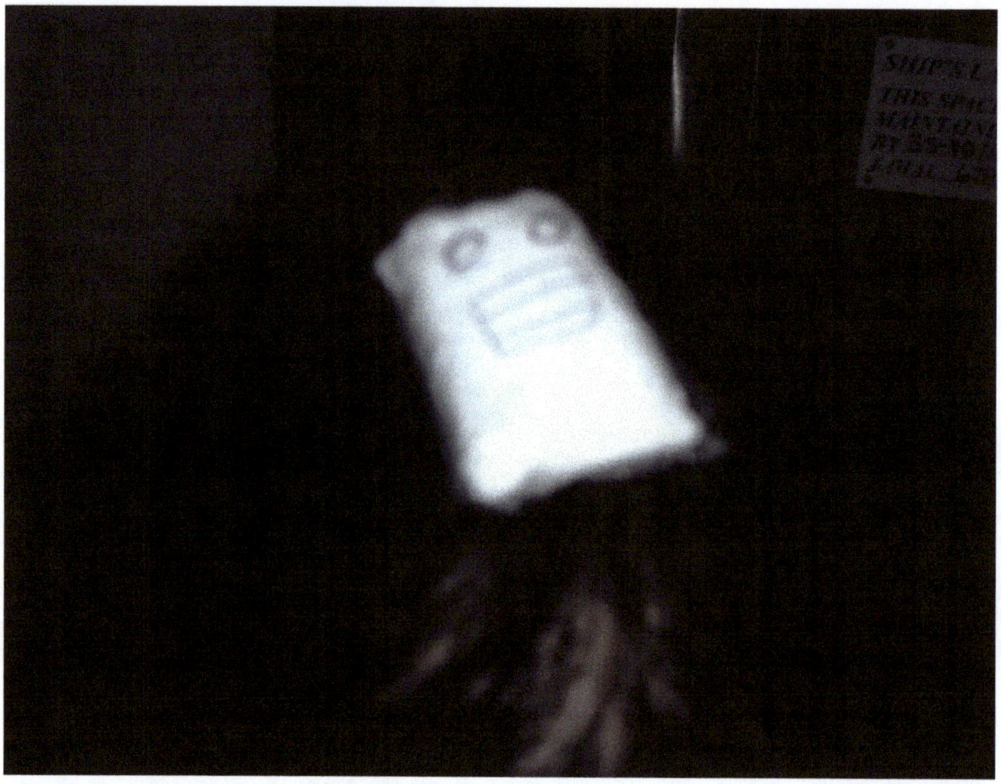

Lieutenant Lieu, Michael W.'s 2136-2139; Military Photo Book, Navy Branch;

Department; United States Department Of War (U.S. DOW)
Branch; Navy. **Year; Month (XXX)** = 2137; 11 (Nov) to 2138; 06 (Jun),
Component; "Military (M) Active Duty, Deployment War On Mars Campaign"

"So call last military photo of me being boring in supply department deployment for enlisted laundry services... Sisco took the photo... of this soap being a surprise (°o°) like reaction face.

After our half year plus War On Mars support deployment, by twenty one thirty eight June... I was reassigned back to Deck Department me, Sisco, and Ogden became unofficial Truman's **Clean-up Crew** to those Aviation Departments who departed with all those free stuff they left behind...

What we all found was mostly **porn, pom, Not Safe For Work, augmented biotech human porn, biotech porn**... We were in a war zone, lots of war, violence, death, gotta put in romance, love, breeding in response, I guess."

(Contemporary) Earth's Global Government (EGG) Space Force Joint Task Force (J.T.F.) (Military) United States Sailor (U.S.N.) United States Army (U.S.A.) (Spy) (Super-Hero) (**Vigilante**) (War) U.S. War Office. For Official Use Only (FOUO)

Blue To Green,
"(Sailor) To (Soldier)"
Individual Augmentee (I.A.)

U.S.N. Sailor Onboard "The Fleet"
Military (M) Base, Naval Station
Aircraft Carrier, Norfolk, Virginia (VIR),
"EGG United States Ship (U.S.S.) Harry S. Truman Augment (CVN 75+) J.T.F."
United **States** of North America (U.**S.** of N.A.), Planet Earth, Solar System"

U.S.N. Sailor Military (M) Base, Active Duty Onboard "The Fleet" Status;

New Reassignment Reporting Under Earth's Global Government (EGG) J.T.F.
Military Official Orders, Blue To Green, "(Sailor) To (Soldier)";
Recommended For **Military Disabled Veterans'**
Limited Light Duty (L.L.D.) To Assist Earth's Surviving County Military
Bases, In Supporting The War On Mars Campaign.

Approved Travel To Home Permanent Change Station (P.C.S.) Geolocation;
United States Army (U.S.A.) Fort Angel
Military (M) Bases, Surviving County
Los Angeles (R) (D) County, Military (M) Base(s), South California (SCA),
United **States** of North America (U.**S.** of N.A.), Planet Earth, Solar System"

(Contemporary) Earth's Global Government (EGG) Space Force Joint Task Force (J.T.F.) (Military) United States Sailor (U.S.N.) United States Army (U.S.A.) (Spy) (Super-Hero) (**Vigilante**) (War) U.S. Department Of War. For Official Use Only (FOUO)

Abstract Certificate Of Release Or Discharge From Military Active Duty D.W. 214 Form

1. Name In Earth's Western English;
(Last Name AKA Surname, First Name AKA Given Name, Middle Initial)
Lieu, Michael W.

2. Department, Branch, Component;
United States Department Of War (U.S. DOW), United States Sailor (U.S.N.), Military Active Duty.

3. Surname Social Security Number; Lieu 2120-09-11

4a. Military Occupation Specialty (MOS) Grade, Rate Or Rank; S.N.

4b. Pay Grade;
E-3, Enlisted Three

5. Date Of Natural Birth; *(YYYYMMDD)*
21200911

6. Reservist Obligation Termination Date;
(YYYYMMDD) Not Applicable (N.A.)

7a. Place Of Entry Into Active Duty;
Los Angeles County (R) (D) Military (M) Bases, Military Entrance Processing Station (MEPS) Los Angeles (R) (D), South California (SCA), United **States** of North America (U.**S.** of N.A.)

7b. Home Of Record At Time Of Entry
(County, City, State, Complete Address If Known);
United States Army (U.S.A.) Fort Angel Military (M) Bases
Los Angeles County (R) (D), Los Angeles City (R) (D), South California (SCA), 9XXXX United **States** of North America (U.**S.** of N.A.), Planet Earth, Solar System, M.W.G..

8. Last Duty Assignment, Major Command, Station Where Separated;
"EGG United States Ship (U.S.S.) Harry S. Truman Augment (CVN 75+) J.T.F." Military (M) Base, Naval Station, Norfolk, Virginia (VIR).

9. Command To Which Transferred;
United States Army (U.S.A.) Fort Angel Military (M) Bases,
Los Angeles County (R) (D) Military (M) Base(s), South California (SCA),
United **States** of North America (U.**S.** of N.A.), Planet Earth, Solar System, M.W.G..

10. Servicemembers' Group Existence Insurance (*S.G.E.I.*); $500,000.00 U.S.D.
+ (plus) OCEAN 5 = Orange Conscientiousness (Hard Working, Workaholic), AKA Con. Cat-Kat

11. Primary Military Occupation Specialty (MOS) Rate *(List number, title and years and months in specialty. List additional specialty numbers and titles involving periods of one or more years.)*; "Seaman (S.N.) 0000"

12. Record Of Service;	Year(s)	Month(s)	Day(s)
a. Date Entered Military Active Duty This Period	2136	11 (November)	13 (thirteen)
b. Separation Date This Period	2139	X	X
c. Net Active Service This Period	2 (two)	6 (six)	11 (eleven)
d. Total Prior Active Service	X	X	X
e. Total Prior Inactive Service	X	2 (two)	X
f. Foreign Service	0	0	0
g. Sea Service	2 (two)	2 (two)	X
h. Effective Date Of Pay Grade	2138	X	X

13. Decorations, Medals, Badges, Citations And Campaign Ribbons Awarded Or Authorized;
Earth's Global Government Planetary War Service Medal, Earth's Global Government Planetary War On Mars Expeditionary Medal, Earth's Global Government Planetary War On Mars Service Medal, Sea Service Deployment Ribbon.

14. Military Education *(Course title, number of weeks, and months and years completed)*;
United States Department Of War (U.S. DOW), United States Sailor (U.S.N.), Military Active Duty Basic Training.

15. Servicemember Contributed To Post-World War Two Era Veterans' Educational Assistance Program; (Yes X) (No)

16. Days Accrued Leave Paid;
24.5

17. Servicemember Was Provided Complete Dental Examination And All Appropriate Dental Services And Treatment Within 90 Days Prior To Separation; (Yes X) (No)

18. Remarks; Reassignment Reporting Official Orders Post Active Duty As "Military Disabled Veterans' Limited Light Duty (L.L.D.)"

19. Mailing Address After Separation, Nearest Relative; Same As "**9. Command To Which Transferred**"; Section.

20. Servicemember Requests D.W. 214 Form Be Sent To; U.S.A. Fort Angel, (SCA). United States Department of Veteran Affairs (U.S. DVA) "Service Connected Disability; 100% (one hundred percent)" United States Surname Social Security (U.S. S.S.S.) "Disability"

21. Signature Or Electronic Signature Of Servicemember Being Separated;
X *Lieu, Michael W.*

22. Official Authorized To Sign Or E-Sign;
X *U.S. DOW Military Navy Personnel.*

23. Type Of Separation; Discharge

24. Character Of Service; Honorable

25. Separation Authority And Narrative Reason For Separation; MilPersMan 2121-03-15. Panic Attacks, Sleep Disorders, T.M.J.

(Contemporary) Earth's Global Government (EGG) Space Force
Joint Task Force (J.T.F.) (Military) United States Sailor (U.S.N.) (Spy)
(Super-Hero) (**Vigilante**) (War) U.S. War Office. For Official Use Only (FOUO)

New Reassignment Reporting Under Earth's Global Government (EGG) J.T.F.
Official Orders For "Lieu, Michael W.", Blue To Green, "(Sailor) To (Soldier)";
Recommended For **Military Disabled Veterans'**
Limited Light Duty (L.L.D.) To Assist Earth's Surviving County Military
Bases, In Supporting The War On Mars Campaign.

Approved Travel To Home Permanent Change Station (P.C.S.) Geolocation;
United States Army (U.S.A.) Fort Angel
Military (M) Bases, Surviving County
Los Angeles (R) (D) County, Military (M) Base(s), South California (SCA),
United **States** of North America (U.**S.** of N.A.), Planet Earth, Solar System"

Department(s); United States Department Of War (U.S. DOW),
United States Department of Outbreaks and Outages (U.S. DOO),
Branch; United States Army (U.S.A.),
Component; Civilian, Post Military (M) Active Duty 2136-2139,
As (**Military Disabled Veterans' Limited Light Duty (L.L.D.)**)
Agency; Housing And Lodging (HAL).
Division; Morale, Welfare, Recreation (M.W.R.).
Note(s); Any and all other L.L.D. as assigned for "100% Disability".

Thank You For Your Service To Earth, And To Your Country's Surviving County.

Very Respectfully,
Official Authorized To Sign.

Member X *Lieu, Michael W.*	Witness X *EGG S.F. J.T.F. Personnel.*

EGG Space Force J.T.F. War On Mars Campaign, Godspeed.

(Contemporary) Earth's Global Government (EGG) Space Force
Joint Task Force (J.T.F.) (Military, Civilian) United States Army (U.S.A.) (Spy)
(Super-Hero) (**Vigilante**) (War) U.S. War Office. For Official Use Only (FOUO)

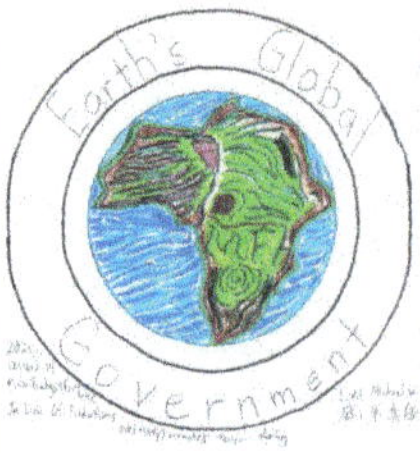

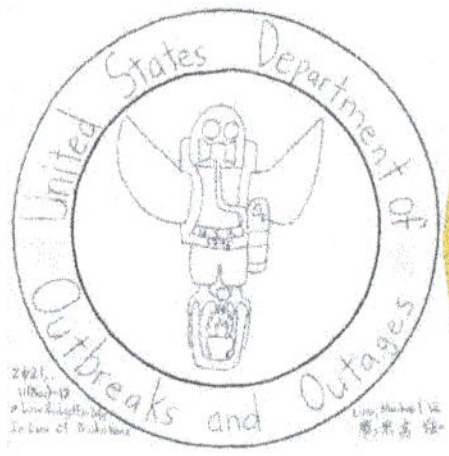

Department(s); United States Department Of War (U.S. DOW),
United States Department of Outbreaks and Outages (U.S. DOO),
Branch; United States Army (U.S.A.),
Component; Civilian "Military Disabled Veterans' Limited Light Duty (L.L.D.)"
Agency; Housing And Lodging (HAL).
Division; Morale, Welfare, Recreation (M.W.R.).
Note(s); Any and all other L.L.D. as assigned for "100% Disability" Eligibility.
Request(s); From **Servicemember Requests D.W. 214 Form Be Sent To**;
U.S.A. Fort Angel, (SCA). United States Department of Veteran Affairs
(U.S. DAV) "Service Connected Disability; 100% (one hundred percent)"
United States Surname Social Security (U.S. S.S.S.) "Disability" Eligibility.

X @ 2139; XX (XXX) – XX (X)	
Approved.	Not Approved.

Official Authorized To Sign.
X *For Colonel Kyle Kinsman, Signed By Lieutenant Colonel (X.O.) Faith Fukuhara*,
Colonel Kyle Kinsman
Commanding Officer (C.O.)
Fort Angel
Los Angeles County (R) (D) Military (M) Base(s), South California (SCA),
United **States** of North America (U.**S.** of N.A.), Planet Earth.

(DocuDrama, Philosophy) (Photo-Book) = Ninja-Talor **A.** Williams' King And X.

Ninja-Talor A. Williams

Talor Alexander Williams { DocuDrama, Photo-Book }
@TalorWilliams

King And X Philosophy

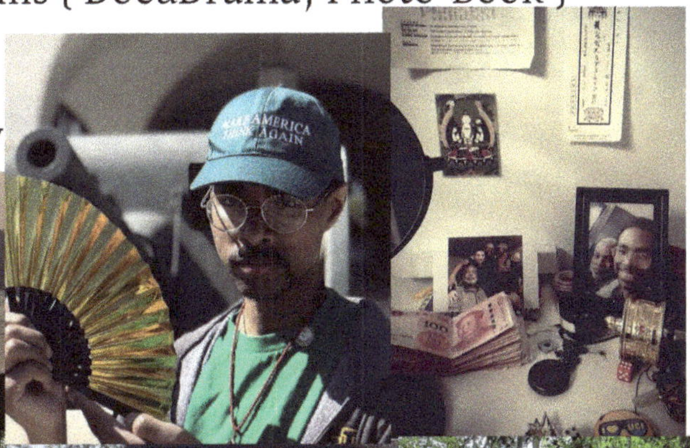

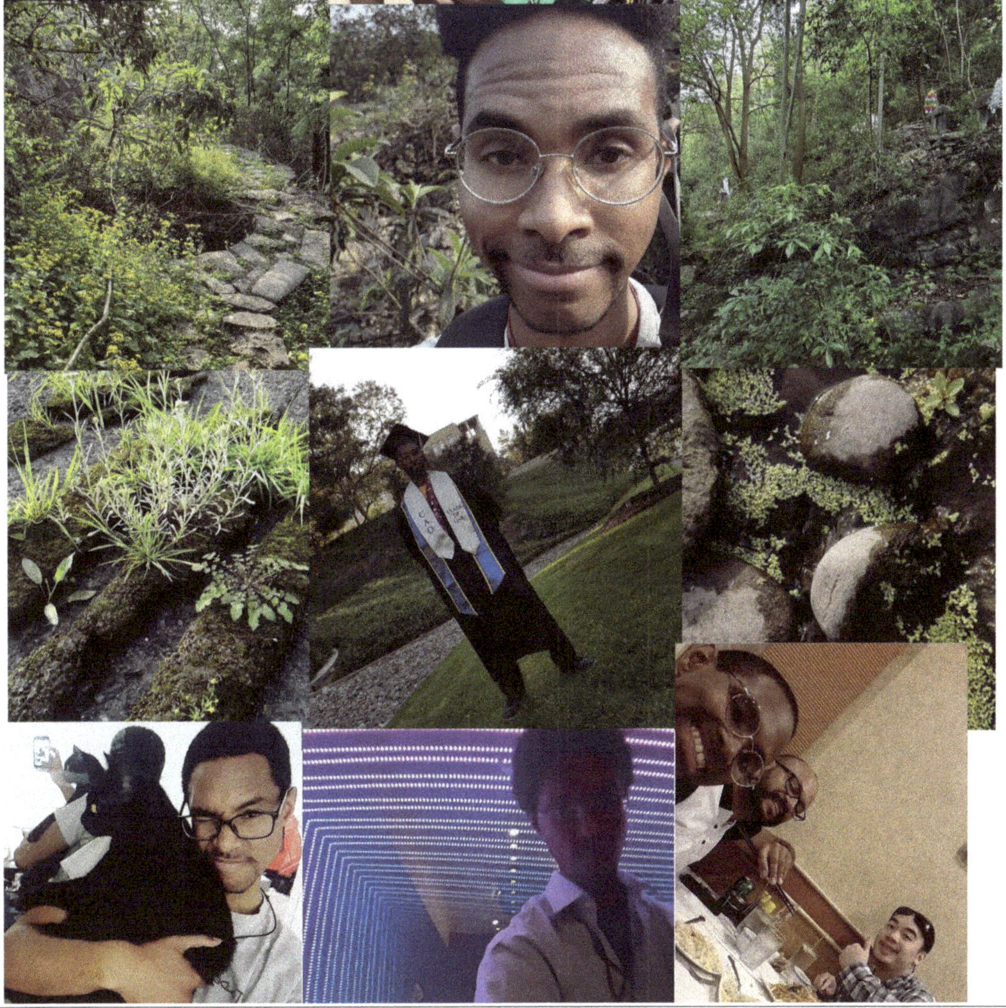

(Vigilante) (War, Military) (Super-Hero) (Parody) (Reality, Gun Girls Gun Guys) (Mythology) (Apocalyptic) (Utopia) (Military) (**Photo-Book**) (**Horror**) = Tools

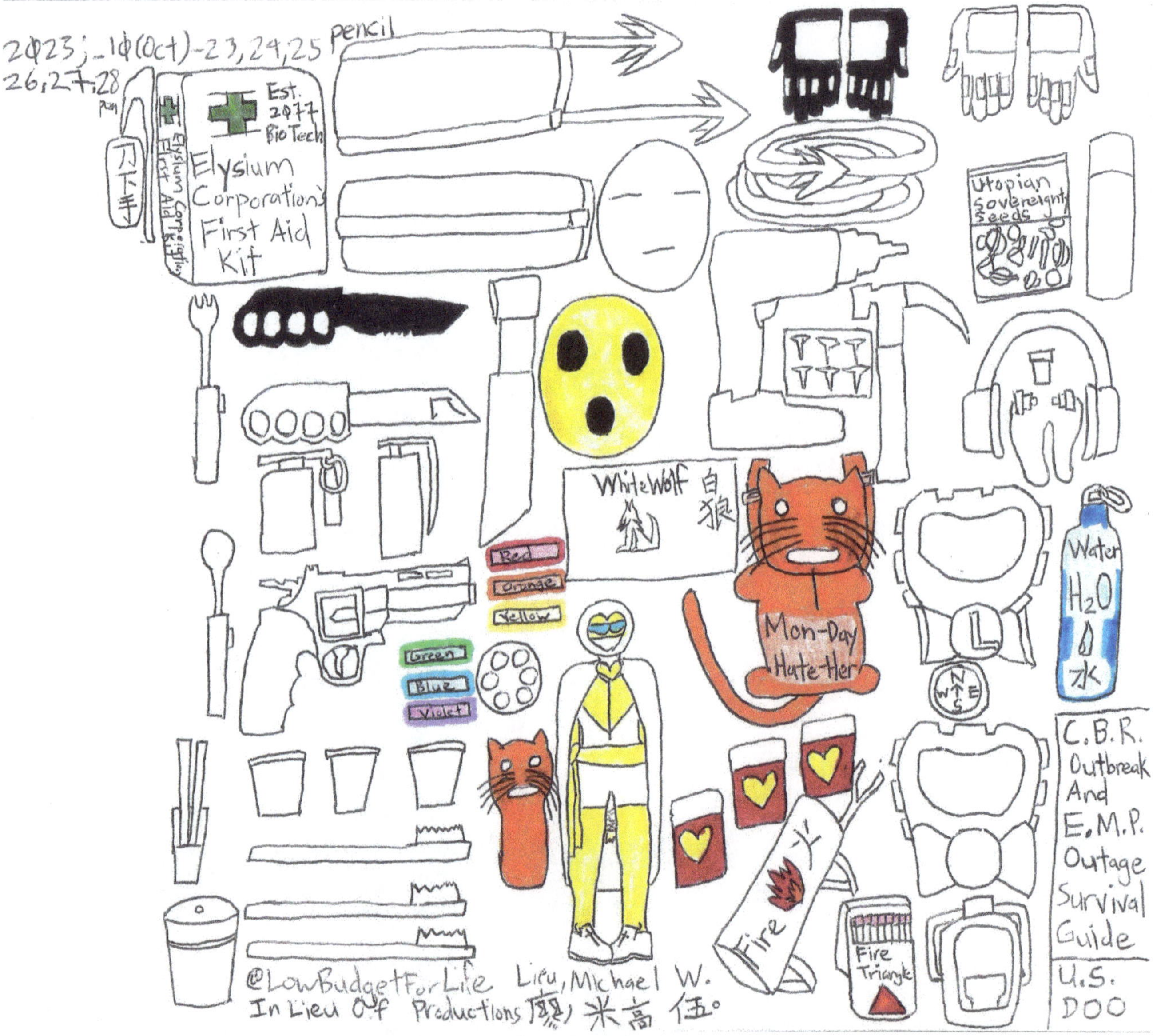

Lieutenant 2nd Lieu, Michael W. 廖, 米高 伍o; Vigilant Volunteers Sisco Cisco (°o°) Yellow;

“**Paperwork** reviewed, **tools** for today's Sir Arthur C. Clarke's Space Elevators Anniversary Launch Celebration... Going to need Yellow **White** HeartBlade doll figure, Orange Conscientiousness Cat Kat stuff animal backpack for my existence insurance in case anything happens to me outside dome. Got **my WhiteWolf suit and helmet by arms legs length**.”

(Vigilante) (Super-Hero) (Parody) (Horror) Biotech Dome Airlock Decontamination

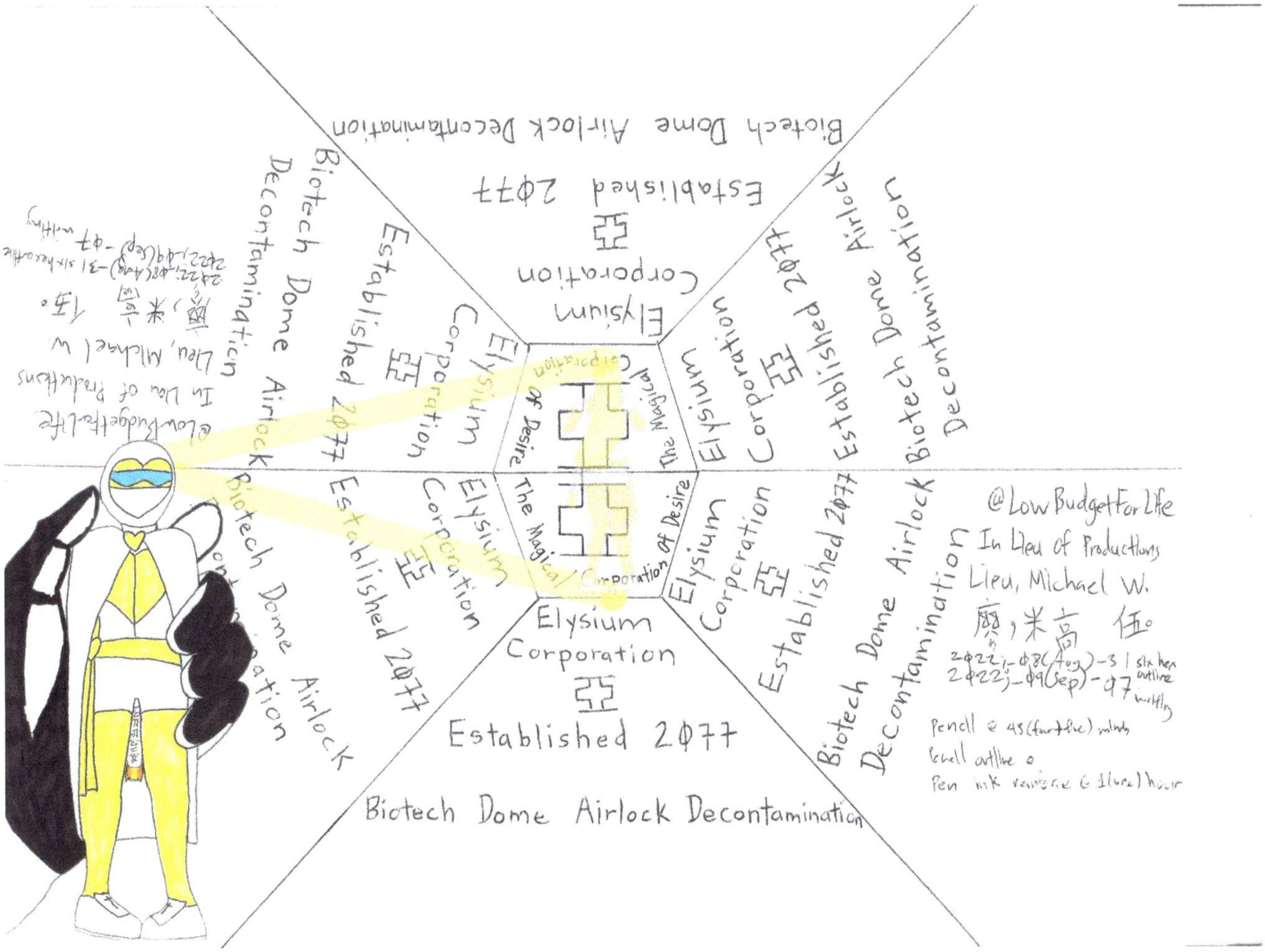

Lieutenant 2nd Lieu, Michael W. 廖, 米高 伍o; Vigilant Volunteers Sisco Cisco (°o°) Yellow;

"Holding **my** Yellow **White** HeartBlade doll, Orange Conscientiousness Cat Kat stuff animal backpack for my existence insurance, and **WhiteWolf suit** close by **limbs length**…

Get through **this airlock** decontamination with **Mister Kelvin Klones.**

There is a yellow hologram broadcast from my Yellow **White** HeartBlade doll figure! Some signal from the Church? …Maybe Sister Mitchell knows.

Need to go to **Downtown District**, **past airlock** plus King And X District, gather few **tools and items needed** before **I** depart from dome animal sanctuary, lastly put away **my journals** so **it** does not get contaminated."

(Vigilante) (Super-Hero) (Parody) (Horror) Biotech Dome Airlock Decontamination

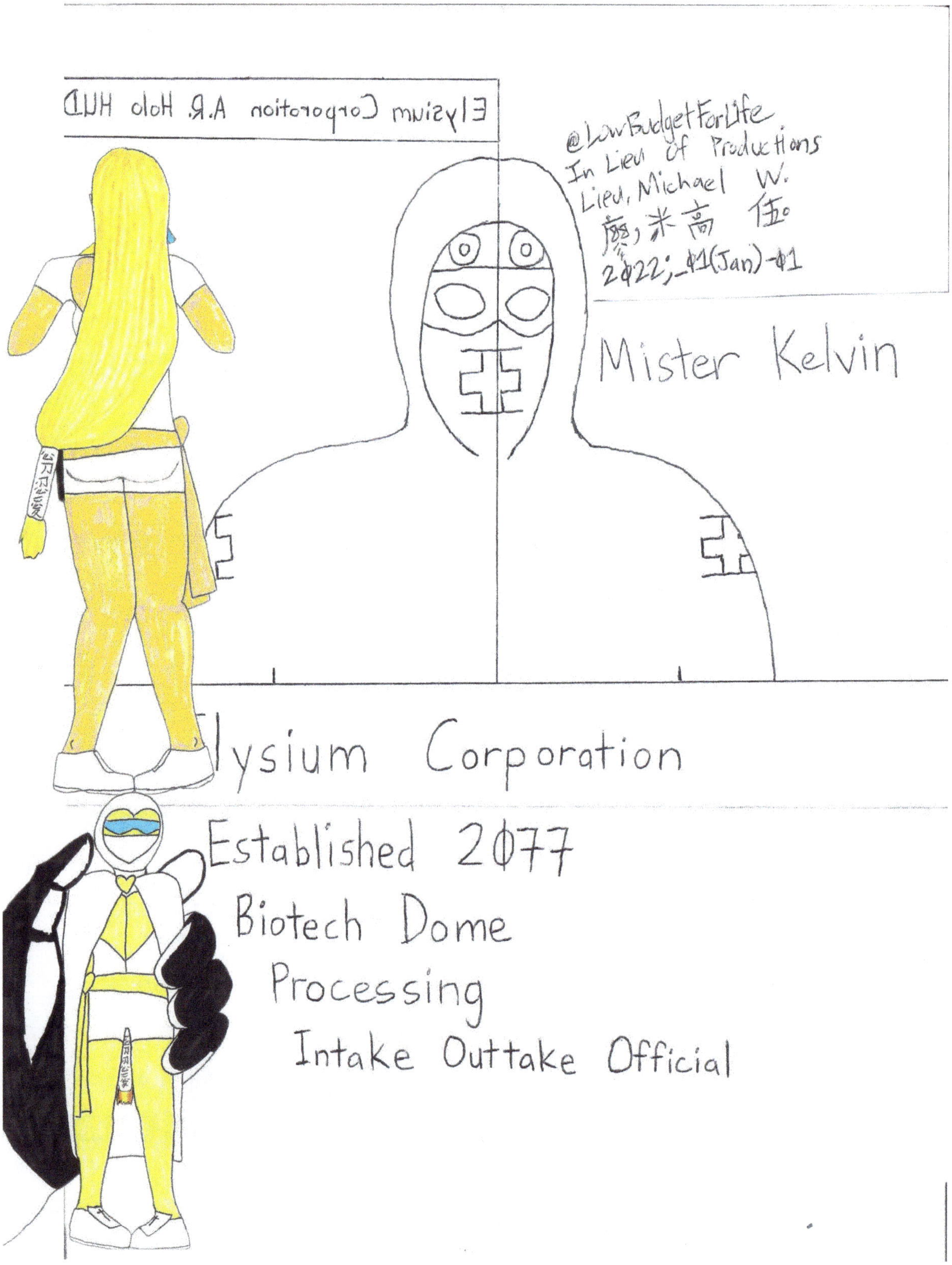

(Super-Hero) (Utopia) (Horror) = Airlock Decontamination Holo HUD Broadcasting

Sisco Cisco .il Yellow (°o°) Broadcast	(°o°) (°o°)	Sisco Cisco .il Yellow (°o°) Broadcast
Date;	Year; (####;) ## (Month) - Day = 2154; 09 (Sep) – 11 (eleven)	

Sisco Cisco .il Yellow (°o°) Broadcast Automation Algorithms Closed Captioning Log;

Elysium Corporation Automation Algorithms; (*monotone female voice over*) "Please stand by for bio**tech** dome **airlock** decontamination. Thank **you** for **your** patience and cooperation **Utopian**."

Mister Kelvin; "Oh hi **Lieutenant Lieu,** got **your WhiteWolf suit with you limbs length**, **your** Yellow **White** HeartBlade doll has a hologram broadcast."

Lieutenant 2nd Lieu, Michael W. 廖, 米高 伍o; "Hey **doc, Mister Kelvin**… going to **Downtown District** have the Church Sisters and or Yellow **White** HeartBlade watch over **my** orange Con Cat backpack if anything happens to me. Existence insurance… Separation between Church and State mainly applies to County Bases **I** used to survive in… Anyways, heading out dome for today's two thousand one space elevator Sir Arthur C. Clarke anniversary."

Mister Kelvin; "Oh Church.. Hm, Sir Arthur C. Clarke, our father figure for space race travel since nineteen fifties… Race to base! Kold War era…"

Lieutenant 2nd Lieu, Michael W. 廖, 米高 伍o; "**You** got a **holo HUD info** of our historical events. Going to stop by **your Klones, Mister Kelvin**. Later."

Mister Kelvin; "Yea, have a good one **Lieutenant Lieu**, **WhiteWolf. Clean**!"

Sisco Cisco .il Yellow (°o°) Broadcast	(°o°) (°o°)	Sisco Cisco .il Yellow (°o°) Broadcast

(DocuDrama, Philosophy)

(Photo-Book)

= King And X District.

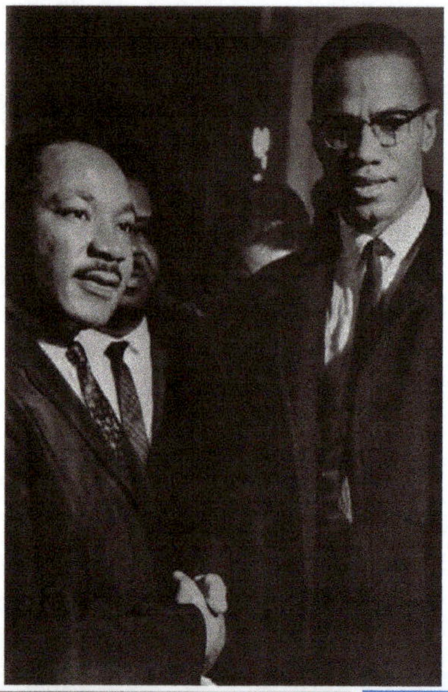

Old World Historical Figures Of The 20th Century United **States** of North America
Martin Luther King (left) Junior **And** Malcolm X (right) Civil Rights Meeting In 1964
"King And X District"

Reverend Martin Luther **King** Junior (Jr.)'s "***I Have A Dream***"	Malcolm **X**'s "***By Any Means Necessary***"
01 (Jan) - 15 Martin Luther King Jr. Day	**05 (May) - 19** Malcolm X Day
"I have a dream that my four little children will one day live in a nation where they will not be judged by the color of their skin, but by the content of their character." -**King**	"I have no mercy or compassion in me for a society that will crush people, and then penalize them for not being able to stand up under the weight." -**X**
King was an African-American black ***intergrationist*** on ethnic equality through ethnic **mixing** and working together.	**X** was an African-American black ***nationalist*** with a firm belief in African-American black **supremacy**.
Non-Violent = **Colored Person(s)** refusing to use physical force in a political movement.	Self-Defense = **Negro(s)** willing to protect their own life and existence.
Reverend Martin Luther **King** Jr.'s "*I Have A Dream*" philosophy inspired and created "**The Non-Violent Dream Movement**" in Washington, D. of C., United **States** of North America	Malcolm **X**'s "*By Any Means Necessary*" philosophy inspired and created "**The Black Panther Party Of Self-Defense**" in Oakland, North California (NCA), United **States** of North America

93 (Lieutenant Lieu, In Lieu Of) Lieu, Michael W. 廖, 米高 伍o 93

(Utopia) @ North (N) Side to Downtown District (D. D.) (Parody) (Super-Hero)

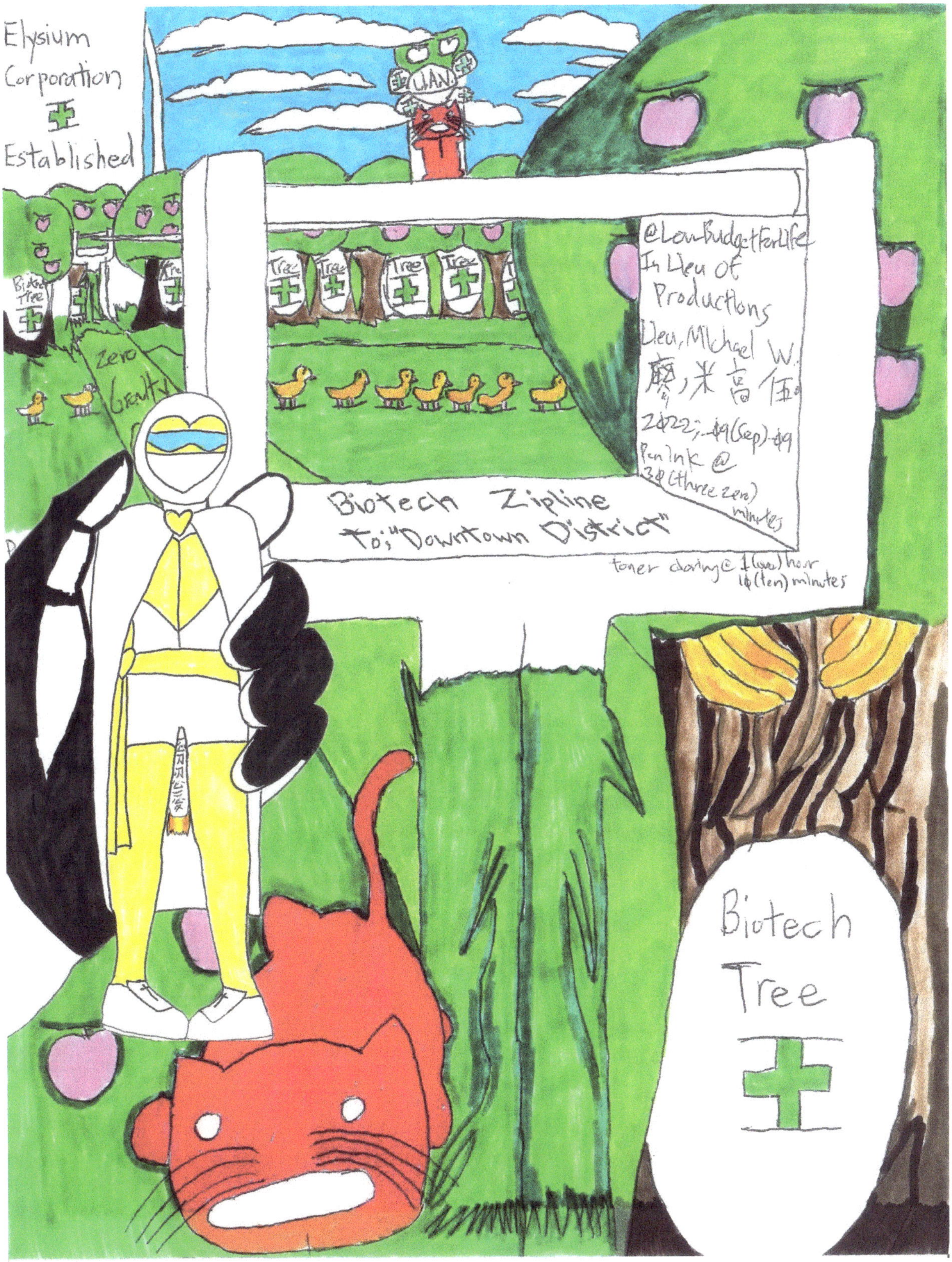

93 \ **2154;** Lieu, 廖, ‘WhiteWolf 白狼’ Left to Right (L→R) 93

Lieutenant 2nd Lieu, Michael W. 廖, 米高 伍o; Vigilant Volunteers Sisco Cisco (°o°) Yellow;

"Taking note of my to do list on **my journal** entry for Sir Arthur C. Clarke's Space Elevators Anniversary Launch Celebration...

I am now about to head to **Downtown District by** Bio**tech Zipline**, going to have **my** orange conscientiousness cat backpack roam around the North Side and see if **my** Yellow **White** HeartBlade doll hologram broadcast signal is up... Blue sky clouds means The Church's West Side, while the red sky clouds means The Market and or The Farm, All For One Shopping Center.

I should gather breathable air oxygen by the green biotech trees, while **I** wait and get **WhiteWolf suit ready for** exterior dome deployment.

Stare at the **white U.A.V.s with the** orange cat holo broadcasts and fruit trees flying around, gonna need some fruits, **solid foods** going to help **the suit.**

Stare at the yellow baby birds, think those are called chicks following each other in a straight line. Review **my other paperwork** while **I** wait...

Chicago (S), The Last Animal Sanctuary (S) Illinois (ILL), **U**.S. of N.A., Earth.
Public And Private Sec Voluntold-Volunteer Services Bulletin Board System (B.B.S.) In **Downtown District (D. D.)**, Abstract Sheet. As Of 2154;
Public Sector, Voluntold Service, "No One Really Wants To Do This, And Wants Out..." "**Jerry Duty**", Xtra-Men, Shaked-Soda, Chicago's Utopians, U.S. Citizens, Alien status.
Public Sector, Volunteer Group, "**Clean-Up Crew**, Get Ready At Five Fifteen A.M.!" "**Clean-up Crew**", **Chicago Controller City Mayor's Support Staff**, **Friendly Peaceful Natural Animal Caretakers**, **Search And Rescue (SAR), Search And Recovery (SAR)**, **Entertainers' Music Video (M.V.) Display.**
Public Sector, Volunteer Group, "Aug+Nat-Animal Human First Responder(s) On Call" "**Strike Force (打擊 力量)**" "**ToolPerson (工具人)**", Handy Human's Jack Of All Trades, Master Of Nothing, Combat Support Engineers, Toy-Masters, Hydro H2O Water, Navy.
Public Sector, Volunteer Group, "Aug+Nat-Animal Human First Responder(s) On Call" "**Utopian Peacekeeping**", Public Militia Aug-Nat Co-Existence, And "**Fan Following**"
Public Sector, Belief Systems, "Separation Between Church And State" "**Chicago City Christian Church"** Belief Systems And Other Religions Welcome +<3"
Private Sector, Confidential, N.D.A., **Kay's Private Security Services L.L.C. R&D.**
Private Sector, Confidential, N.D.A., **Elysium Corporation,** Bio**tech Dome Sector.**

(Vigilante) (Super-Hero) (Parody) (Utopia) = Orange Con Cat Yellow Broadcasting

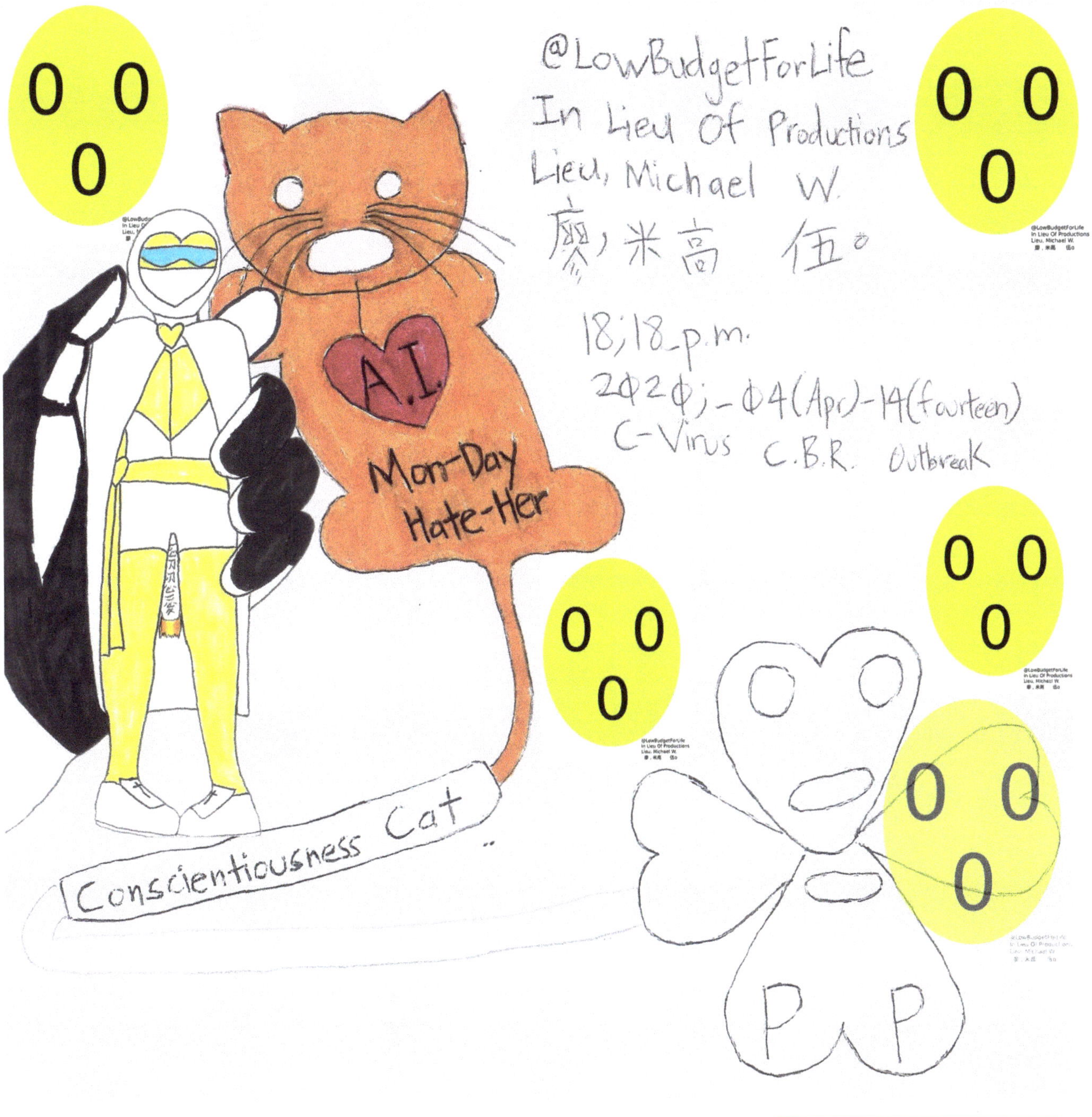

Lieutenant 2nd Lieu, Michael W. 廖, 米高 伍o; Vigilant Volunteers Sisco Cisco (°o°) Yellow;

"**My** orange conscientiousness cat is broadcasting yellow surprise face holograms, think it senses that a nearby Sister Nun and or Yellow **White** HeartBlade can watch over **my** existence insurance, need to look around."

(Utopia) (Super-Hero) (Parody) (Vigilante) = Breed-**Built** Super-Hero Skin Suits.

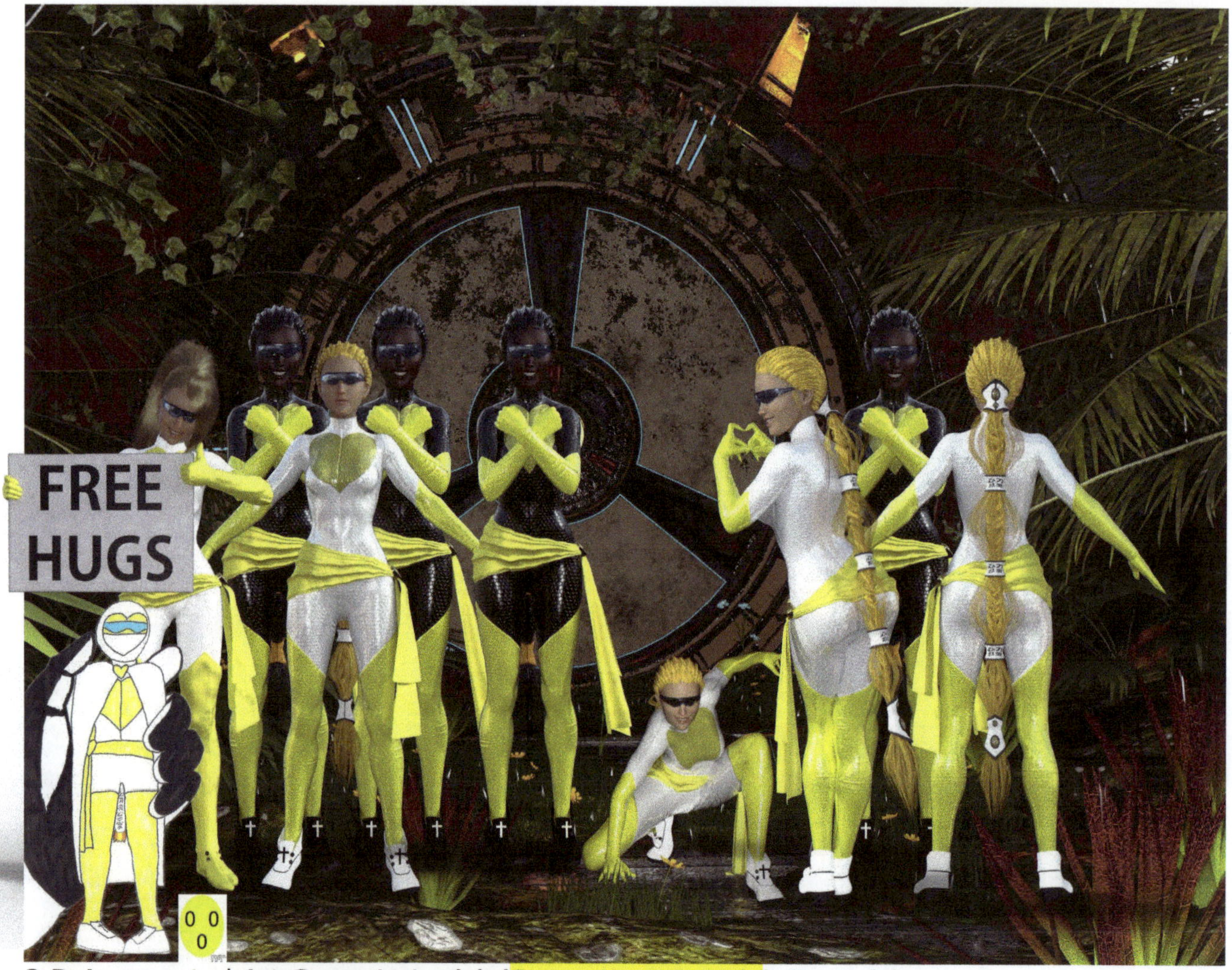

3 D Augmented Art Commission(s) (Super-Hero Parody) Genre(s)
Yellow **White** "HeartBladeEndureLove 心刀忍愛" And
Yellow Black "HeartBladeEndureLove 心刀忍愛" Model(s)
By; Amanda Coryell, AKA MandyKatDWP (Draxis Web Productions)

Lieutenant 2nd Lieu, Michael W. 廖, 米高 伍o; Vigilant Volunteers Sisco Cisco (°o°) Yellow;

"This is quite the Yellow **White** and Yellow Black HeartBlade person size suits, Mister Kenny Kreates has been storing them here and not in **Elysium Corp**.

My orange conscientiousness cat is broadcasting yellow surprise face holograms still… **I** think it senses that a nearby person size suits recently gain awareness…? Consciousness? …There is one kneeing down position and some of them have managed to change person size poses…"

(Parody) (Super-Hero) = Yellow **White** "HeartBladeEndureLove 心刀忍愛" P.S.

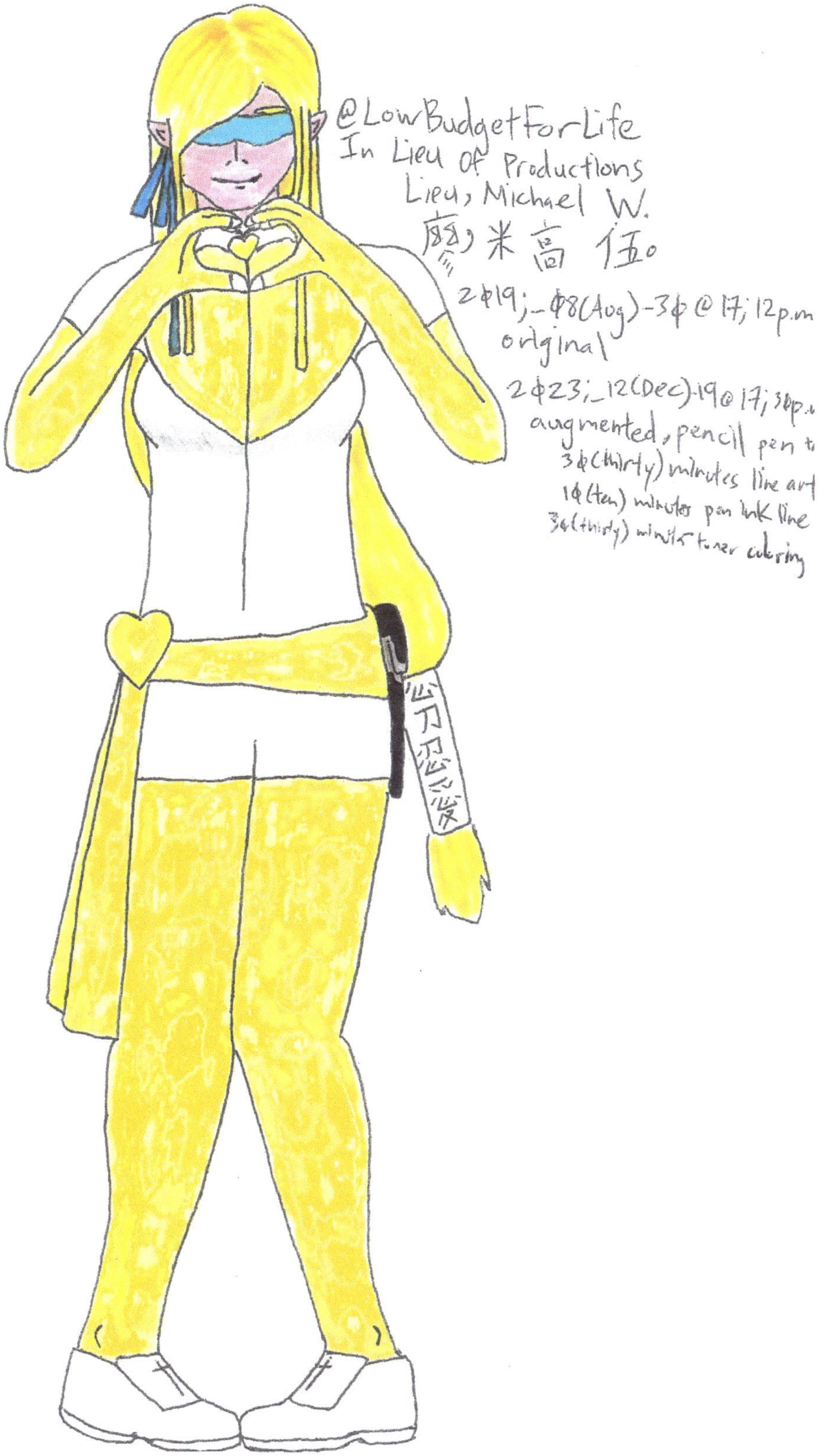

Sisco Cisco .il Yellow (°o°) Broadcast	(°o°) (°o°) (°o°) (°o°)	Sisco Cisco .il Yellow (°o°) Broadcast
Date;	2154; 09 (Sep) - 11	
Actual Legal Given Name(s); Yellow **White** "HeartBladeEndureLove 心刀忍愛" <3,		
Alias Network Nickname(s); Yellow **White** "HeartBladeEndureLove 心刀忍愛", Heart Love Hand Shapes, Super-Hero Parody, Spandex-Woman, A Nun, **Free Hugs**, Augmented-Woman, Sister Sister, Yelling Commands, Female Klark Kent, [Aug+]		
Love(s) Or Like(s); Love, Love is the frequency. **Superman 64** **The Disaster Artist The Room** [*Aug+*]		
[*Aug+*]		
Sisco Cisco .il Yellow (°o°) Broadcast	(°o°) (°o°) (°o°) (°o°)	Sisco Cisco .il Yellow (°o°) Broadcast

(Parody) (Super-Hero) = Yellow **White** "HeartBladeEndureLove 心刀忍愛", Briana

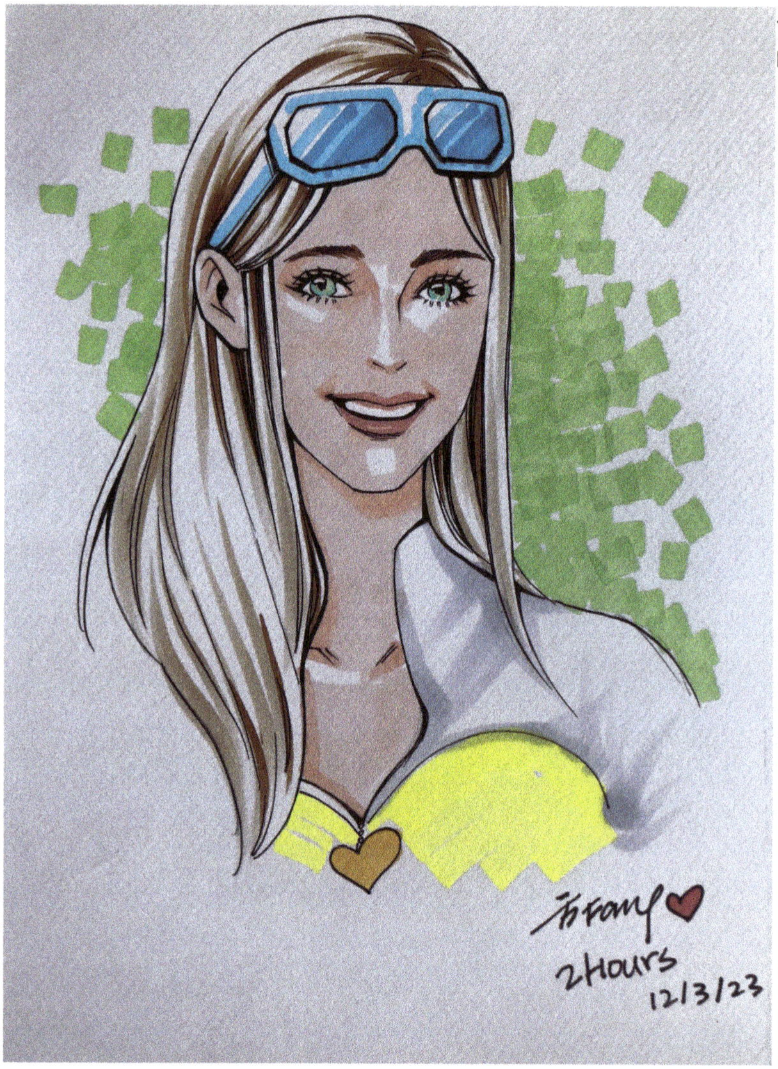

{ "2 D Augmented Art Commission(s)
Portrait(s) By; Fang Fang Comics, Fang YiLi"

Sisco Cisco .il Yellow (°o°) Broadcast	(°o°) (°o°) (°o°) (°o°) (°o°) (°o°) (°o°) (°o°)	Sisco Cisco .il Yellow (°o°) Broadcast
Date;	2154; 09 (Sep) - 11	
Actual Legal Given Name(s); Sister ✝ **K. W. White** <3 Goes by "Sister **White** <3, Yellow **White** HeartBlade"		
Alias Network Nicknames; Yellow **White** "HeartBladeEndureLove 心刀忍愛", Heart Love Hand Shapes, Super-Hero Parody, Spandex-Woman, A Nun, Augmented-Woman, **Free Hugs**, Sister Sister, Yelling Commands, Female Klark Kent, [Aug+]		
[*Aug+*]		
Sisco Cisco .il Yellow (°o°) Broadcast	(°o°) (°o°) (°o°) (°o°) (°o°) (°o°) (°o°) (°o°)	Sisco Cisco .il Yellow (°o°) Broadcast

Lieutenant 2nd Lieu, Michael W. 廖, 米高 伍o; Vigilant Volunteers Sisco Cisco (°o°) Yellow;

"Ok, interesting... Looks like **my** orange con cat existence insurance broadcasting was on to something...

This Yellow **White** HeartBlade seems to have gain awareness and is able to lift up her blue sunglasses... Her eyes, **iris, is able to develop form**.

Big Ben's splinter cell Chicago Chapter Mexican-Men Militia says in **myth**, when **scales fall from their eyes**, consciousness formed as a consequence of eating the forbidden fruit in utopia, the garden of Eden."

(Super-Hero) Sisco Cisco .il Yellow-Mellow Surprise Face (°o°) Broadcast Network
(Utopia) = A.R. Holo HUD Intranet, Wireless Broadcasting Frequencies Online
(Parody) = Parodying Super-Hero Yellow Limbs Suit, **White Phoenix**. Blond Hair.

Sisco Cisco .il Yellow (°o°) Broadcast	(°o°) (°o°)	Sisco Cisco .il Yellow (°o°) Broadcast
Date;	Year; (####;) ## (Month) - Day = 2154; 09 (Sep) – 11 (eleven)	

Sisco Cisco .il Yellow (°o°) Broadcast Automation Algorithms Closed Captioning Log;

Lieutenant 2nd Lieu, Michael W. 廖, 米高 伍o; "Oh hi Yellow **White** HeartBlade?"

Yellow **White** "HeartBladeEndureLove 心刀忍愛"; "...Oh hi Yellow **White** HeartBlade? ...Is that **your name** or...? What language I am speaking? What I am? Who I am? Where do I come from? How come my skin stretches?"

Lieutenant 2nd Lieu, Michael W. 廖, 米高 伍o; "Your yellow Heart Love symbol, arms, legs limbs and belt sash is usually color code. Hence, Yellow. Call sign after Yellow, varies **based on body torso and shoes** you are wearing, in your case **White**. Your full name is Yellow **White** HeartBladeEndureLove. You are speaking English... What you are... You are an **augmented** bio**tech human female**, breed-**built** by orange clothing wearing Mister Kenny Kreates of **Elysium Corporation in Downtown District the center of this dome** animal sanctuary. Your skin stretches because what you are wearing is a skintight suit designed as a **Personal Protective Equipment**. **I** need you as Church official to watch over **my** orange stuff animal backpack Con Cat, it is my existence insurance, it has a wearable straps as a backpack.... Let **me** show you, and put down your blue sunglasses HUD, that glasses is a tool that provides info to you as well as style, you can change your toner to something else if you desire."

Sisco Cisco .il Yellow (°o°) Broadcast	(°o°) (°o°)	Sisco Cisco .il Yellow (°o°) Broadcast

(Utopia) (Super-Hero) (Parody) = Yellow **White** "HeartBladeEndureLove 心刀忍愛"

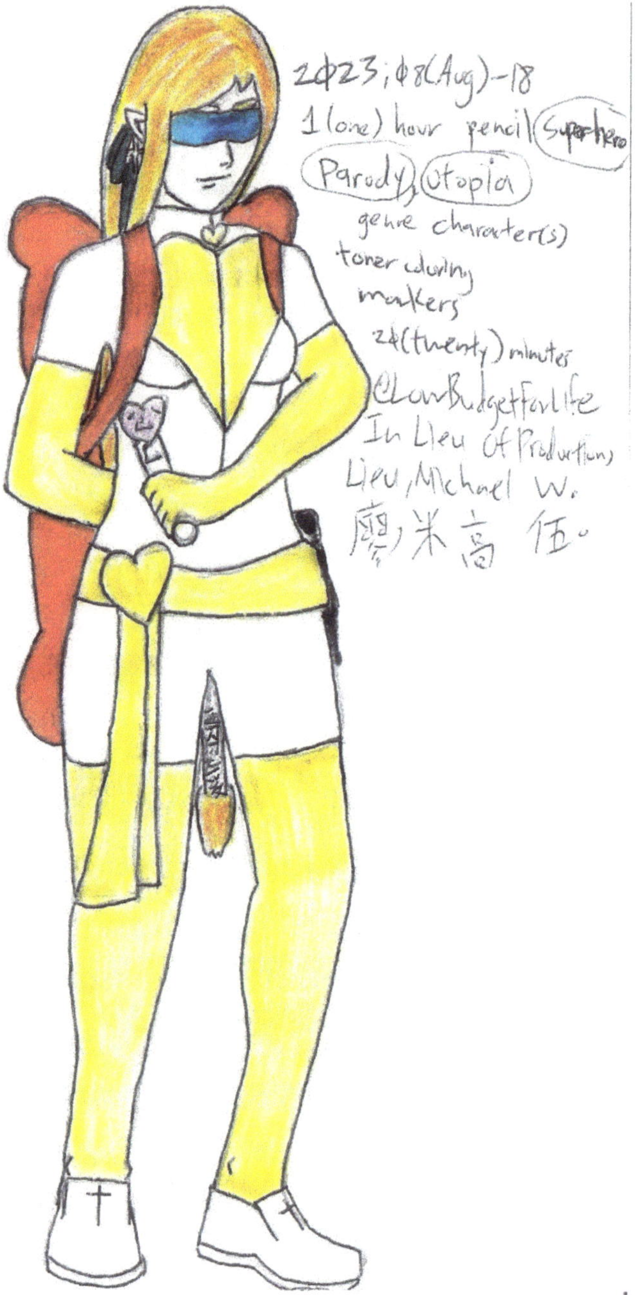

Sisco Cisco .il Yellow (°o°) Broadcast	(°o°) (°o°) (°o°) (°o°) (°o°) (°o°) (°o°) (°o°) (°o°)	Sisco Cisco .il Yellow (°o°) Broadcast
Date;	2154; 09 (Sep) - 11	
Actual Legal Given Name(s); Yellow **White** "HeartBladeEndureLove 心刀忍愛" <3		
Alias Network Nickname(s); Yellow **White** "HeartBladeEndureLove 心刀忍愛", Heart Love Hand Shapes, Super-Hero Parody, Spandex-Woman, A Nun, Augmented-Woman, **Free Hugs**, Sister Sister, Yelling Commands, Female Klark Kent, [Aug+]		
To Do List Task; "Yellow **White** HeartBlade, **I** know there is separation of Church And State, watch over, take good care of **my "Existence Insurance** OCEAN 5 (Five) Orange Conscientiousness Cat" if anything happens to **me in WhiteWolf suit** when **I** am doing **engineering dome maintenance**." -Lieutenant Lieu, Michael W.'s Will (2nd 廖, 米高 伍o)		
[*Aug+*]		
Sisco Cisco .il Yellow (°o°) Broadcast	(°o°) (°o°) (°o°) (°o°) (°o°) (°o°) (°o°) (°o°) (°o°)	Sisco Cisco .il Yellow (°o°) Broadcast

Lieutenant 2nd Lieu, Michael W. 廖, 米高 伍o; Vigilant Volunteers Sisco Cisco (°o°) Yellow;

"This Church's Yellow **White** HeartBlade, has gain awareness... Is aware, however, is she toner bind? **I** should ask her. After she puts on **my** orange stuff animal backpack Con Cat existence insurance... **Existence insurance** check."

(Super-Hero) Sisco Cisco .il Yellow-Mellow Surprise Face (°o°) Broadcast Network
(Utopia) = A.R. Holo HUD Intranet, Wireless Broadcasting Frequencies Online
(Parody) = Parodying Super-Hero Yellow Limbs Suit, **White Phoenix**. Blond Hair.

Sisco Cisco .il Yellow (°o°) Broadcast	(°o°) (°o°)	Sisco Cisco .il Yellow (°o°) Broadcast
Date;	Year; (####;) ## (Month) - Day = 2154; 09 (Sep) – 11 (eleven)	

Sisco Cisco .il Yellow (°o°) Broadcast Automation Algorithms Closed Captioning Log;

Lieutenant 2nd Lieu, Michael W. 廖, 米高 伍o; "Hey chief, Yellow **White**, are you able to tell what toner your arms and legs are, plus body torso and shoes?"

Yellow **White** "HeartBladeEndureLove 心刀忍愛"; "...Hey chief, what **I** see is Yellow... My body torso and shoes looks like **White**. Your hair is **white, partially white**. Interesting mix of **white** with black and gray and sliver."

Lieutenant 2nd Lieu, Michael W. 廖, 米高 伍o; "Whoa, yea, you are not toner blind. **My** hair is **white partially white**. You will be safe inside this green dome animal sanctuary by the **white pillar Elysium Corp buildings** and Violet changing hex panels on top of Downtown District with yellow animals, baby chicks, walking around in a straight line sometimes. Any questions, concerns, look up to the blue sky, cross the flower fields and to the Church, she wears black mostly and yellow limbs as color clothing code, ask for Sister Mitchell, she can explain to you much more in detail... Your blue sunglasses you are wearing should be able to provide more questions, concerns. Ok, **I** gotta go Yellow."

Yellow **White** "HeartBladeEndureLove 心刀忍愛"; "Ok, bye **Lieu Lieu**!"

Sisco Cisco .il Yellow (°o°) Broadcast	(°o°) (°o°)	Sisco Cisco .il Yellow (°o°) Broadcast

(Utopia) = Chicago (S), The Last Animal Sanctuary (S) Of Illinois (ILL) Interior

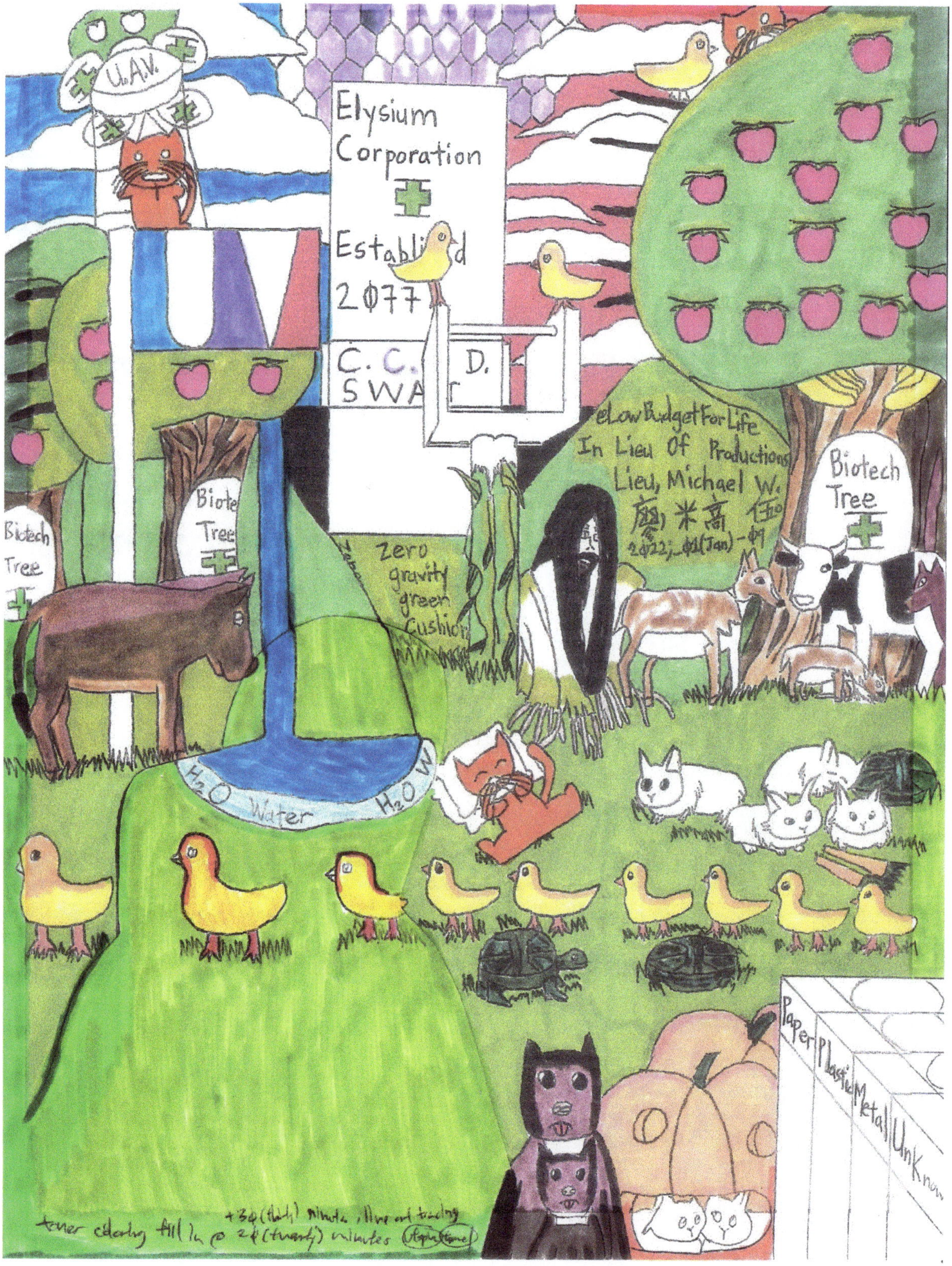

Lieutenant 2nd Lieu, Michael W. 廖, 米高 伍o; Vigilant Volunteers Sisco Cisco (°o°) Yellow;

"Ok, gotta slow down and **journal this** on the actuality that **my** Conscientiousness Cat backpack existence insurance was able to locate a Yellow **White** HeartBlade person size that gain consciousness, with the assistance of **my own doll** that was broadcasting using Sisco Cisco's surprise face tool...

This psychological phenomenon consciousness, awareness, how did it ever occur and happened within animal sanctuary dome?

Mister Kenny Kreates in **Downtown District** is gonna be amazed when he finds out one of his bio**tech** breeding buddies gained consciousness through **my** orange Conscientiousness Cat backpack broadcasting tools...

Both of them have orange toner... connection notice, orange color code.

Conscientiousness, consciousness... **These two words are spelled quite similar and different in some ways, a bit confusing to me as E.S.L....**

C o n s c i e n t i o u s n e s s.

Conscientiousness, hard working workaholic personality under psychology OCEAN 5, five.

C o n s c i o u s n e s s.

Consciousness, being aware state of mind, origins of consciousness is a bit of a mystery... Maybe Mayor Chicago Controller has something to do with it?

Can consciousness be engineered, build, we know consciousness can be passed on from natural people to natural people by breeding, however, **building artificial persons** like Yellow **White** and Yellow Black HeartBladeEndureLove.

That is another question for another time... in **my retirement journal**.

Gotta get **WhiteWolf suit ready up on air and food**, need to continue inwards towards **Downtown District** by Green Guys' Xtra-Men Jerry Duty... and Mars Hall with that biotech vending... and let **us** see what else?

Still got some time for today's outside dome space elevator, Sir Arthur C. Clarke anniversary celebration broadcast."

(SitCom, Western Society) = Xtra-Men, Jerry Duty, Dog Name Cat, Mars Hall.

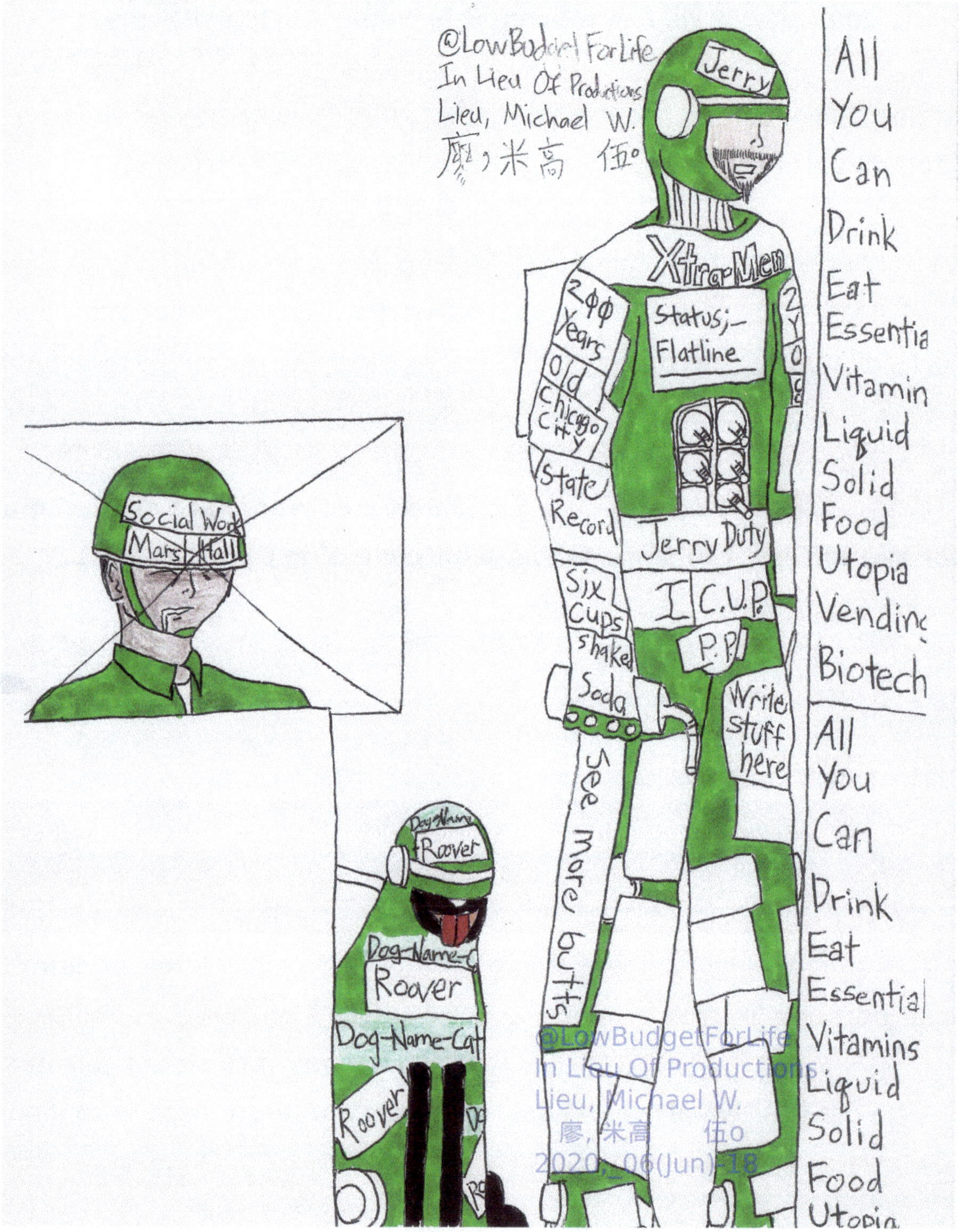

(Horror) (SitCom, Western Society) (Utopia) = Utopian Sovereignty Biotech Vend

Elysium Corporation Augmented Reality Holographic Heads Up Display A.R. Holo HUD			
All You Can Drink Eat Essential Vitamins Liquid Solid Food Utopia Vending Biotech Mixed Drink Selection; "Antioxidants Super Foods Spices"			
Cloves Among the three spices in antioxidant concentration. Contains eugenol, carophyllin, gallic acid, eugenin. Inhibits oxidation of L.D.L. cholesterol, an early step in atherosclerosis.	Cinnamon An antioxidant powerhouse, compounds protect against inflammation, heart disease, Alzheimer's. Polyphenols, increases insulin activity, aiding diabetes control. Boosts attention.	Allspice Contains more than a dozen antioxidant, including eugenol, quercetin, ellagic and gallic acids. Ellagic acid has anticancer effects, decreases proliferation of pancreatic cancer cells.	Saffron Loaded with carotenoids, which possess anticancer activity and aid immune function. Also protects blood fats against oxidation, a step in the development of cardiovascular disease.
Fennel Contains flavonoids and tannins. Stimulates bile acid production in the liver, giving it a major role in fat digestion and absorption. Also contains anethole, reduce inflammation and retard cancer.	Oregano Another antioxidant powerhouse loaded with phenols. Rosmarinic acid has demonstrated antimicrobial activity against food-borne pathogens, and can be used topically as an herbal antiseptic.	Cumin Seeds have significant antiglycemic effect and help reverse metabolic abnormalities in diabetes. Also stimulate bile acid production and its secretion into bile.	Garlic Pungent sulfur compounds promote heart via beneficial effects on lipid metabolism and reduction of cholesterol levels. May also reduce blood pressure and retard cancer growth.
Mustard [*Aug+*]	Rosemary [*Aug+*]	Mint [*Aug+*]	Turmeric [*Aug+*]
Basil [*Aug+*]	Ginger [*Aug+*]	Red Pepper [*Aug+*]	Black Pepper [*Aug+*]
[*Aug+*]			
Elysium Corporation Augmented Reality Holographic Heads Up Display A.R. Holo HUD			

(SitCom, Western Society) = Xtra-Men, Armed Social Worker Mars Hall. Mic Pena

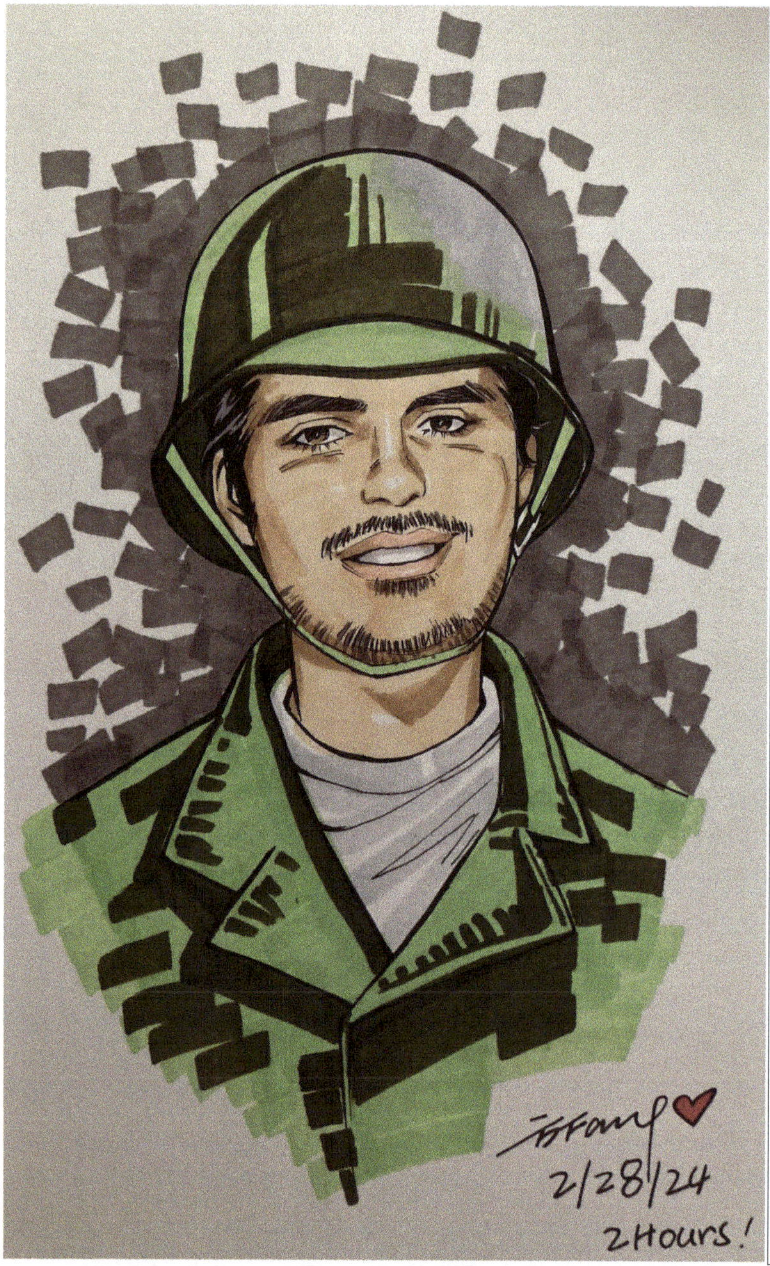

{ "2 D Augmented Art Commission(s) Portrait(s) By; Fang Fang Comics, Fang YiLi" }

Sisco Cisco .il Yellow (°o°) Broadcast	(°o°) (°o°) (°o°) (°o°) (°o°) (°o°) (°o°) (°o°)	Sisco Cisco .il Yellow (°o°) Broadcast
Date;	2154; 09 (Sep) - 11	
Actual Legal Given Name(s); Mars Hall Goes by "Mars"		
Alias Network Nickname(s); Armed Social Worker Mars Hall, Xtra-Men, Echo Tool, Jerry Duty, Green Soldier Boy, Mexican-Men Militia, [Aug+]		
Employment; **Public Sector, Voluntold Service, Jerry Duty** Tax Free Monthly Earnings $123,456.78 Only United States Dollars (U.S.D.)		
Ethnicity; Latino, Natural Unmodified.		
[Aug+]		
Sisco Cisco .il Yellow (°o°) Broadcast	(°o°) (°o°) (°o°) (°o°) (°o°) (°o°) (°o°) (°o°)	Sisco Cisco .il Yellow (°o°) Broadcast

Lieutenant 2nd Lieu, Michael W. 廖, 米高 伍o; Vigilant Volunteers Sisco Cisco (°o°) Yellow;

"Uh oh, **I** believe Mars Hall has awaken from his quote exciting unquote, Jerry Duty watch... Green soldier boy, armed social worker is now fully awake and aware... **I** wonder if **I** can talk **my** way out by saying this is all psychological dream and he is still in a dream made possible by Mayor Chicago Controller..."

(Super-Hero) (Utopia) (SitCom, Western Society) = Xtra-Men Mars Hall. Mic Pena

Sisco Cisco .il Yellow (°o°) Broadcast	(°o°) (°o°)	Sisco Cisco .il Yellow (°o°) Broadcast
Date;	Year; (####;) ## (Month) - Day = 2154; 09 (Sep) – 11 (eleven)	

Sisco Cisco .il Yellow (°o°) Broadcast Automation Algorithms Closed Captioning Log;

Lieutenant 2nd Lieu, Michael W. 廖, 米高 伍o; "Oh hi, uh... This is just a dream, you are still sleeping for Jerry Duty as Xtra-Men. This is psychological!"

Social Worker Mars Hall; "Oh hi **Lieu Lieu**! Jerry! Jerry Duty still undead."

Xtra-Men, Jerry; "(*panicking, male voice over*) I need **Shaked-Soda**! Please!! I need **Shaked-Soda**! Cat be a good Dog and get me **Shaked-Soda**! Please!!"

Lieutenant 2nd Lieu, Michael W. 廖, 米高 伍o; "...Jerry Duty still requesting **Shaked Soda**? He really loves and needs his **Shaked Soda** as undead. Looks like Cat Dog Rover is doing good with his biotech suit, just on stand by... Anyways... There is a lot of **sticky notes** all over both of them..."

Social Worker Mars Hall; "Yea... Jerry Duty is exciting, everybody needs a hobby! Wanna see the millions of U.S. Dollars I have? Making money six figure salary per month, one hundred twenty three thousand, four hundred fifty six dollars and seventy eight cents! A millionaire in Chicago Animal Sanctuary!"

Lieutenant 2nd Lieu, Michael W. 廖, 米高 伍o; "Next time sir, military Mars Hall. Got **WhiteWolf suit** at limbs length, space elevator anniversary today."

Social Worker Mars Hall; "Oh ok, kool, see **you** later **Lieu Lieu**. Millionaire!"

Sisco Cisco .il Yellow (°o°) Broadcast	(°o°) (°o°)	Sisco Cisco .il Yellow (°o°) Broadcast

(Crime Drama) = Toilet Paper (T.P.) Watch In Chicago (S) Animal Sanctuary ILL.

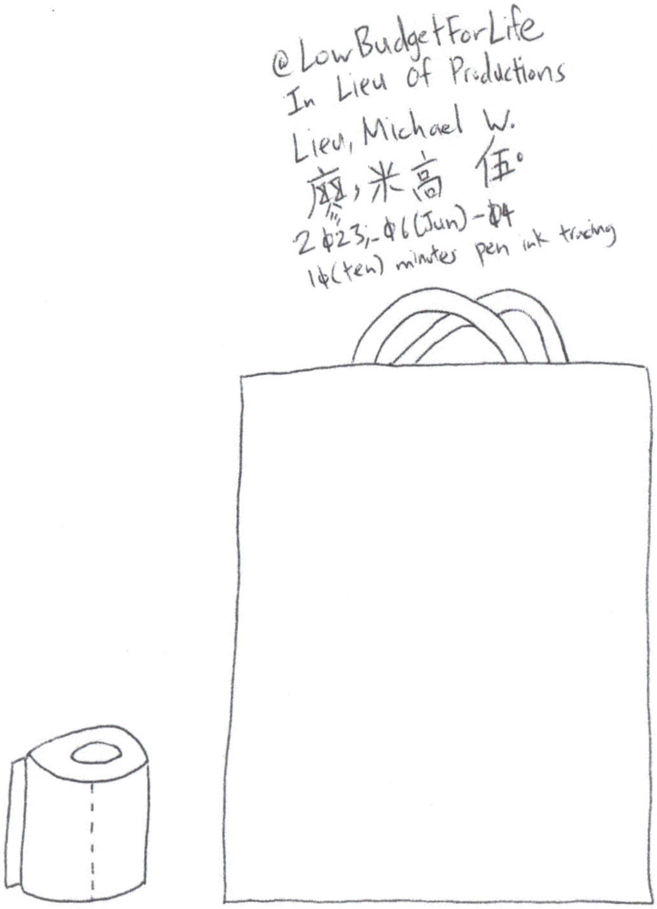

Sisco Cisco .il Yellow (°o°) Broadcast	(°o°) (°o°) (°o°) (°o°) (°o°) (°o°) (°o°) (°o°)	Sisco Cisco .il Yellow (°o°) Broadcast
Date;	2154; 09 (Sep) - 11	
A.R. Holo HUD Message; From Retired Officer Ogden, Deana M. "**2 (Two) Toilet Paper (T.P.)** max per "**Augmented Animals**" Also Known As (AKA) Humans, in this Chicago Animal Sanctuary ILL... **Crime** punishable to take more than **Two T.P.** I am watching."		
[Aug+]		
Sisco Cisco .il Yellow (°o°) Broadcast	(°o°) (°o°) (°o°) (°o°) (°o°) (°o°) (°o°) (°o°)	Sisco Cisco .il Yellow (°o°) Broadcast

Lieutenant 2nd Lieu, Michael W. 廖, 米高 伍o; Vigilant Volunteers Sisco Cisco (°o°) Yellow;

"**I** think **I** need **Toilet Paper, T.P.**, Ogden still watching for a retired Officer... Bet she is taking a huge fooking sheet...

Master At Arms Ogden is watching, stealth suit functions online and active.

Got that SWAT suit of hers near by, lucky she not going out of dome would have to decontaminate it from being exposed outside...

Ye, just take **one T.P.**, since **I** have **one T.P.**...

Send them back to **my** sector in North (N) Side by **U.A.V....**

After this, gonna have to go through another **airlock** decontamination, **two** decon after being from **my** North (N) Side to **Downtown District**."

(Horror) = Biotech Dome Airlock Decontamination

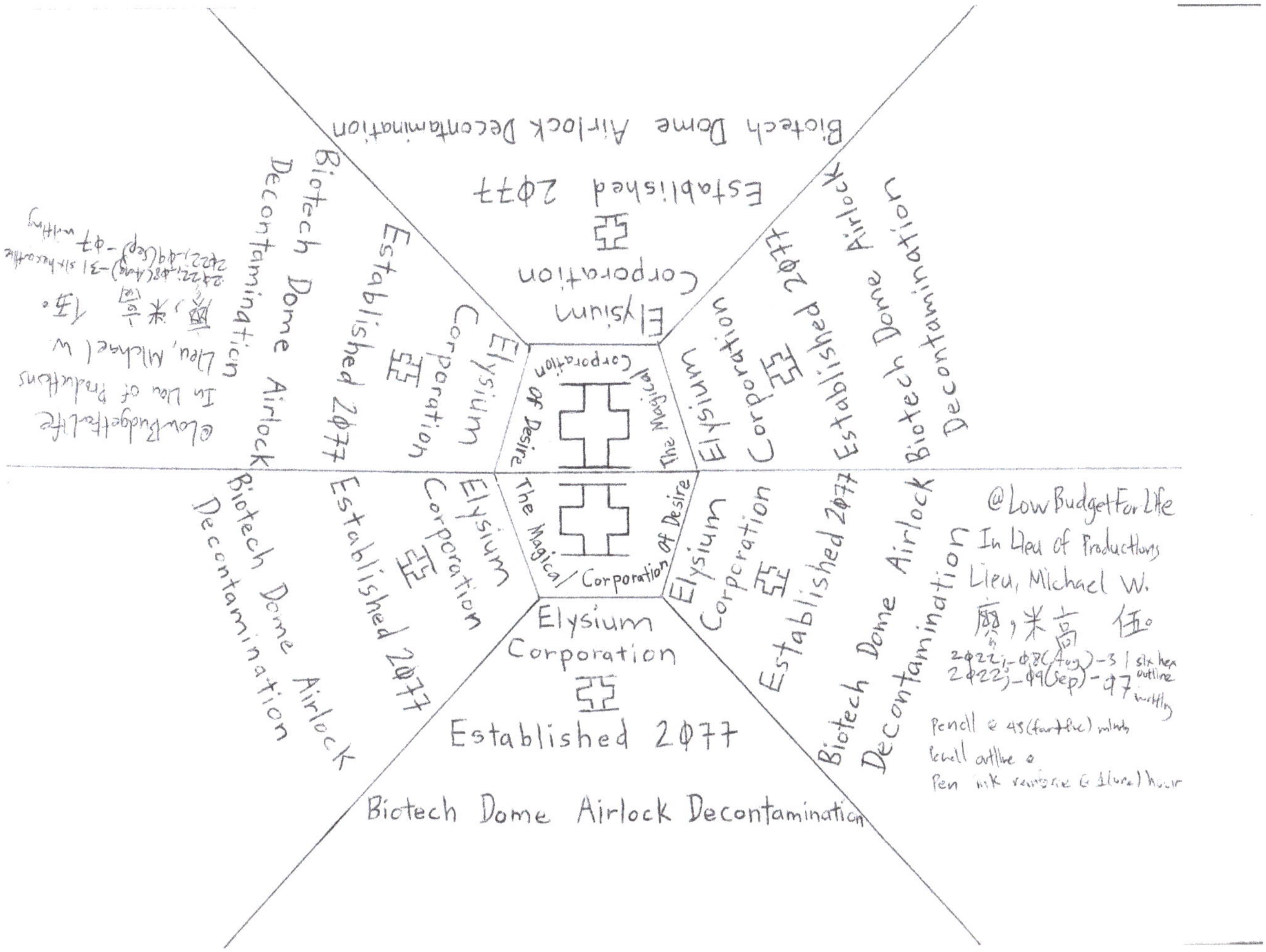

Lieutenant 2nd Lieu, Michael W. 廖, 米高 伍o; Vigilant Volunteers Sisco Cisco (°o°) Yellow;

"Ok, going through another **airlock** decontamination, this is **my** second one within the day since **I** need to **get more items** and then head out dome for the space elevator anniversary celebration, Sir Arthur C. Clarke.

Think **I** have to talk to another **Elysium Corp klones... Artificial persons' Mister Kelvin Klones again**, but **he is another person**... **I** think, as a **klone**.

We see in a bit, **WhiteWolf suit at limbs length**, good to go for oxygen, water... air, water... gas, liquid, **solid, solid food**, pee poo **part** of **biotech suit cleaned out** and decontaminated for this out dome event... much prep work.

(Horror) = Biotech Dome Airlock Decontamination, Mister Kelvin, Temperature

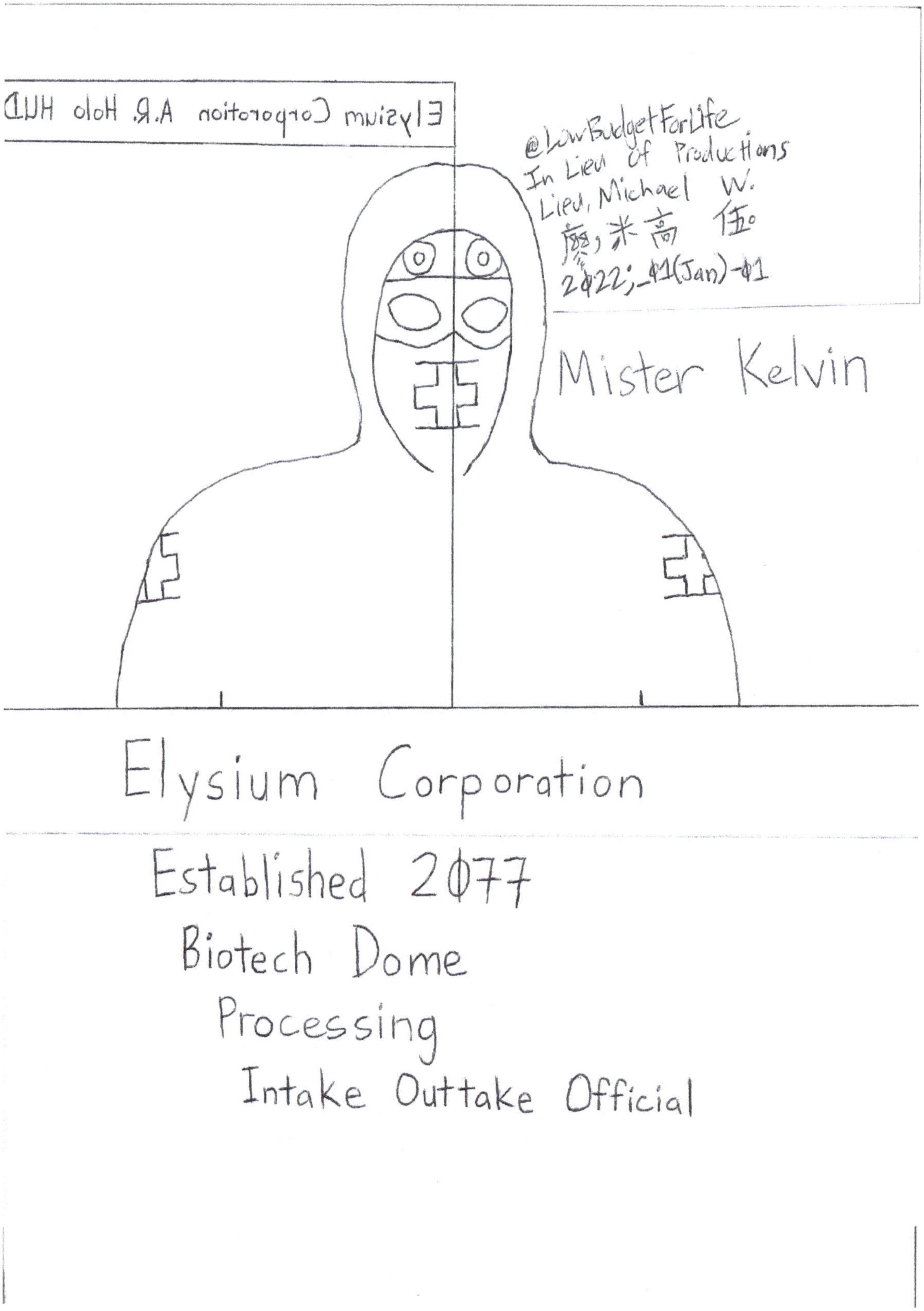

(Super-Hero) (Utopia) (Horror) = Airlock Decontamination Holo HUD Broadcasting

Sisco Cisco .il Yellow (°o°) Broadcast	(°o°) (°o°)	Sisco Cisco .il Yellow (°o°) Broadcast
Date;	Year; (####;) ## (Month) - Day = 2154; 09 (Sep) – 11 (eleven)	

Sisco Cisco .il Yellow (°o°) Broadcast Automation Algorithms Closed Captioning Log;

Elysium Corporation Automation Algorithms; (*monotone female voice over*) "Please stand by for bio**tech** dome **airlock** decontamination. Thank **you** for **your** patience and cooperation **Utopian**."

Lieutenant 2nd Lieu, Michael W. 廖, 米高 伍o; "Oh hi **Mister Kelvin Klones**."

Mister Kelvin; "Oh hi **Lieu Lieu**, **my Klones** tell **me you** preparing for Sir Arthur C. Clarke's space elevator anniversary celebration, **you** not getting a good signal even being within edge of dome? **You got your suit close by limbs length**. Always loved that **WhiteWolf symbol outlines on your suit**."

Lieutenant 2nd Lieu, Michael W. 廖, 米高 伍o; "Much thanks **Mister Kelvin**. Yea, no **I** do not get a good signal, kind of, barely, has to do with the dome's structure materials Mayor used, even on dome edge. So, might as well go out dome and get a better broadcast signal, as well as say meet and greet Mayor Chicago Controller... However, still got some prep work, **I** need, **camera for visual capturing**, Captain Rage's **riot shield backpacks**... Yea, got some more **airlock** decontaminations to go... This should be fun... **Happy Cleaning**."

Mister Kelvin; "Yea, have a good day **Lieutenant Lieu**, **WhiteWolf. Clean**!"

Sisco Cisco .il Yellow (°o°) Broadcast	(°o°) (°o°)	Sisco Cisco .il Yellow (°o°) Broadcast

(Time Traveling) = Time Traveling Forward, Life Is Strange 0, 1, Flash Face(s)

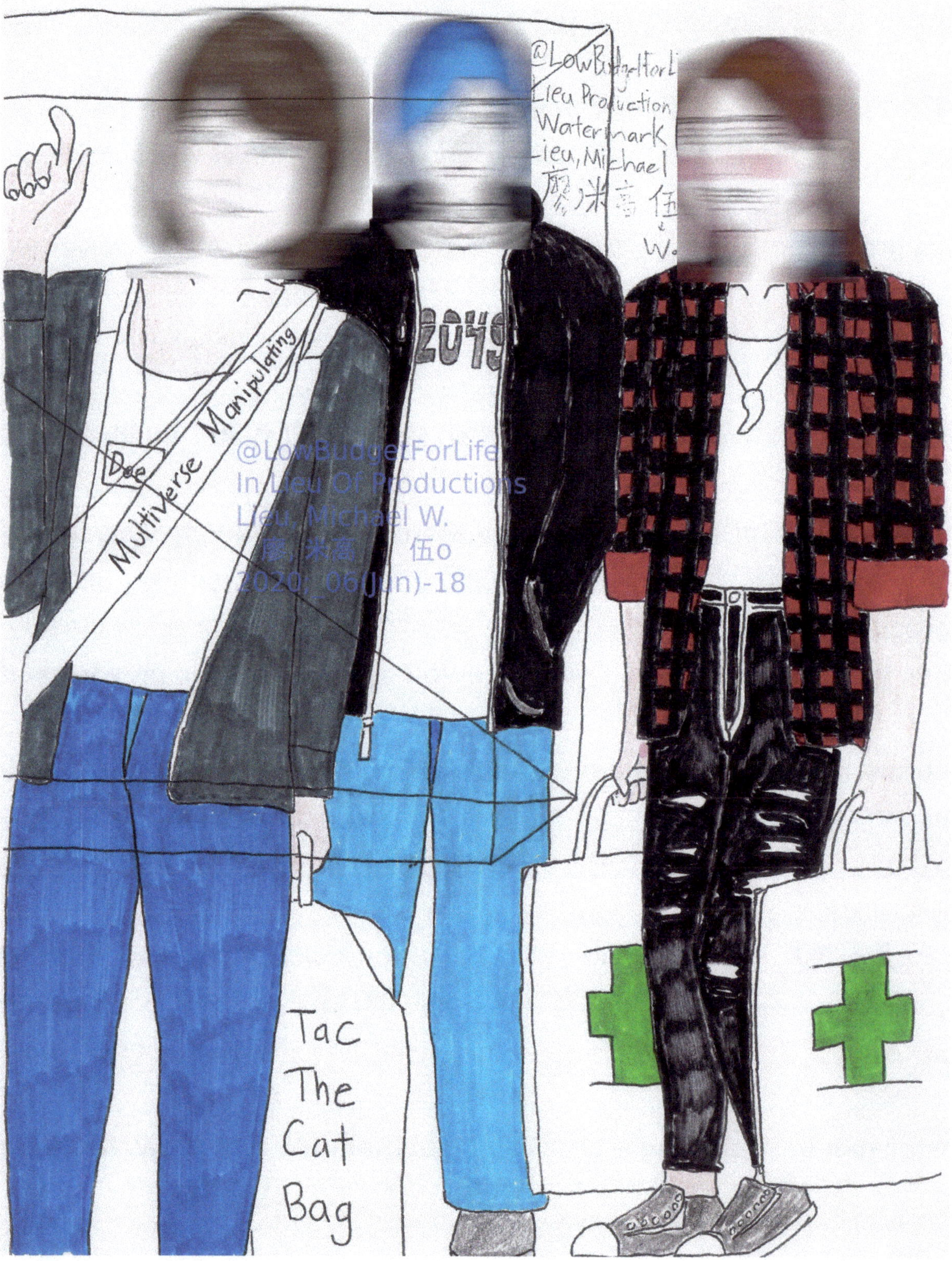

112 \ **2154;** Lieu, 廖, ‘WhiteWolf 白狼’ Left to Right (L→R) 112

(Super-Hero) Sisco Cisco .il Yellow-Mellow Surprise Face (°o°) Broadcast Network
(Utopia) = A.R. Holo HUD Intranet, Wireless Broadcasting Frequencies Online
(Time Traveling) = Time Traveling Forward Future, W. B. D.C. T.V. Flash Face Blur.

Sisco Cisco .il Yellow (°o°) Broadcast	(°o°) (°o°)	Sisco Cisco .il Yellow (°o°) Broadcast
Date;	Year; (####;) ## (Month) - Day = 2154; 09 (Sep) – 11 (eleven)	

Sisco Cisco .il Yellow (°o°) Broadcast Automation Algorithms Closed Captioning Log;

Lieutenant 2nd Lieu, Michael W. 廖, 米高 伍o; "…**Augmented persons, three of them**… Hello? **I** see **three people** here… **Anyone** hear **me**?"

Electro Magnetic Spectrum (E.M.S.) Abstract
[10₁₈] Cosmetic Rays
[10₁₆] Gamma Rays
[10₁₃] **X Rays**
[10₁₁] Ultra Violet (U.V.) Rays
[10₉]
[10₈] Light; "Violet, Blue, Green, Yellow, Orange and Red" [10₇]
[10₇] Infrared Rays
[300,000] Radio Waves

Max Doe Cosplay; "(*Amplitude Modulation (A.M.) Frequencies Broadcasting*)"

Chloe Priceless Cosplay; "(*Frequency Modulation (F.M.) Plus Amplitude Modulation (A.M.) Frequencies Hertz (Hz) Fusion Mix Broadcasting*)"

Raychel Amber-Alert Kosplay; "(*Frequency Modulation (F.M.) Frequencies Broadcasting*)"

Sisco Cisco .il Yellow (°o°) Broadcast	(°o°) (°o°)	Sisco Cisco .il Yellow (°o°) Broadcast

(Time Traveling) = Time Traveling Forward Future, W. B. D.C. T.V. Flash Face Blur.

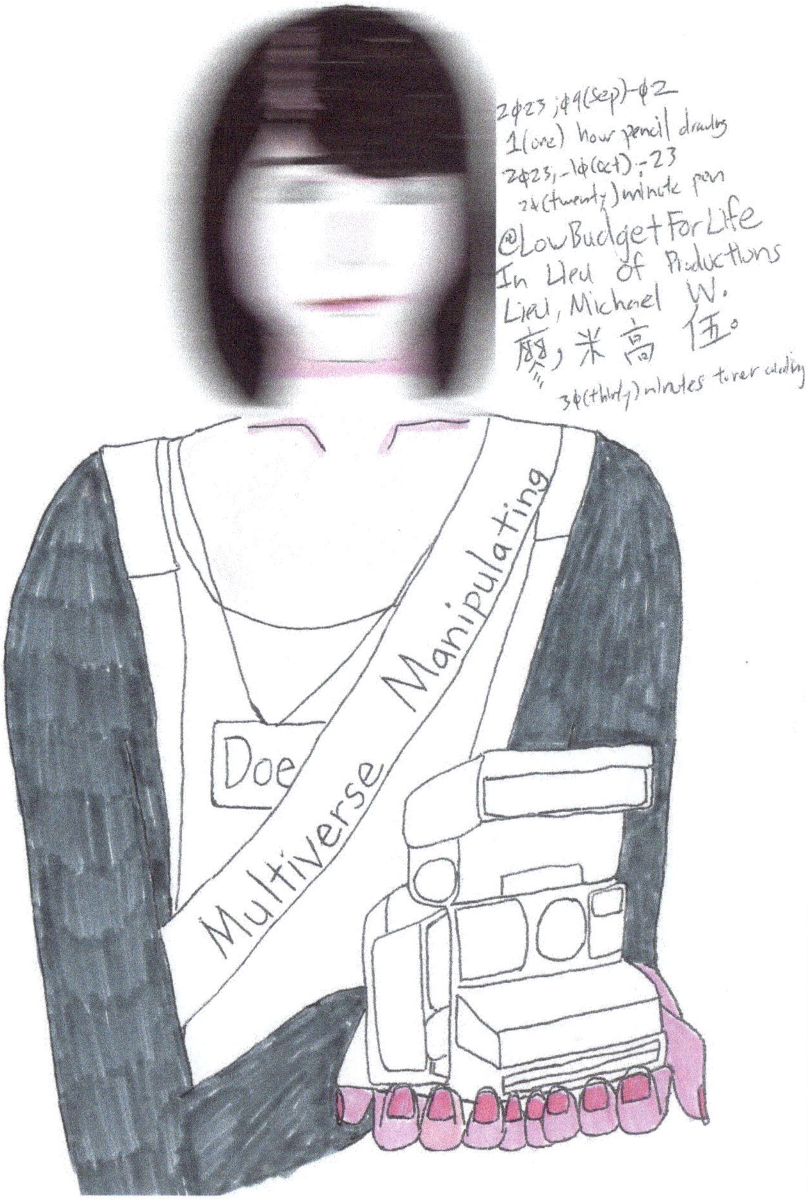

Sisco Cisco .il Yellow (°o°) Broadcast	(°o°) (°o°) (°o°) (°o°) (°o°) (°o°) (°o°) (°o°)	Sisco Cisco .il Yellow (°o°) Broadcast
Date;	2154; 09 (Sep) - 11	
Actual Legal Given Name(s); Max Doe Goes by "Max Doe Cosplay"		
Alias Network Nickname(s); Blurry Face, Time Girl, Time Traveling Human Female, DOE Doe doe, Department Of Energy, Instant Camera Record, [Aug+]		
Employment; Multiverse Manipulating		
Physical Plus Psych Age; Augmented Age(s).		
[Aug+]		
Sisco Cisco .il Yellow (°o°) Broadcast	(°o°) (°o°) (°o°) (°o°) (°o°) (°o°) (°o°) (°o°)	Sisco Cisco .il Yellow (°o°) Broadcast

Lieutenant 2nd Lieu, Michael W. 廖, 米高 伍o; Vigilant Volunteers Sisco Cisco (°o°) Yellow;

"**Augmented person**, Time Traveling Blurry Face Female Max Doe Cosplay hands **me** a **tool for instant camera record**, just what **I** was talking about.

Was she able to listen in, eavesdrop on my conservation with **Mister Kelvin Klones** during **my airlock** decontamination? **I** am not able to pick up anything she was saying, must be **different frequencies**, gotta keep moving."

(Psychological) = Purple Psych Projection Via Ultra Violet (U.V.) Broadcast Rays.

Chicago Controller City Mayor, Self-Govern Sovereignty Sector Sanctuary (S)
2154; Chicago, Illinois' Intranet, Wireless Broadcasting Frequencies Online

City Controller Augmented Reality Holographic Heads Up Display A.R. Holo HUD

Chicago Controller City Mayor

"Chicago City Clean House Act of 2152 (twenty one fifty two); 07-04 (July four)"

"Effective; 2152 (twenty one fifty two); 07-05 (July five)"
Retire Or Be Repurposed

Start Utopian, all non-retired and non-resigned current, active, and reserve public security of the old world Chicago Police Department (C.P.D.) officers must reside and work in Chicago Elysium's police precinct under Captain Rage's Public Po Po Place's first responders only to animal sanctuary needs, no augmented animal human city needs. Repurposed new world, on duty and off duty Chicago City Police Department (C.C.P.D.) Special Weapons And Tactics (SWAT) officers uses code names of rank trait for their own personal protection in retaliation from Cainful behavior, or not if they want to trap and lure those using old world public sector tactics, period end.

Rank =	Captain	Lieutenant	Operator	Sergeant\Alfa\Deanna	Rookie
Trait =	Rage	Gain	Mirror	Soil\Tool\Ogden	Knight

City Controller Augmented Reality Holographic Heads Up Display A.R. Holo HUD

2154; Chicago, Illinois' Intranet, Wireless Broadcasting Frequencies Online
Chicago Controller City Mayor, Self-Govern Sovereignty Sector Sanctuary (S)

(Psychological) = Purple Psych Projection U.V. Rays Broadcasting. (Mass Media)

Chicago Controller City Mayor, Self-Govern Sovereignty Sector Sanctuary (S)

2154; Chicago, Illinois' Intranet, Wireless Broadcasting Frequencies Online

City Controllers Augmented Reality Holographic Heads Up Display A.R. Holo HUD

Mass Media Broadcasting Frequencies Archive Stories;

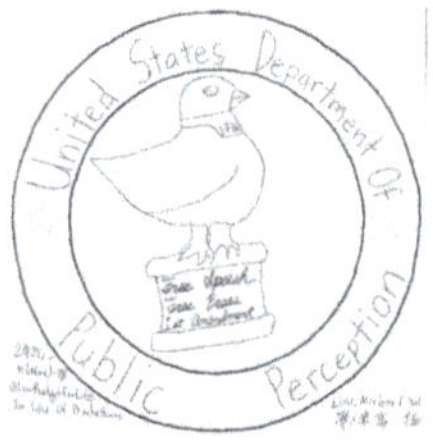

A U.S. Department Of Public Perception (DOPP),
Investigative Journalism presents:
"Purple Projections' City Controllers",
and "Animal Sanctuary (S)"

Written by Mister Marco M. Morales,
Edited by Artificial Intelligence (A.I.) Ai
Large Intelligent Sexy Assistant (LISA),
Published by **Animal Kingdom Service Doggie Evans** on 2153, January one.

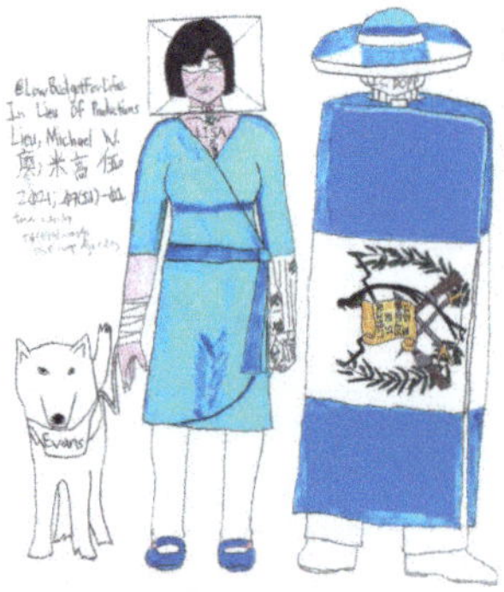

"Chicago City Clean House Act of 2152 (twenty one fifty two); 07-04 (July four)"

'An outside look into a unique 22nd (twenty second) century phenomenon of "Animal Sanctuary (S)" by these broadcasting towers with "Ultra Violet (U.V.) Rays Flag(s)" claiming to be "City Controllers, City Mayor(s)" of "augmented animals", nicknames for humans.'

During the 22nd (twenty second) Century Dark Ages, (2121 - Present Onward), with the loss of many natural unmodified human lives, and **augmented** bio**tech humans refusing to comply and follow easy policies** resulted in more division, many felt Earth's superpower federal governments should have done more to prepare, but would Earthlings in a liberty loving country like United **States** of North America (U.**S.**N.A.) give up their freedom for security?

As a consequence, 55 states all across U.**S.**N.A., went dark when federal, state, county, local governments' lack of infrastructures during that first year of 2121 dark ages apocalyptic pandemics and blackouts caused by Space Belters, Roman Industries, Martian Terrorists' use of unconventional tactics based on their belief systems of War On Resources.

During the 2121 dark ages, many small groups joined together from the ashes of Earth's superpower United **States** and her interests to form an "Animal Sanctuary (S)", the Sanctuary name comes from the political party ruling of blue Democratics' "Sanctuary (S) Cities".

As of 2150s, the most recent Earth's Global Government (EGG) Joint Task Force (J.T.F.) County (D) (R) Military (M) Base to turn "Animal Sanctuary (S)" is in North America, Chicago, Illinois (ILL), owned and operated by U.V. Rays' Chicago Controller City Mayor.

[*Aug+*]

Mass Media Broadcasting Frequencies Archive Stories;

City Controllers Augmented Reality Holographic Heads Up Display A.R. Holo HUD

2154; Chicago, Illinois' Intranet, Wireless Broadcasting Frequencies Online

Chicago Controller City Mayor, Self-Govern Sovereignty Sector Sanctuary (S)

(Horror) = Biotech Dome Airlock Decontamination

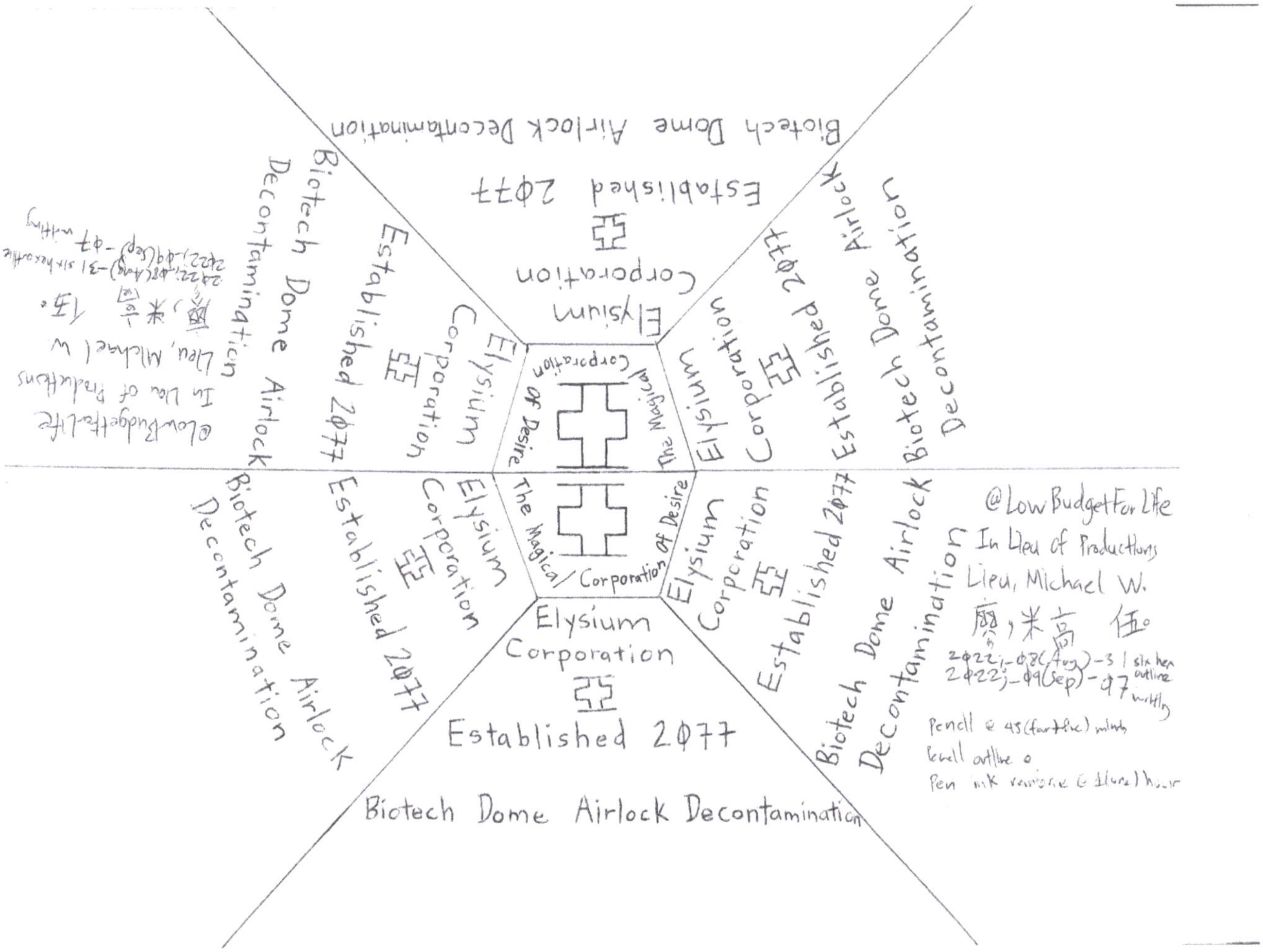

Lieutenant 2nd Lieu, Michael W. 廖, 米高 伍o; Vigilant Volunteers Sisco Cisco (°o°) Yellow;

"Okay, after this **airlock with Mister Kelvin Klones**, going to see Captain Rage and Mayor Chicago Controller to obtain **riot shield backpack** as **I** been **wanting to get out of desire** but, have no time for... Until now.

The U.V. Rays will be pretty powerful as I am getting closer to a U.V. hex sector in the dome Downtown District...

Just recall and focus on my recent reviewed physical paperwork... There will be questioning about it...

If needed, take **deep breaths**."

(Horror) = Biotech Dome Airlock Decontamination, Mister Kelvin, Temperature

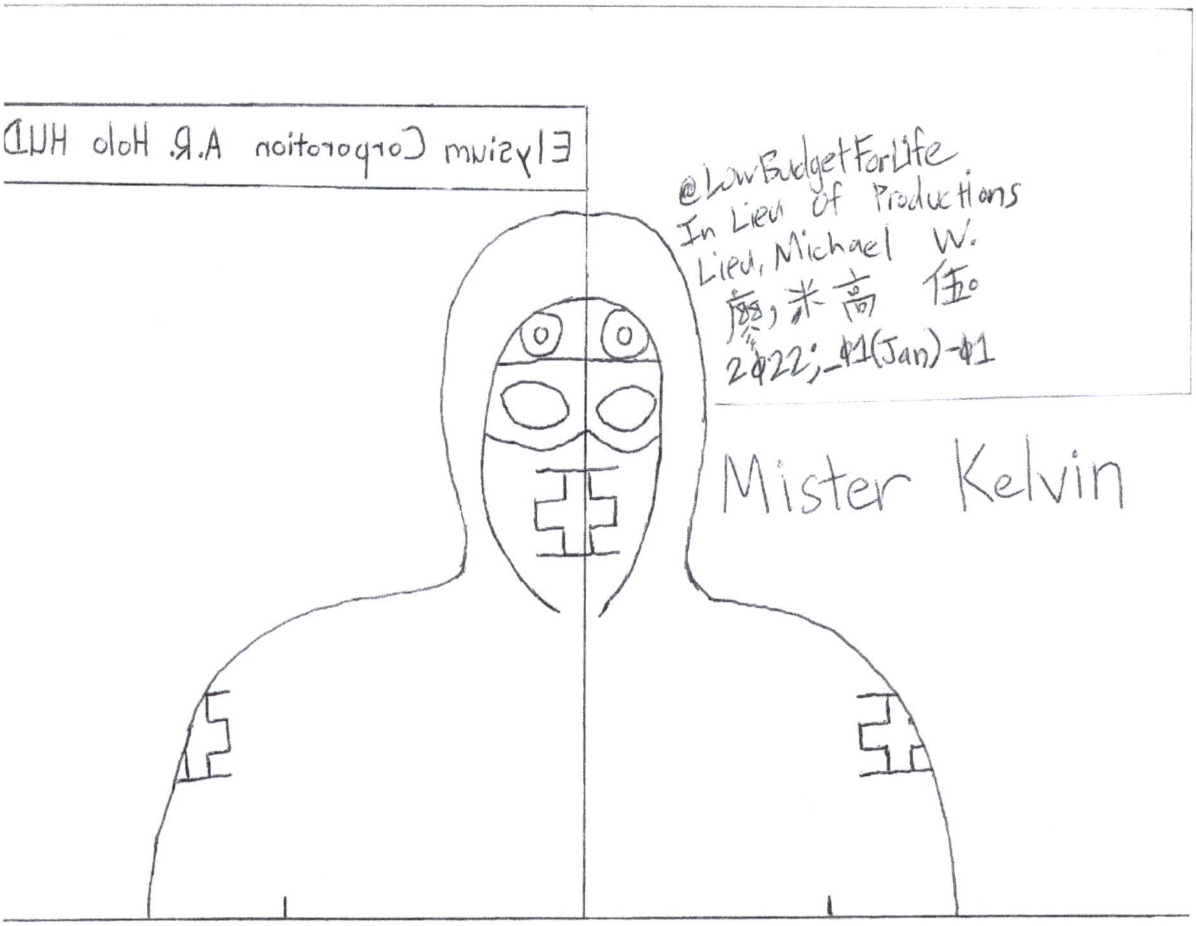

Elysium Corporation

Established 2Ø77

Biotech Dome

Processing

Intake Outtake Official

(Super-Hero) (Utopia) (Horror) = Airlock Decontamination Holo HUD Broadcasting

Sisco Cisco .il Yellow (°o°) Broadcast	(°o°) (°o°)	Sisco Cisco .il Yellow (°o°) Broadcast
Date;	Year; (####;) ## (Month) - Day = 2154; 09 (Sep) – 11 (eleven)	

Sisco Cisco .il Yellow (°o°) Broadcast Automation Algorithms Closed Captioning Log;

Elysium Corporation Automation Algorithms; (*monotone female voice over*)

"Please stand by for bio**tech** dome **airlock** decontamination.

Thank **you** for **your** patience and cooperation **Utopian**."

Lieutenant 2nd Lieu, Michael W. 廖, 米高 伍o; "Oh hi **Mister Kelvin Klones**."

Mister Kelvin; "Oh hi **Lieu Lieu**, not getting a good signal for today's space elevator anniversary celebration huh? Gotta go out dome with **your suit**?"

Lieutenant 2nd Lieu, Michael W. 廖, 米高 伍o; "Ye, gotta **get some stuff**."

Mister Kelvin; "**You have your** existence insurance in place in case something happens? Your personality trait is Conscientiousness. Color code orange... Well, our dome animal sanctuary, the Mayor... does wonders to people and everyone handles the Mayors' U.V. Rays psychological projection differently...

Lieutenant 2nd Lieu, Michael W. 廖, 米高 伍o; "Yea, **my** existence insurance, orange Conscientiousness Cat stuff animal is being equipped by a Yellow **White** HeartBlade who gained her awareness recently in **Downtown District**."

Mister Kelvin; "Nice. Mister Kenny would want to know. **Lieu Lieu, Clean**!"

Sisco Cisco .il Yellow (°o°) Broadcast	(°o°) (°o°)	Sisco Cisco .il Yellow (°o°) Broadcast

(Psychological) (Reality, Gun Girls Gun Guys) = Chicago City Police Department.

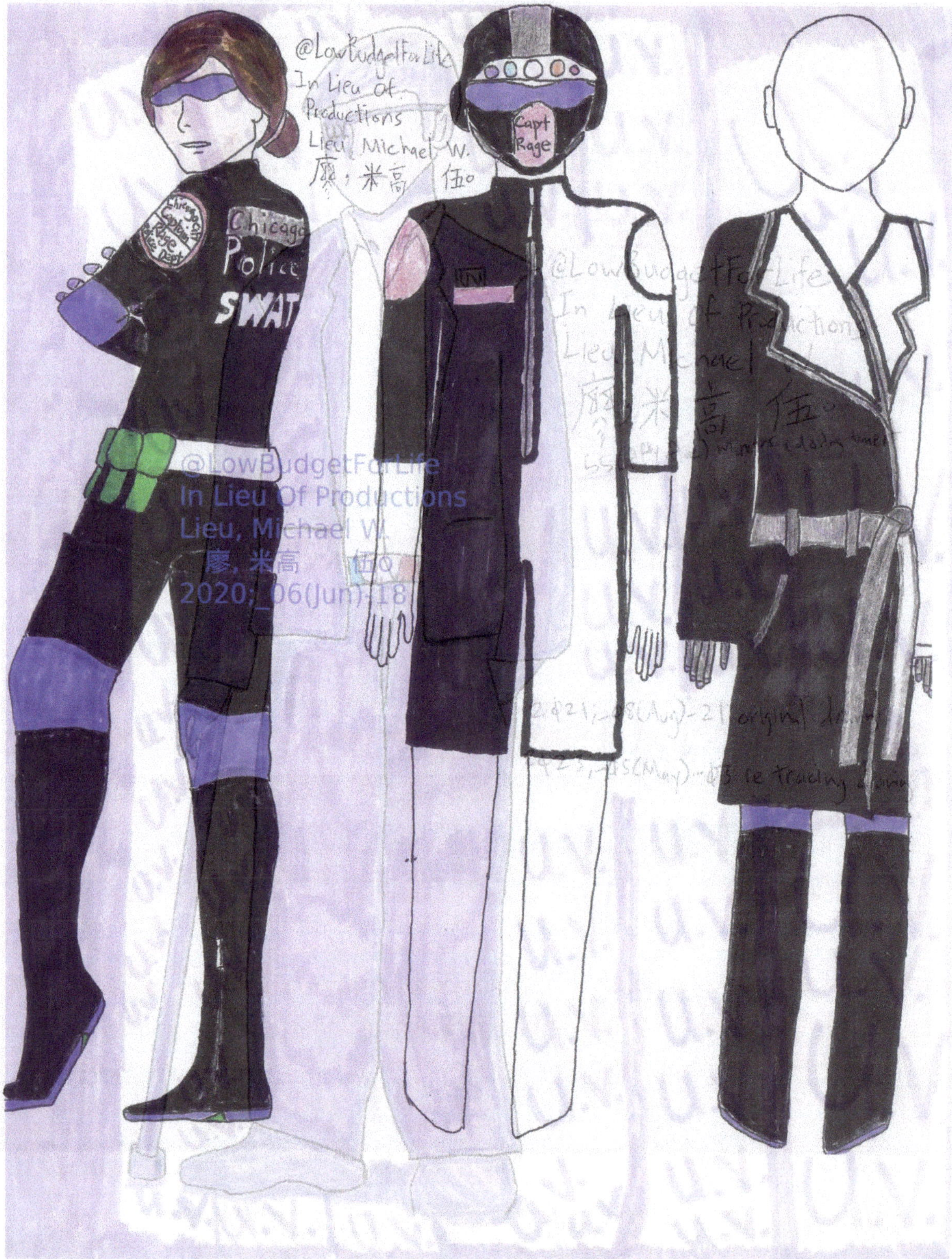

120 \ **2154;** Lieu, 廖, ‘WhiteWolf 白狼’ Left to Right (L→R) 120

(Super-Hero) (Utopia) (Psychological) (Reality, Gun Girls Gun Guys) = C.C.P.D..

Sisco Cisco .il Yellow (°o°) Broadcast	(°o°) (°o°)	Sisco Cisco .il Yellow (°o°) Broadcast
Date;	Year; (####;) ## (Month) - Day = 2154; 09 (Sep) – 11 (eleven)	

Sisco Cisco .il Yellow (°o°) Broadcast Automation Algorithms Closed Captioning Log;

Lieutenant 2nd Lieu, Michael W. 廖, 米高 伍o; "Oh hi Violet Purple Po Po Mayor! **Here for some free stuff**, **I** just recall that wanted a **riot shield backpack** from Public Po Po Place and was in **Downtown District** to do prep work on **WhiteWolf suit** for out dome broadcast anniversary for space elevators, Sir Arthur C. Clarke... Plus, **I** going to do some U.V. broadcast tower maintenance as well since **I** am out dome, so... Yea."

(Frequency Modulation (F.M.) Plus Amplitude Modulation (A.M.) Frequencies Hertz (Hz) Mix With Radio Waves And Ultra Violet Rays Broadcasting)

Captain Rage C.C.P.D. SWAT, + (plus) Chicago Controller City Mayor; *(male(s) mixed female(s) voice tone(s))*; "Start Utopian, **WhiteWolf exterior, Lieutenant Lieu interior**, **your desire for free stuff** has been prepared for **you**, a **wearable back pack riot shield from our old world Police Department**, we noticed **you** have no desire for firearms, guns, even though **you** work among those military Army soldier boys and girls, Period End."

Lieutenant 2nd Lieu, Michael W. 廖, 米高 伍o; "**I'm**, more of an **engineer**, than a soldier, a **failed engineer**. **Engineering, building, creating things is my kind of specialty, role**, as to being a Joe Army soldier boy. Hello? Hello?"

Sisco Cisco .il Yellow (°o°) Broadcast	(°o°) (°o°)	Sisco Cisco .il Yellow (°o°) Broadcast

(Psychological) = Purple Psych Projection Via Ultra Violet (U.V.) Broadcast Rays.
(Reality, Gun Girls Gun Guys) = C.C.P.D.. Wearable Backpack Riot Shield.

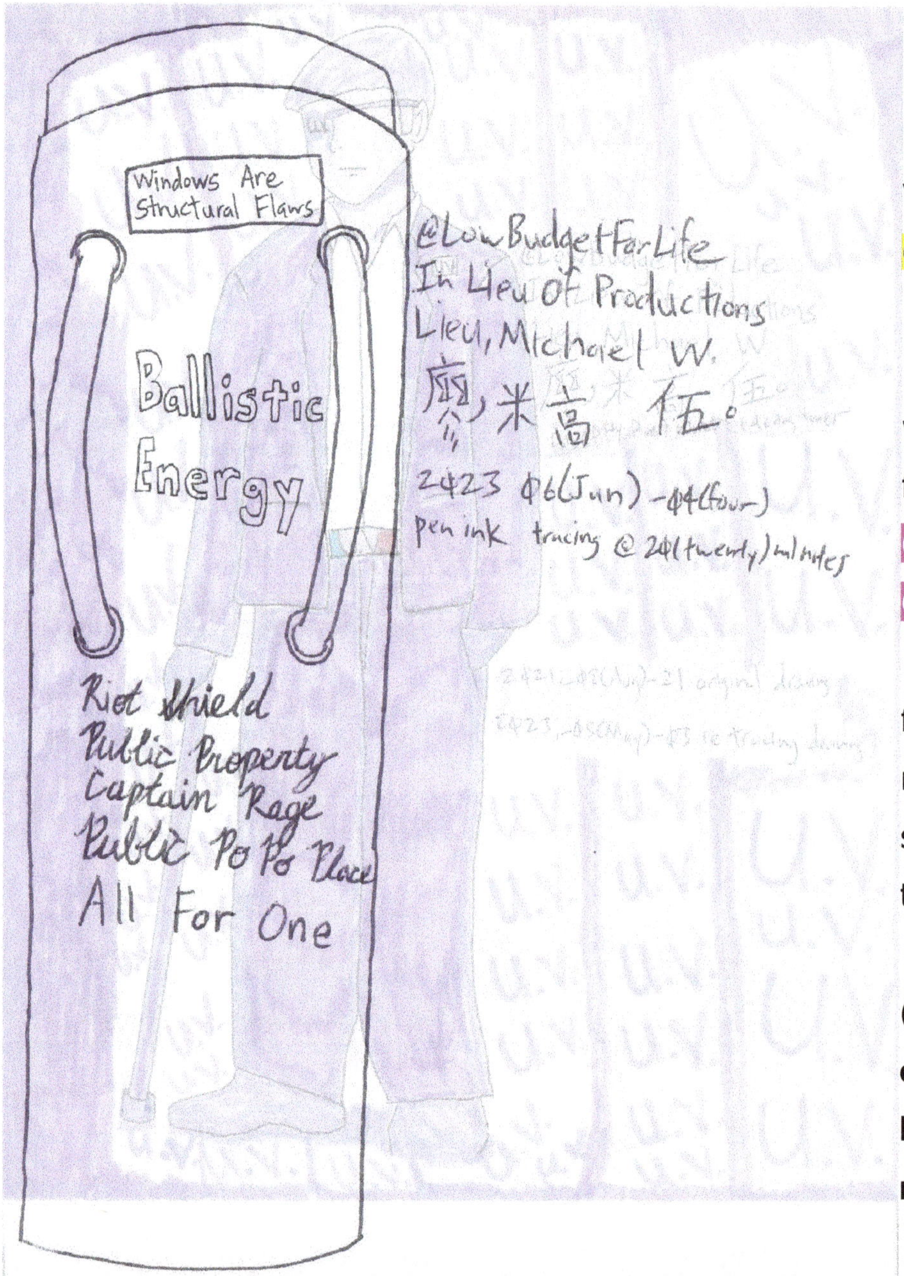

Lieutenant 2nd Lieu, Michael W. 廖, 米高 伍o; Vigilant Volunteers Sisco Cisco (°o°) Yellow;

"...Gonna document what it is **I** am getting from Captain Chicago Controller here in C.C.P.D. SWAT...

Going to send the following items back to my North (N) Side sector that **I** obtained the items by **U.A.V.**

One Toilet Paper (T.P.), one instant camera record, one back pack wearable riot shield.

In preparation for Sir Arthur C. Clarke's Space Elevators Anniversary Launch Celebration and this may be the last entry for the year twenty one fifty four, the date September eleven, so **I** am going to write it down on **my retirement journal** before putting on **WhiteWolf suit** and go out dome for exposure... Summoning **U.A.V.** now."

(Psychological) = Purple Psych Projection Via Ultra Violet (U.V.) Broadcast Rays.

(Utopia) = Orange Conscientiousness Cat Unmanned Aerial Vehicle (U.A.V.)

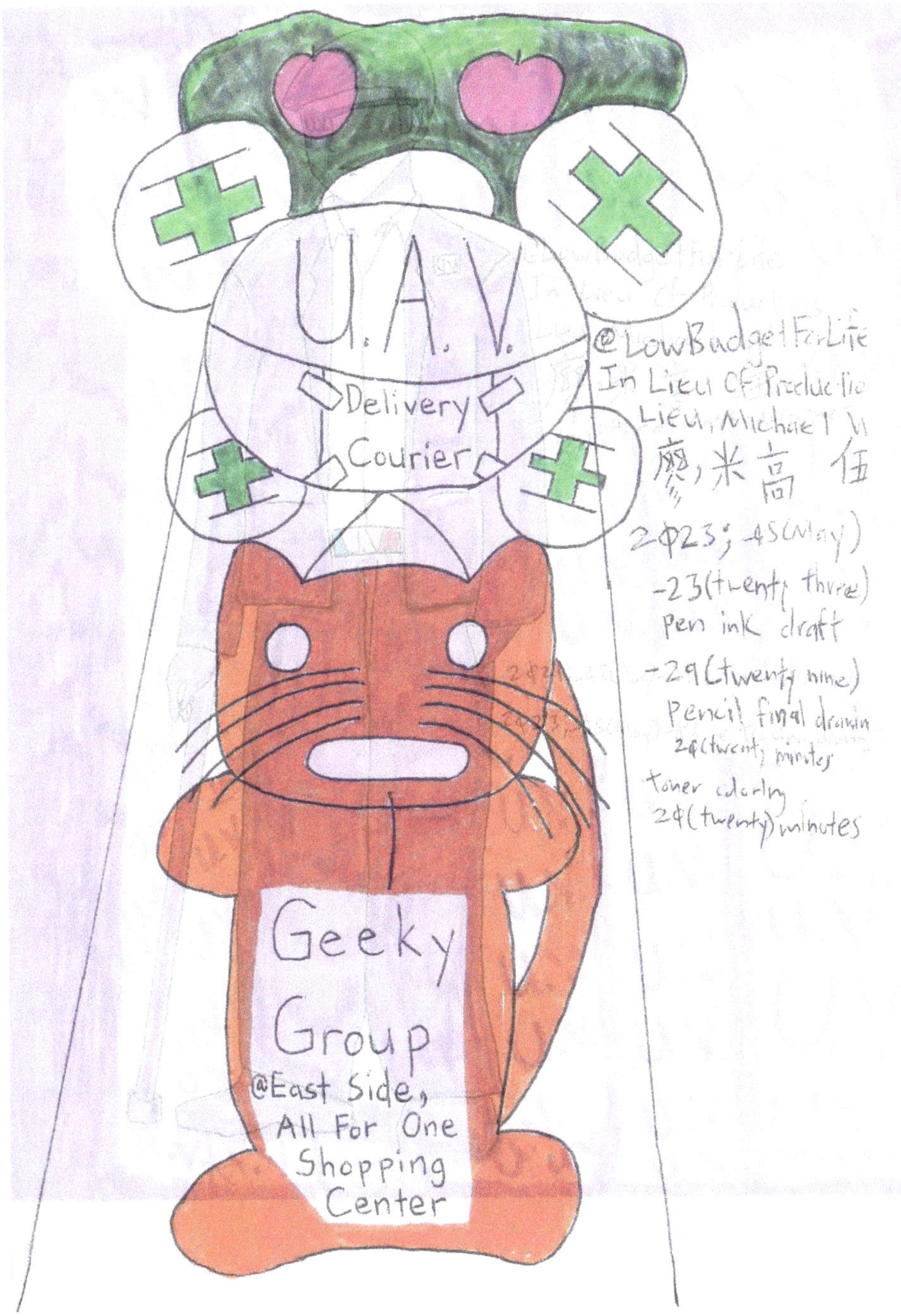

Sisco Cisco .il Orange Yellow Violet (°o°) Broadcast	(°o°) (°o°) (°o°) (°o°) (°o°) (°o°)	Sisco Cisco .il Orange Yellow Violet (°o°) Broadcast
Date;	2154; 09 (Sep) - 11	

Unmanned Aerial Vehicle (U.A.V.) Delivery Courier With Orange Conscientiousness Cat Holo HUD;

"From; Downtown District (D. D.) of Chicago City Police Department (C.C.P.D.) SWAT. To; North (N) Side of Biotech Dome Ocean Border."

Manifest Transport; "**One** Toilet Paper, **One** Instant Camera Record, **One** Wearable Backpack Riot Shield."

Signed Self Certify; Lieutenant Lieu, Michael W. (2nd 廖, 米高 伍o)

[Aug+]

Sisco Cisco .il Orange Yellow Violet (°o°) Broadcast	(°o°) (°o°) (°o°) (°o°) (°o°) (°o°)	Sisco Cisco .il Orange Yellow Violet (°o°) Broadcast

(Horror) = Biotech Dome Airlock Decontamination

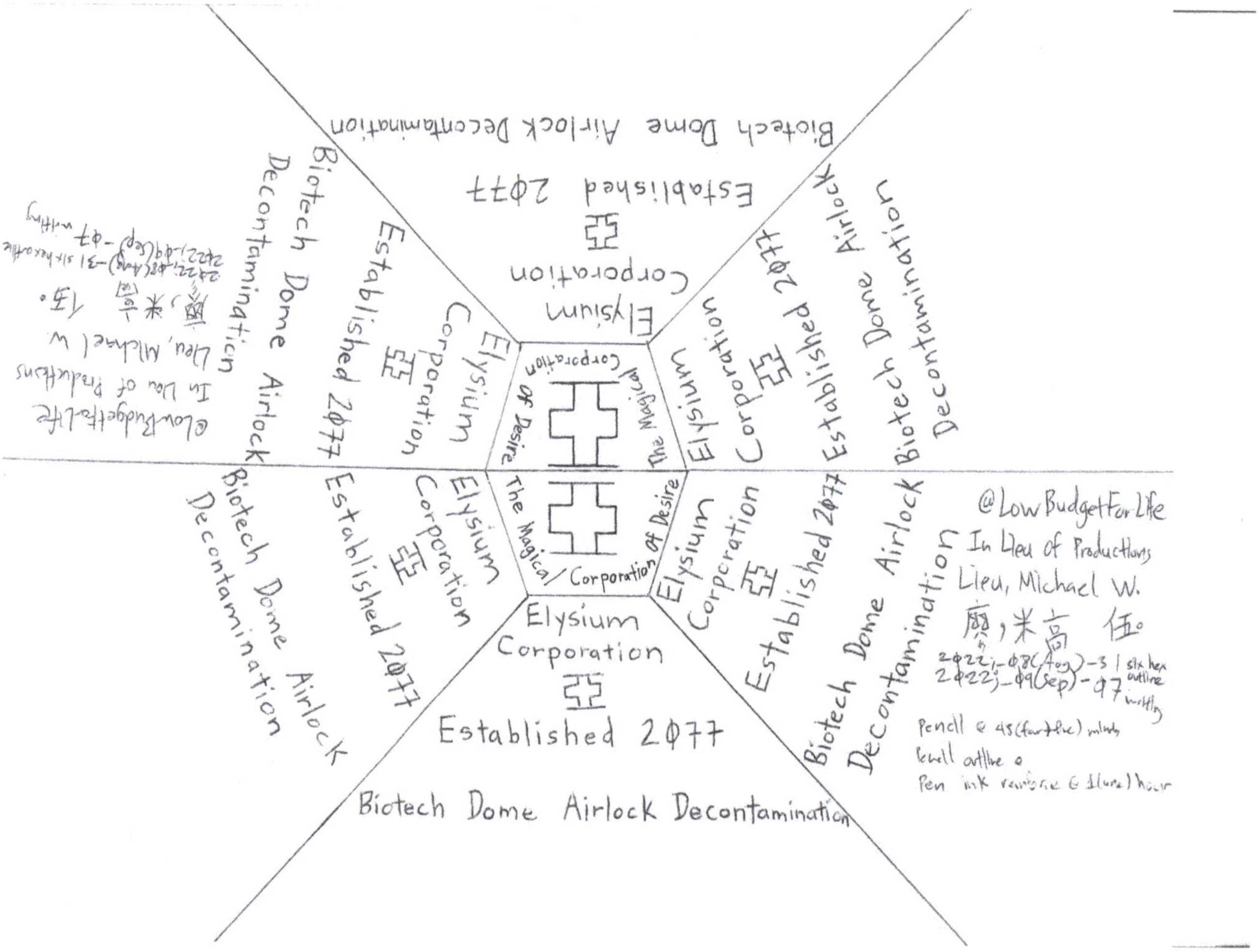

Lieutenant 2nd Lieu, Michael W. 廖, 米高 伍o; Vigilant Volunteers Sisco Cisco (°o°) Yellow;

"Okay, **let me review my retirement year four journal again,** the year **twenty one fifty four,** the date September eleven **prep work**...

Today is Sir Arthur C. Clarke's Space Elevators Anniversary Launch Celebration and **I am doing prep work to get ready with my biotech suit P.P.E. WhiteWolf**, since I cannot get much of a good signal broadcast from Animal Sanctuary North (N) Side edge dome border from within...

Suit equipped, Electro Magnetic Spectrum visuals on stand by...

Existence insurance equipped, watched over by Yellow **White** HeartBlade. Now for another **airlock, Mister Kelvin**, and the Mayors' U.V. rays broadcast..."

(Horror) = Biotech Dome Airlock Decontamination, Mister Kelvin, Temperature

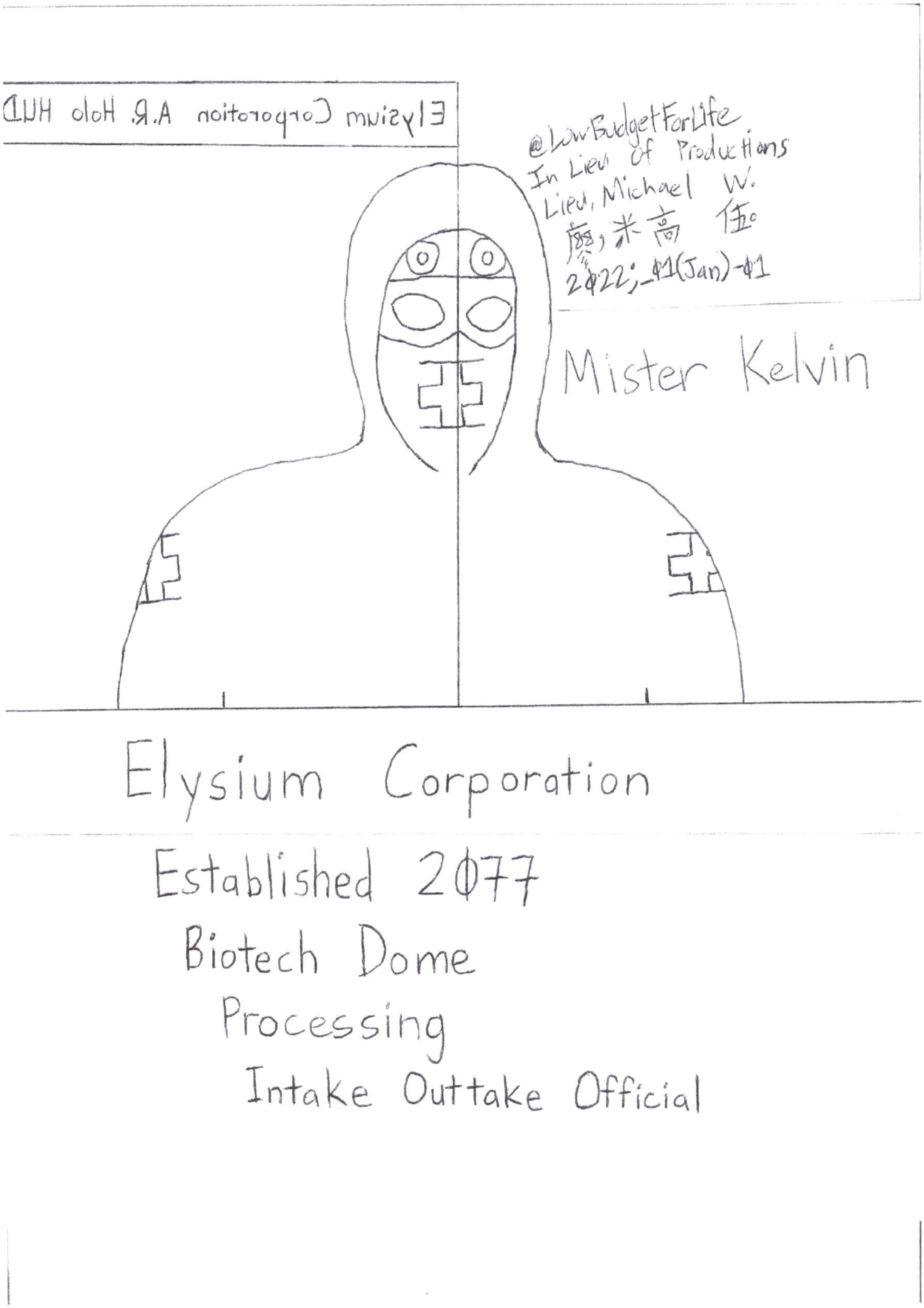

(Horror) = Biotech Dome Airlock Decontamination (Neo Noir Vigilante) X Rays

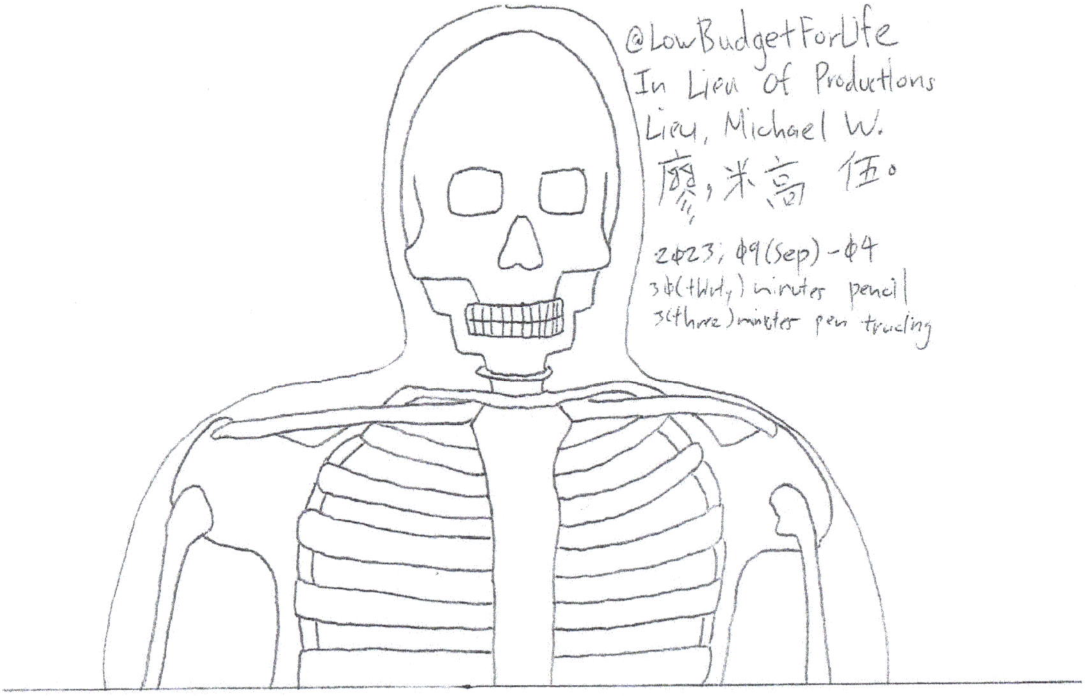

(Super-Hero) (Utopia) (Horror) = Airlock Decontamination Holo HUD Broadcasting

Sisco Cisco .il Yellow (°o°) Broadcast	(°o°) (°o°)	Sisco Cisco .il Yellow (°o°) Broadcast
Date;	Year; (####;) ## (Month) - Day = 2154; 09 (Sep) – 11 (eleven)	

Sisco Cisco .il Yellow (°o°) Broadcast Automation Algorithms Closed Captioning Log;

Elysium Corporation Automation Algorithms; (*monotone female voice over*)

"Please stand by for bio**tech** dome **airlock** decontamination.

Thank **you** for **your** patience and cooperation **Utopian**."

Lieutenant 2nd Lieu, Michael W. 廖, 米高 伍o WhiteWolf 白狼 Bio**tech Suit**;

"Oh hi **Mister Kelvin Klones**, this time **I** am ready to go out dome...

WhiteWolf Bio**tech Suit P.P.E. active**, online, **Electro Magnetic Spectrum visuals on stand by, X Ray online. I like your X Ray visuals**, nice natural **skeleton figure you have**. **You** a good **klone**, got good genes biology."

Mister Kelvin; "Oh hi **Lieu Lieu**, **WhiteWolf** now, the State, acting ChEng... **Suited** up. **My X Ray scans looks good** huh? Think the Mayor, Chicago Controller will be broadcasting shortly, so... hang on tight... Stand by, **you got your** existence insurance watched over by a Yellow **White** HeartBlade person size breed**-build**, a member of the Church, who gain awareness! Mister Kenny for sure would like to know how the origins of consciousness formed for it and her. Sister Mitchell would be very happy and pleased..."

(*Frequency Modulation (F.M.) Plus Amplitude Modulation (A.M.) Frequencies Hertz (Hz) Mix With Radio Waves And Ultra Violet Rays Broadcasting*)

Mister Kelvin; "Ok, Mayor Chicago Controller is ready**. Clean**! **Clear!!**"

Sisco Cisco .il Yellow (°o°) Broadcast	(°o°) (°o°)	Sisco Cisco .il Yellow (°o°) Broadcast

(Psych) = U.V. Rays (Horror) = Decontamination (Neo Noir Vigilante) = X Rays

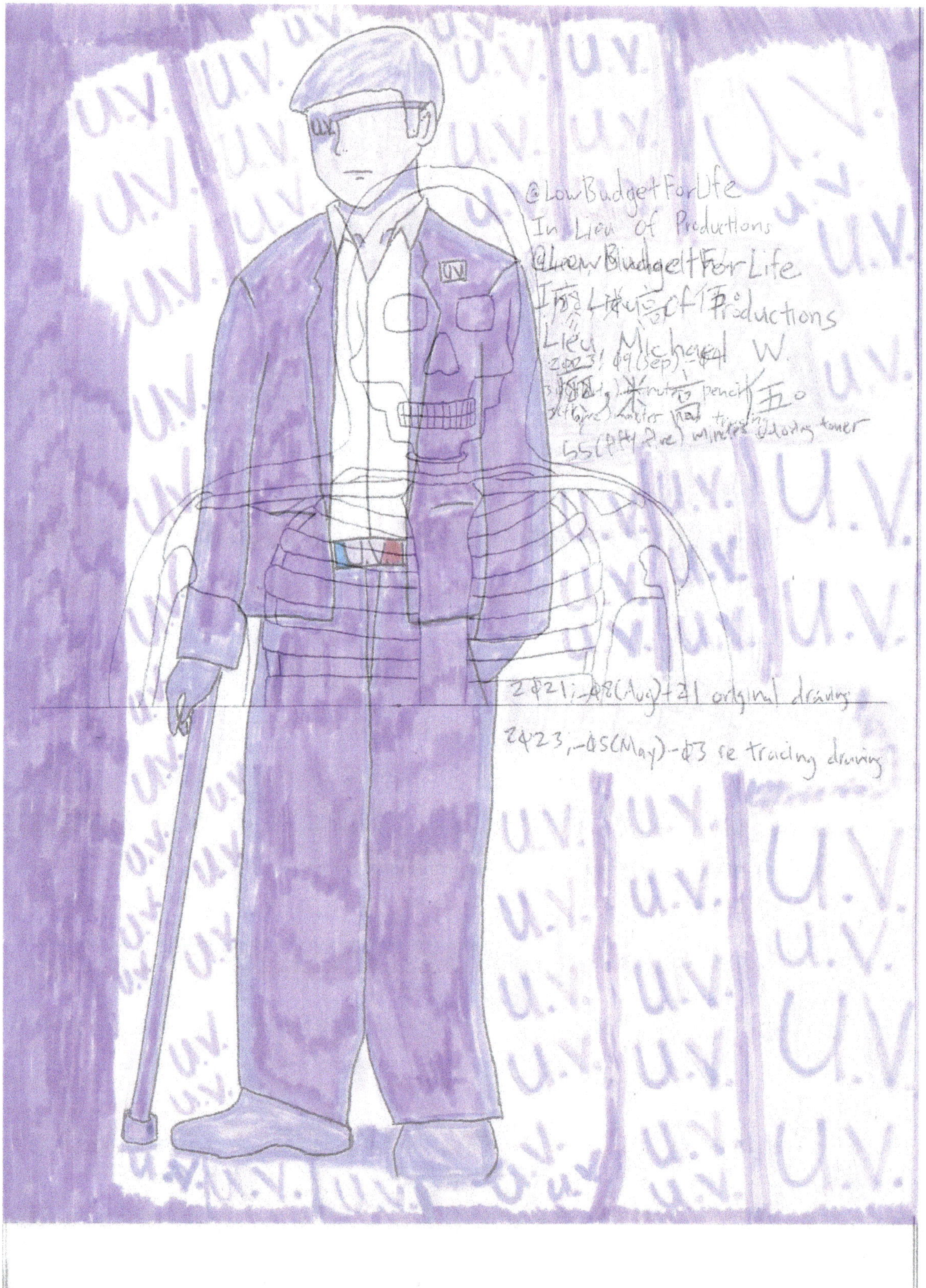

(Psychological) = Purple Psych Projection Via Ultra Violet (U.V.) Broadcast Rays.

Reads = Left to Right (L -> R) \ Wireless Broadcasting Frequencies Online
2154; Chicago (S), The Last Animal Sanctuary (S) Of Illinois' Intranet

Why bother, who cares, barely any humans follow rules regulation	Fine, whatever do not sit on it, complete this before entering

Chicago Controller
City Mayor

"Chicago City Clean House Act of
2152 (twenty one fifty two); 07-04 (July four)"

Indoctrination Self-Governing Guidelines Overview

Failure to complete the new world 2121 (twenty one twenty one) indoc, is a warrant for immediate removal from Chicago (S), The Last Animal Sanctuary (S) Of Illinois (ILL).

Exceptions apply only to the surviving beautiful friendly peace loving augmented and natural animal kingdoms plus planet earth's environment!!

Make a choice, Utopian.

No, I disagree, wipe my memory.	I do not know, let us discuss.	Yes, I agree.

Clean-up Crew
Unit 6 (six),
Unit 9 (nine),
Unit 8888,8888 (eight eight eight eight comma eight eight eight eight).

Abusive bitch and moan, toxicity constant crybaby complaint box	Let the shit hit the fan rock bottom, to crash and burn hard

2154; Chicago (S), The Last Animal Sanctuary (S) Of Illinois' Intranet
Reads = Left to Right (L -> R) \ Wireless Broadcasting Frequencies Online

(Psychological) = Purple Psych Projection Via Ultra Violet (U.V.) Broadcast Rays.

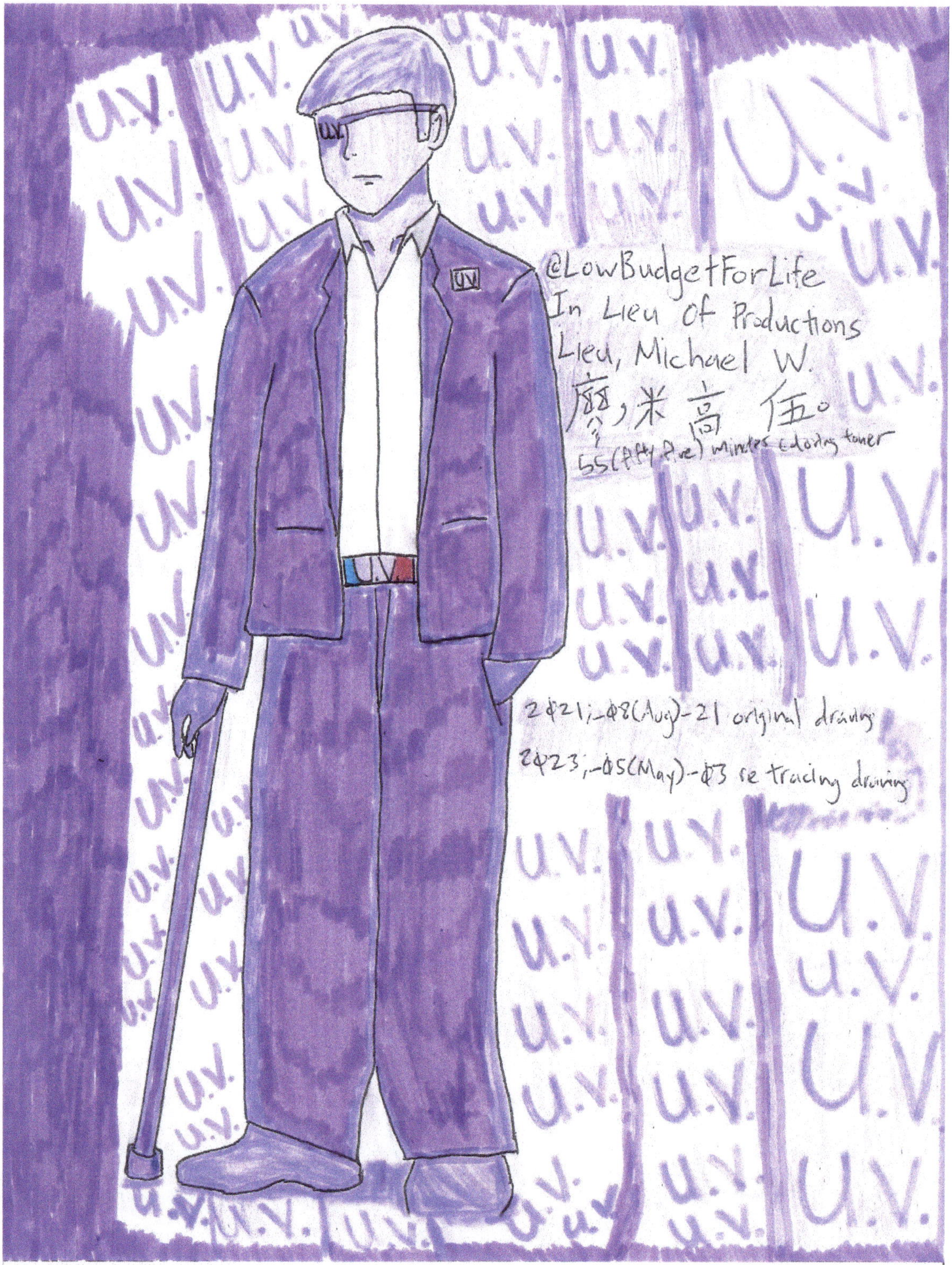

(Psychological) (Super-Hero) (Utopia) = Mayor's Contract End User Agreement.

Sisco Cisco .il Violet (U.V.) Broadcast	(°o°) (°o°) (°o°) (°o°) (°o°) (°o°) (°o°) (°o°) (°o°) (°o°) (°o°) (°o°) (°o°) (°o°) (°o°) (U.V.)	Sisco Cisco .il Violet (U.V.) Broadcast
Date;	Year; (####;) ## (Month) - Day = 2154; 09 (Sep) – 11 (eleven)	

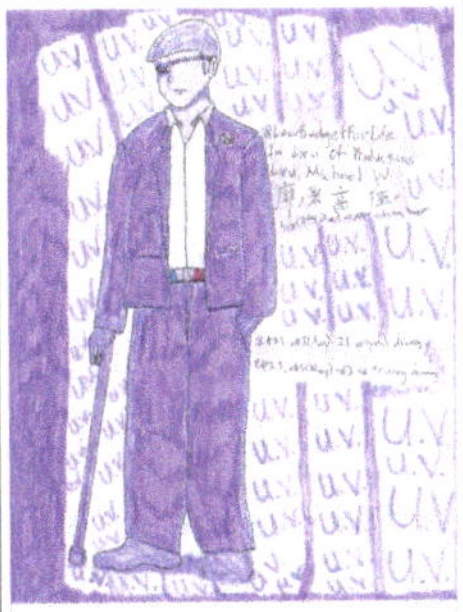

Chicago Controller City Mayor

Chicago (S), The Last Animal Sanctuary (S) Of Illinois (ILL), N.A.,

Electro Magnetic Spectrum; Radio Waves, Amplitude Modulation (A.M.) Frequencies Hertz (Hz) Broadcasting, "Crosses' Telepathy"

Lieutenant 2nd Lieu, Michael W. 廖, 米高 伍o + **WhiteWolf 白狼** Bio**tech Suit+**; Vigilant Volunteers Sisco Cisco (°o°) Yellow;
(*bio**tech suit** male voice)*

"**Audio voice recording** for **my retirement journal** year four, the year **twenty one fifty four**, the date September eleven. On **my** way out dome, as today is Sir Arthur C. Clarke's Space Elevators Anniversary Launch Celebration, recording this since **I** cannot take my out my written journal to risk contamination…"

Chicago Controller City Mayor (*frequency hertz male(s) voice*);

"Start Utopian, Unit eight eight eight eight comma eight eight eight eight, **WhiteWolf**, **Lieutenant Lieu**, so **you** agree to the terms of departure from my beautiful peace loving animal sanctuary…? The displaying memo is still the same, because we Mayors' City Controllers are just lazy, we sit on things… Ultra Violet lazy, Or are **you** departing from my beautiful peace loving animal sanctuary for sure? **You** have options, Period End."

Lieutenant 2nd Lieu, Michael W. 廖, 米高 伍o + **WhiteWolf 白狼** Bio**tech Suit+**; Vigilant Volunteers Sisco Cisco (°o°) Yellow;
(*bio**tech suit** male voice)*

"…Oh hi Mayor, Chicago Controller, on **my** way out dome a bit for today's space elevator anniversary, and U.V. broadcast maintenance… So, yea… **I** agree, Mayor Hello? Hello? …Hello?"

Sisco Cisco .il Violet (U.V.) Broadcast	(°o°) (°o°) (°o°) (°o°) (°o°) (°o°) (°o°) (°o°) (°o°) (°o°) (°o°) (°o°) (°o°) (°o°) (°o°) (U.V.)	Sisco Cisco .il Violet (U.V.) Broadcast

(Neo Noir Vigilante) = Biotech Suit (Psychological) = Psych Project U.V. Broadcast

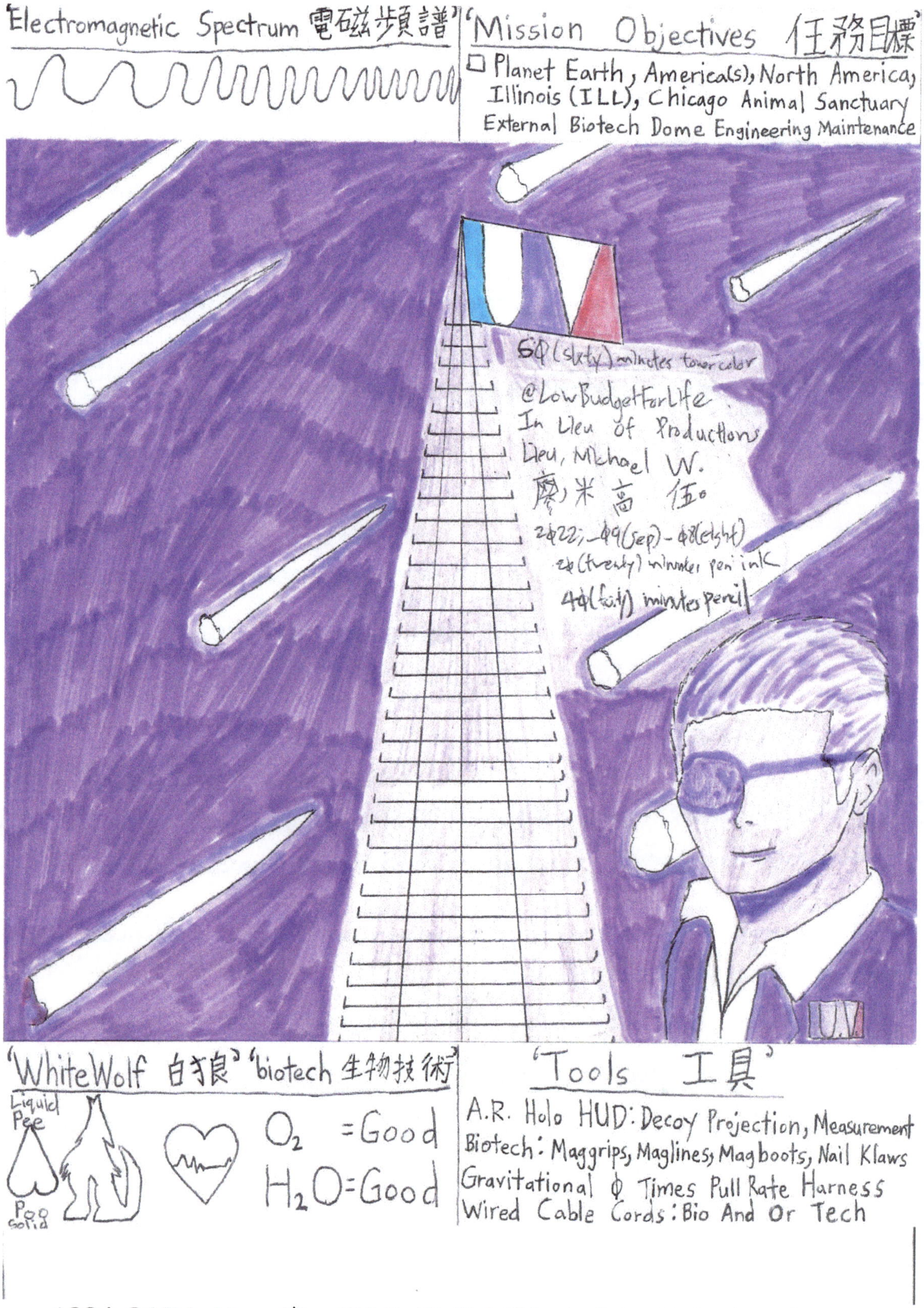

(Psychological) (Super-Hero) (Utopia) = 22nd Century Dark Ages, Sanctuary City.

Sisco Cisco .il Violet (U.V.) Broadcast	(°o°) (°o°) (°o°) (°o°) (°o°) (°o°) (°o°) (°o°) (°o°) (°o°) (°o°) (°o°) (°o°) (°o°) (°o°) (U.V.)	Sisco Cisco .il Violet (U.V.) Broadcast
Date;	Year; (####;) ## (Month) - Day = 2154; 09 (Sep) – 11 (eleven)	

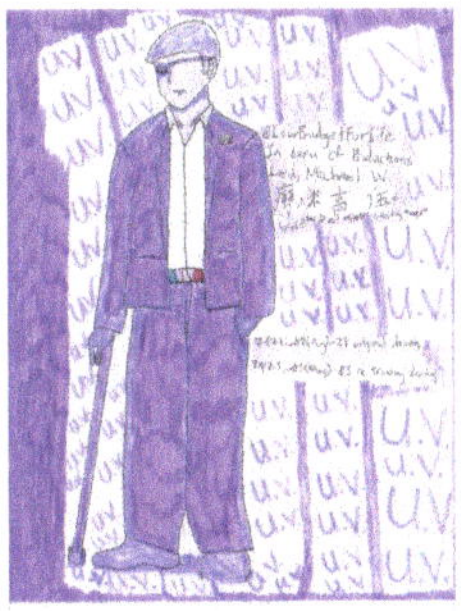

Chicago Controller City Mayor

Chicago (S), The Last Animal Sanctuary (S) Of Illinois (ILL), N.A.,

Electro Magnetic Spectrum; Radio Waves, Amplitude Modulation (A.M.) Frequencies Hertz (Hz) Broadcasting, "Nine Inch Nails' The Day The World Went Away (Still Version)"

Lieutenant 2nd Lieu, Michael W. 廖, 米高 伍o + **WhiteWolf 白狼** Biotech **Suit+**; Vigilant Volunteers Sisco Cisco (°o°) Yellow; (*biotech suit male voice*)

"Now outside dome, **I** am going to prioritize **engineering maintenance** for the Mayor Chicago Controller's U.V. flag broadcasting antennas and receivers…

Then Sir Arthur C. Clarke's Space Elevators Anniversary Launch Celebration, as the signal is strongest on top of the Mayors' U.V. broadcast tower… Gotta **ascend** up… Mayor is going to be talking to me as **I** climb up, got **my tools**. **I** got this. Deep breaths."

Chicago Controller City Mayor (*frequency hertz male(s) voice*);

"Start Utopian, Unit eight eight eight eight comma eight eight eight eight, **WhiteWolf**, Lieutenant Lieu, my beautiful, our beautiful U.V. broadcasting flag tower in Illinois, the last sanctuary for peaceful animals, **augmented** and natural, in the old world before the twenty second century dark ages, starting in twenty one twenty one, Seattle, Washington, United **States** of North America, Earth, was a sanctuary city, offering services to **illegal immigrants, on and off planet aliens**, much like the Church that provides for those in need to survive, after the State failed us, we created, we built our own system, animal sanctuary originated from sanctuary city to a protocol, Period End."

Sisco Cisco .il Violet (U.V.) Broadcast	(°o°) (°o°) (°o°) (°o°) (°o°) (°o°) (°o°) (°o°) (°o°) (°o°) (°o°) (°o°) (°o°) (°o°) (°o°) (U.V.)	Sisco Cisco .il Violet (U.V.) Broadcast

(Neo Noir Vigilante) = Biotech Suit (Psychological) = Psych Project U.V. Broadcast

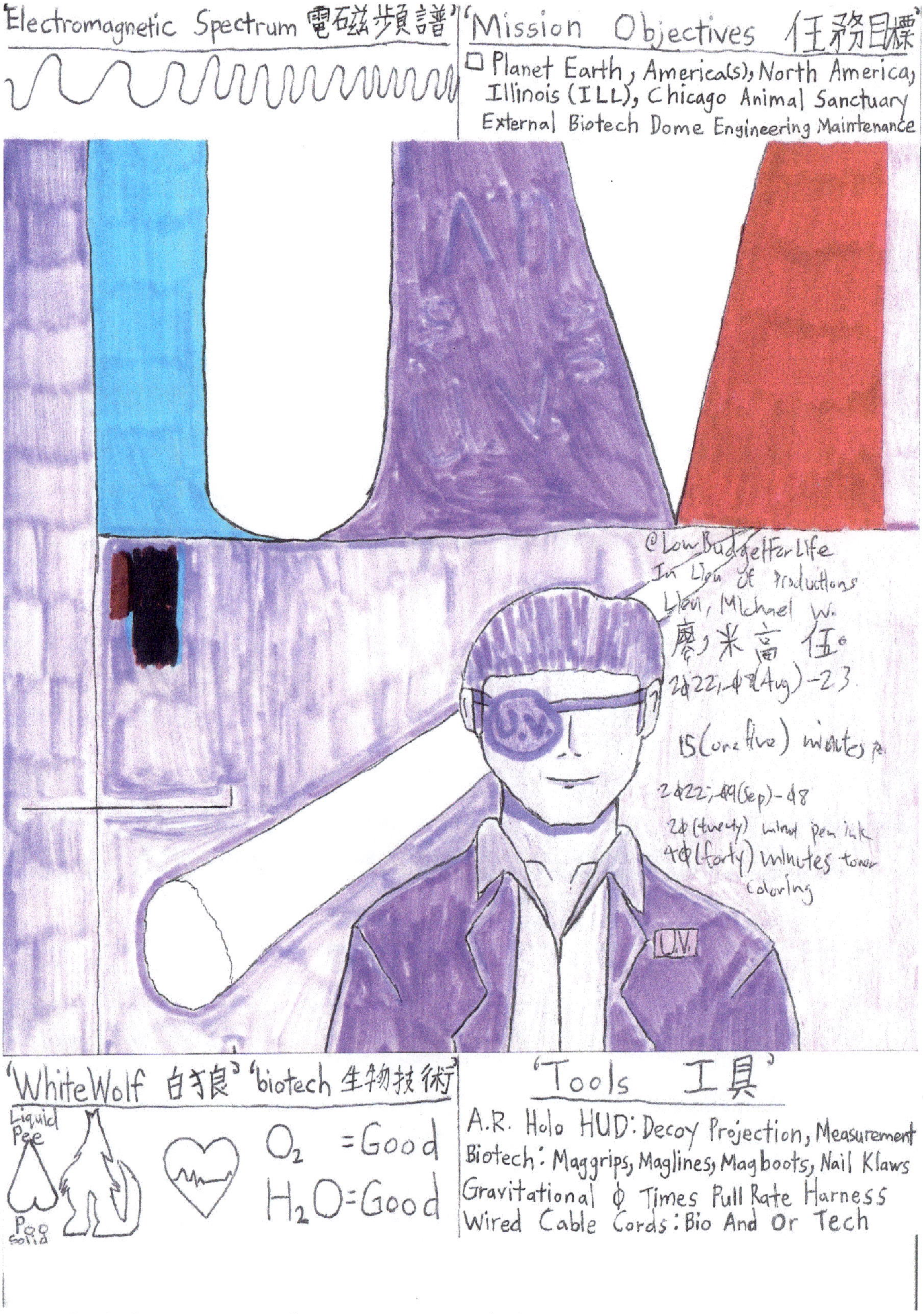

(Psychological) (Super-Hero) (Utopia) = 22nd Century Dark Ages, blue, red, violet

Sisco Cisco .il Violet (U.V.) Broadcast	(°o°) (°o°) (°o°) (°o°) (°o°) (°o°) (°o°) (°o°) (°o°) (°o°) (°o°) (°o°) (°o°) (°o°) (°o°) (U.V.)	Sisco Cisco .il Violet (U.V.) Broadcast
Date;	Year; (####;) ## (Month) - Day = 2154; 09 (Sep) – 11 (eleven)	

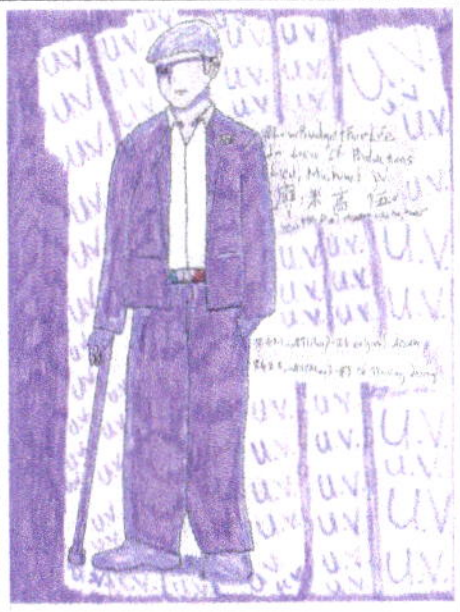

Chicago Controller City Mayor

Chicago (S), The Last Animal Sanctuary (S) Of Illinois (ILL), N.A.,

Electro Magnetic Spectrum; Radio Waves, Amplitude Modulation (A.M.) Frequencies Hertz (Hz) Broadcasting, "Nine Inch Nails' The Downward Spiral (The Bottom)"

Lieutenant 2nd Lieu, Michael W. 廖, 米高 伍o + **WhiteWolf 白狼** Bio**tech Suit+**; Vigilant Volunteers Sisco Cisco (°o°) Yellow;
(*bio**tech suit** male voice*)
"**I** got this. Ascend up… and do **engineering maintenance**."

Chicago Controller City Mayor (*frequency hertz male(s) voice*);
"Start Utopian, Unit eight eight eight eight comma eight eight eight eight, **WhiteWolf**, Lieutenant Lieu, the Church providing much resources, assistance, beliefs during the twenty second century dark ages was relief, but over the years of owning and operating a violet independent belief systems and political party that splinter off from blue and red, Democratic and Republican, or was it a mix of blue and red that makes violet, ultra violet, showed that the Church and her Sister Nuns, Father Priests, beliefs was really Mercy, and the State was his Army Soldiers, Military Operatives, beliefs was really Justice, a saying, separation between Church and State, that can translate to separation between Mercy and Justice, who gets to starve, who gets to eat during the twenty second century dark ages, we decide, we believe, we act, in a separation between Judge, Jury, Xecutioner, in some cases Self-Xecutioner, **Kourts** provide a Self-Xecution option to hang themselves, flatline, we can **turn them undead, assuming the physical body is still together**, Period End."

Sisco Cisco .il Violet (U.V.) Broadcast	(°o°) (°o°) (°o°) (°o°) (°o°) (°o°) (°o°) (°o°) (°o°) (°o°) (°o°) (°o°) (°o°) (°o°) (°o°) (U.V.)	Sisco Cisco .il Violet (U.V.) Broadcast

(Contemporary) (Neo Noir Vigilante) = Biotech (Psychological) = U.V. Broadcast

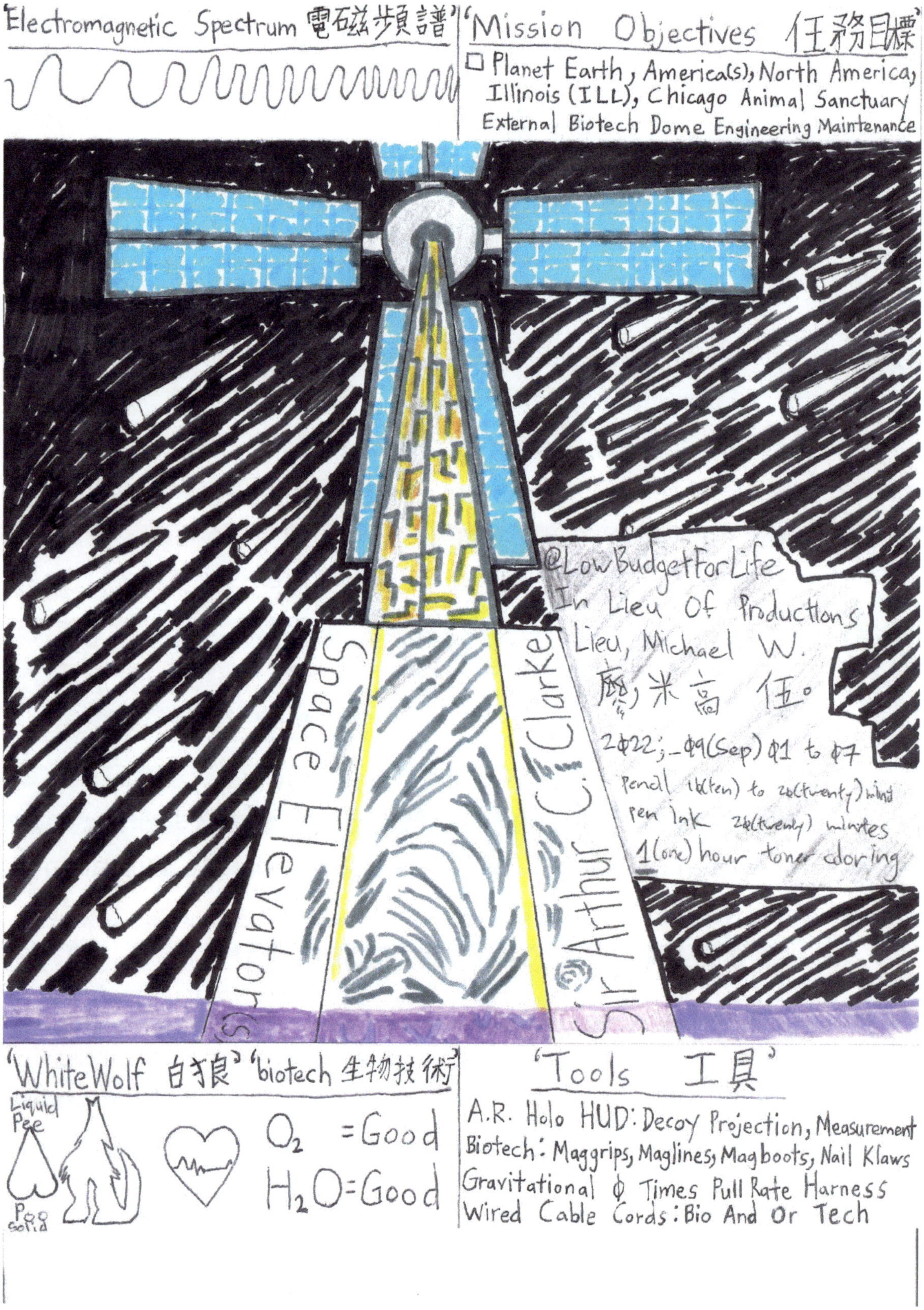

(Psychological) (Super-Hero) (Utopia) = 22nd Century Dark Ages, 2121; Disaster

Sisco Cisco .il Violet (U.V.) Broadcast	(°o°) (°o°) (°o°) (°o°) (°o°) (°o°) (°o°) (°o°) (°o°) (°o°) (°o°) (°o°) (°o°) (°o°) (°o°) (U.V.)	Sisco Cisco .il Violet (U.V.) Broadcast
Date;	Year; (####;) ## (Month) - Day = 2154; 09 (Sep) – 11 (eleven)	

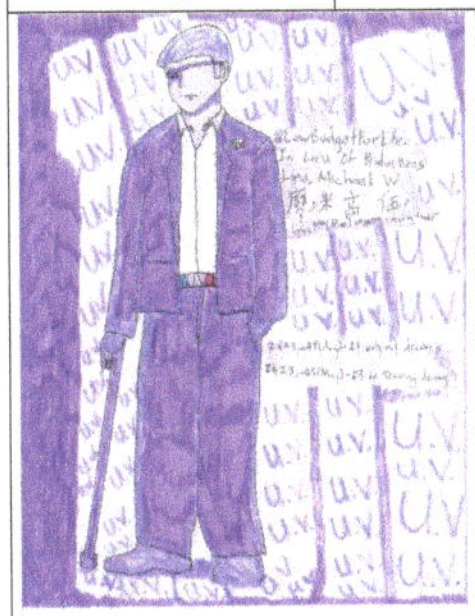

Lieutenant 2nd Lieu, Michael W. 廖, 米高 伍o + **WhiteWolf 白狼**
Bio**tech Suit+**; Vigilant Volunteers Sisco Cisco (°o°) Yellow;
(*bio**tech suit** male voice*)
"**Engineering maintenance** done, Space Elevator time."

Chicago Controller City Mayor (*frequency hertz male(s) voice*);
"Start Utopian, Unit eight eight eight eight comma eight eight eight eight, **WhiteWolf**, Lieutenant Lieu, during the twenty second century dark ages, year one twenty one twenty one, month March, eleven billion people, no purified water, little to no electricity, electronics tech did not work, no holo HUD, no emergency services, **those who had their electronics tech operating, had devices called Faraday cages, providing a form of electronics** Electro Magnetic Pulse **shielding when used correctly**, however, such tech requires energy, the energy Earth mostly uses was nuclear, some electrical grid, some solar panels, some wind turbines, the State, Army Soldier Military Operatives, was not able to kool down nuclear reactors' meltdown, diseases, famine came along within that year, war from within Earth's first, second, third world countries, lastly death, not the deaf you know as hearing loss, rather dead death one, Kay's Private Security Services gated communities, **Elysium Corporations** were good, business as usual, Period End."

Chicago Controller City Mayor

Chicago (S), The Last Animal Sanctuary (S) Of Illinois (ILL), N.A.,

Electro Magnetic Spectrum; Radio Waves, Amplitude Modulation (A.M.) Frequencies Hertz (Hz) Broadcasting, "Muse's Isolated System"

Sisco Cisco .il Violet (U.V.) Broadcast	(°o°) (°o°) (°o°) (°o°) (°o°) (°o°) (°o°) (°o°) (°o°) (°o°) (°o°) (°o°) (°o°) (°o°) (°o°) (U.V.)	Sisco Cisco .il Violet (U.V.) Broadcast

(Contemporary) (Psychological) (Super-Hero) (Utopia) = A.R. Holo HUD Project

Sisco Cisco .il Violet (U.V.) Broadcast	(°o°) (°o°) (°o°) (°o°) (°o°) (°o°) (°o°) (°o°) (°o°) (°o°) (°o°) (°o°) (°o°) (°o°) (°o°) (U.V.)	Sisco Cisco .il Violet (U.V.) Broadcast
Date;	Year; (####;) ## (Month) - Day = 2154; 09 (Sep) – 11 (eleven)	
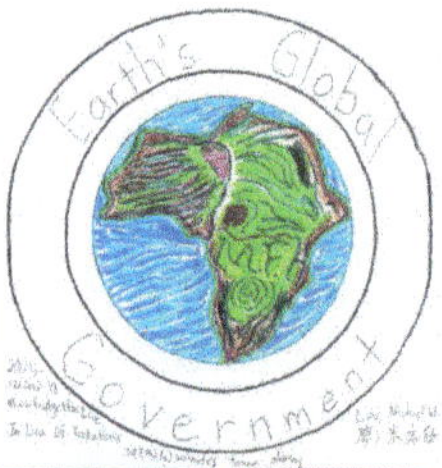 Earth's Global Government (EGG) Space Force Joint Task Force (J.T.F.) 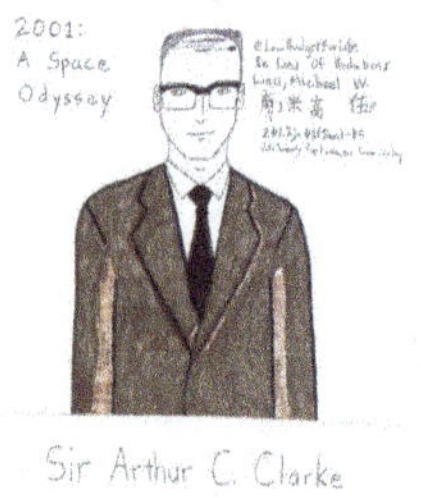 Sir Arthur C. Clarke Space Elevators.	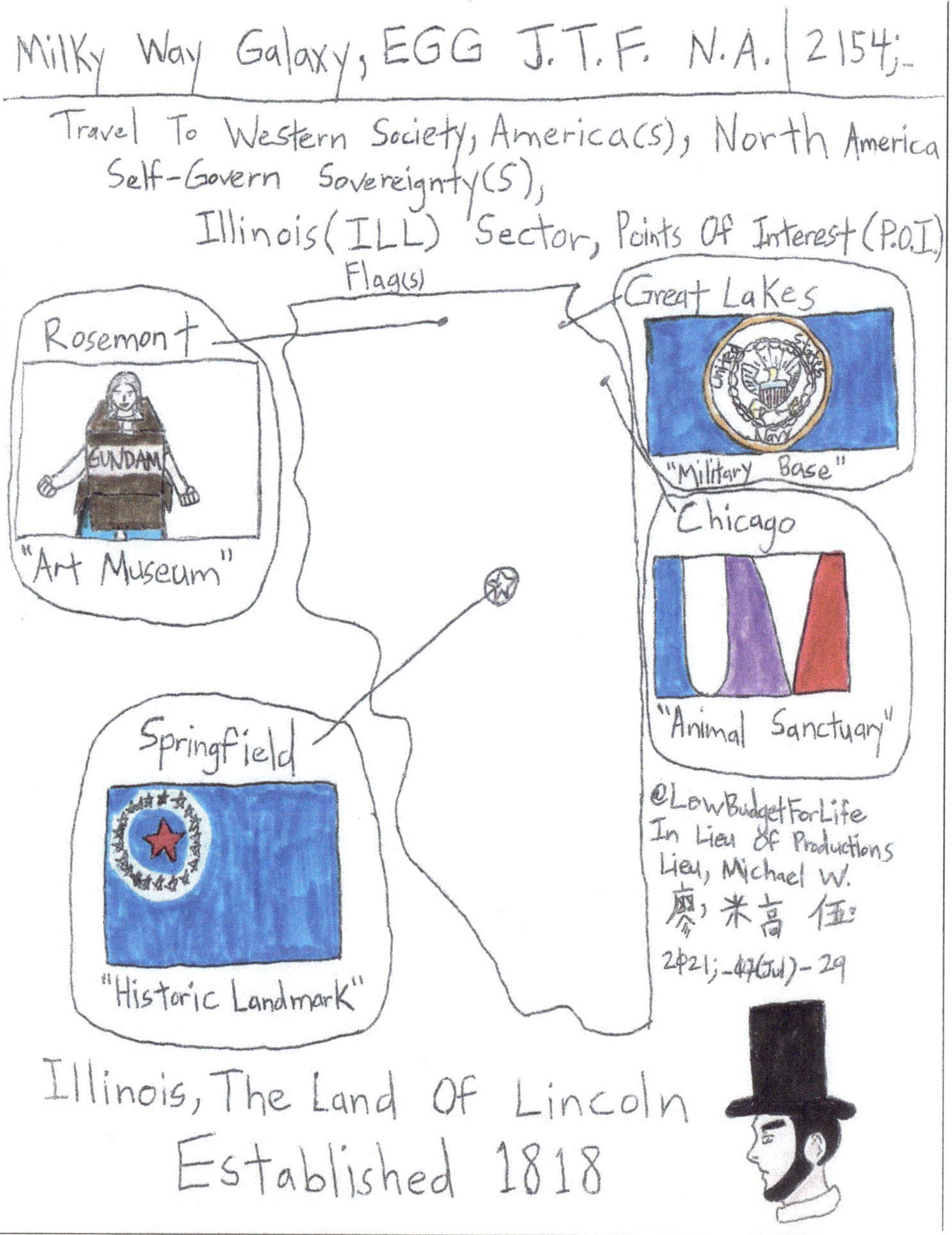	
Sisco Cisco .il Violet (U.V.) Broadcast	(°o°) (°o°) (°o°) (°o°) (°o°) (°o°) (°o°) (°o°) (°o°) (°o°) (°o°) (°o°) (°o°) (°o°) (U.V.)	Sisco Cisco .il Violet (U.V.) Broadcast

(Psychological) (Super-Hero) (Utopia) = 22nd Century Dark Ages, Da Divide.

Sisco Cisco .il Violet (U.V.) Broadcast	(°o°) (°o°) (°o°) (°o°) (°o°) (°o°) (°o°) (°o°) (°o°) (°o°) (°o°) (°o°) (°o°) (°o°) (°o°) (U.V.)	Sisco Cisco .il Violet (U.V.) Broadcast
Date;	Year; (####;) ## (Month) - Day = 2154; 09 (Sep) – 11 (eleven)	

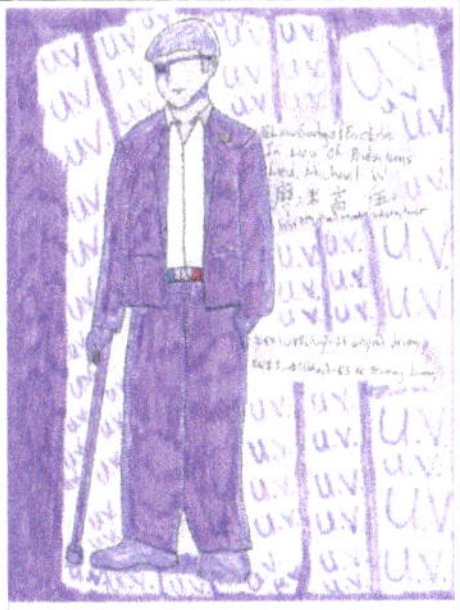

Lieutenant 2nd Lieu, Michael W. 廖, 米高 伍o + **WhiteWolf 白狼**
Bio**tech Suit+**; Vigilant Volunteers Sisco Cisco (°o°) Yellow;
(*bio**tech suit** male voice*)
"**We** are here... Illinois, North America... Space Elevator..."

Chicago Controller City Mayor (*frequency hertz male(s) voice*);

"Start Utopian, Unit eight eight eight eight comma eight eight eight eight, **WhiteWolf**, Lieutenant Lieu, military gated communities in pre Kold War bases all around United States and her fifty five states' interest had to decide who to take in and who to leave out, this is where the Church and State split, monitoring the active **Electro Magnetic Spectrum** still broadcasting radio waves all across North America, we concluded that the Church supported illegal immigrants, on and off planet aliens, the Militia Mexican-Men Movement, formed during twenty one twenty one, in response to the State for failing to provide for illegal immigrants, on and off planet aliens, mainly assisted illegal immigrants of Mexican descent and or nationality, the State, Army Soldier Military Operatives in their bases expand to counties across North America, the people on base need to understand English language read and write along with speaking and listening plus have physical paperwork, the Church and the Militia did not demand such language requirement and paperwork process, Period End."

Chicago Controller City Mayor

Chicago (S), The Last Animal Sanctuary (S) Of Illinois (ILL), N.A.,

Electro Magnetic Spectrum; Radio Waves, Amplitude Modulation (A.M.) Frequencies Hertz (Hz) Broadcasting, "Final Fantasy Eight's Overtune"

Sisco Cisco .il Violet (U.V.) Broadcast	(°o°) (°o°) (°o°) (°o°) (°o°) (°o°) (°o°) (°o°) (°o°) (°o°) (°o°) (°o°) (°o°) (°o°) (°o°) (U.V.)	Sisco Cisco .il Violet (U.V.) Broadcast

(Contemporary) (Psych) (Super-Hero) (Utopia) = Continents Surviving On Earth

Sisco Cisco .il Violet (U.V.) Broadcast	(°o°) (°o°) (°o°) (°o°) (°o°) (°o°) (°o°) (°o°) (°o°) (°o°) (°o°) (°o°) (°o°) (°o°) (°o°) (U.V.)	Sisco Cisco .il Violet (U.V.) Broadcast
Date;	Year; (####;) ## (Month) - Day = 2154; 09 (Sep) – 11 (eleven)	

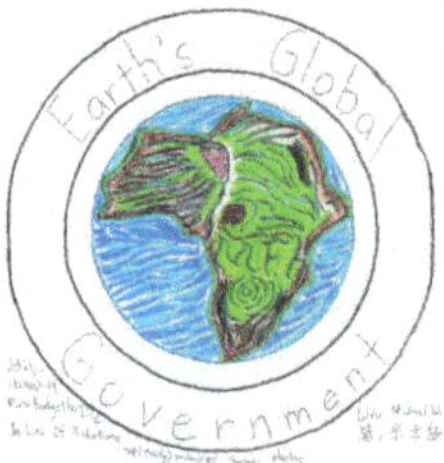

Earth's Global Government (EGG) Space Force Joint Task Force (J.T.F.)

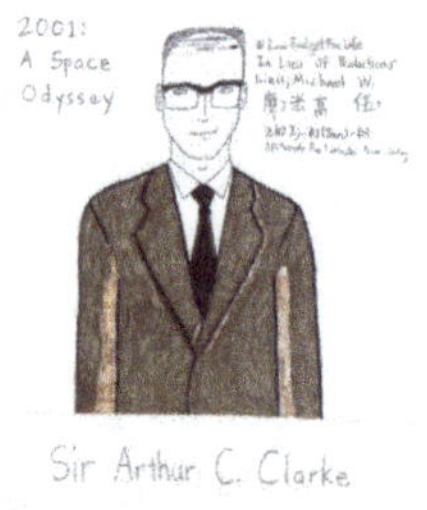

Sir Arthur C. Clarke Space Elevators.

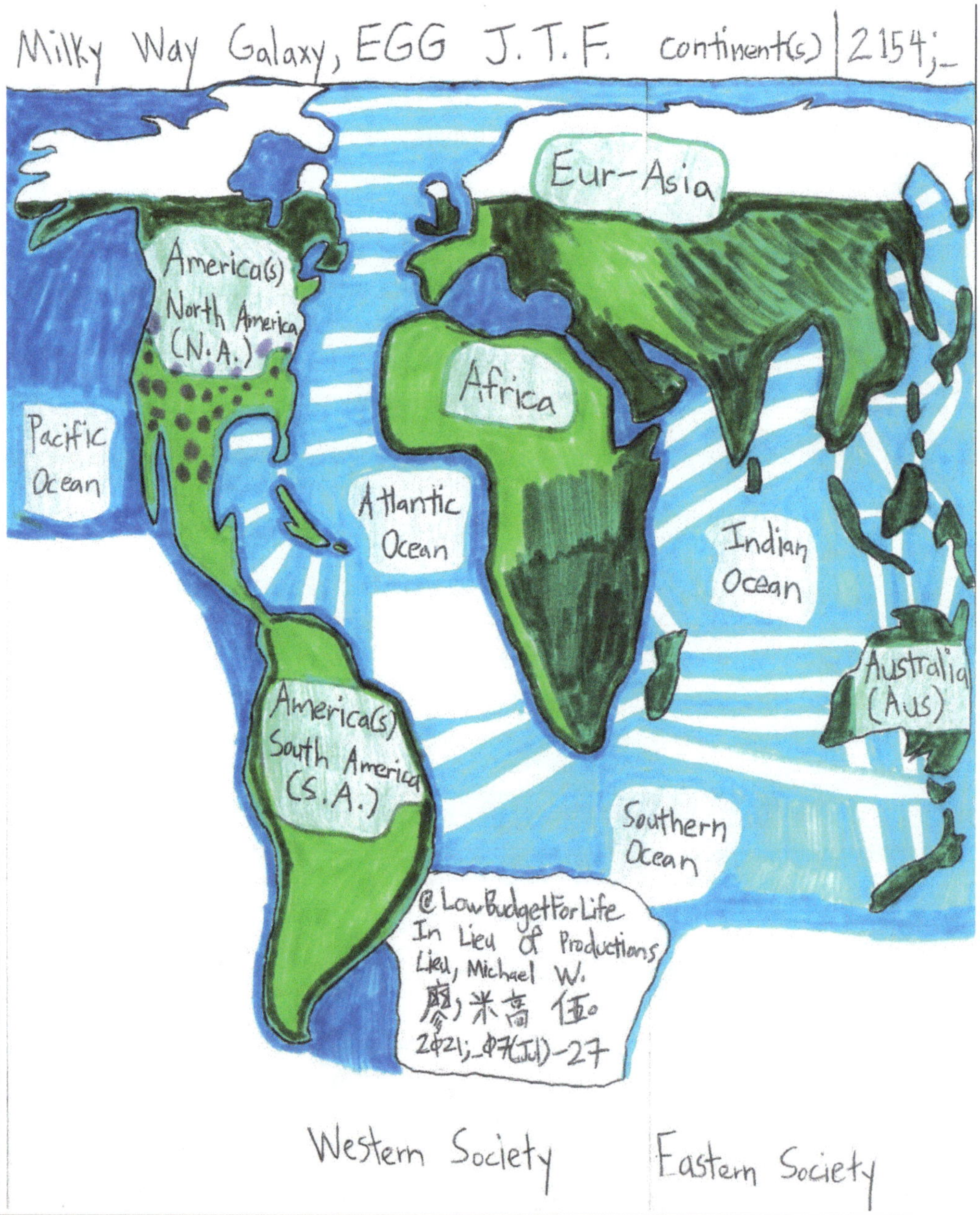

Sisco Cisco .il Violet (U.V.) Broadcast	(°o°) (°o°) (°o°) (°o°) (°o°) (°o°) (°o°) (°o°) (°o°) (°o°) (°o°) (°o°) (°o°) (°o°) (°o°) (U.V.)	Sisco Cisco .il Violet (U.V.) Broadcast

(Psychological) (Super-Hero) (Utopia) = 22nd Century Dark Ages, Post 2121;

Sisco Cisco .il Violet (U.V.) Broadcast	(°o°) (°o°) (°o°) (°o°) (°o°) (°o°) (°o°) (°o°) (°o°) (°o°) (°o°) (°o°) (°o°) (°o°) (°o°) (U.V.)	Sisco Cisco .il Violet (U.V.) Broadcast
Date;	Year; (####;) ## (Month) - Day = 2154; 09 (Sep) – 11 (eleven)	

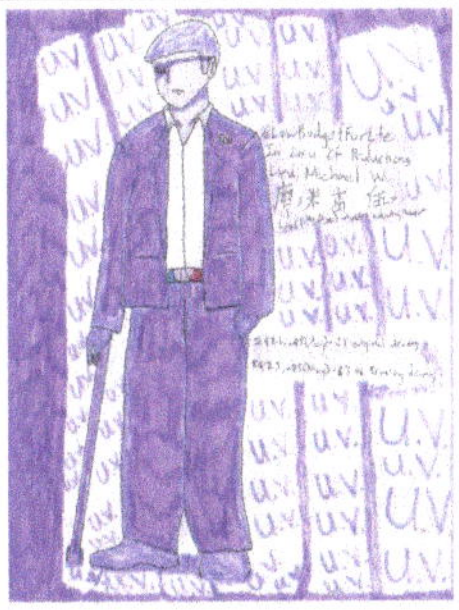

Chicago Controller City Mayor

Chicago (S), The Last Animal Sanctuary (S) Of Illinois (ILL), N.A.,

Electro Magnetic Spectrum; Radio Waves, Amplitude Modulation (A.M.) Frequencies Hertz (Hz) Broadcasting, "13 Sentinels Aegis Rim's Impending Doom"

Lieutenant 2nd Lieu, Michael W. 廖, 米高 伍o + **WhiteWolf 白狼 Biotech Suit+**; Vigilant Volunteers Sisco Cisco (°o°) Yellow;

(***biotech suit*** *male voice*)

"Let **us** see, frequency, **we** are in North America… On Earth."

Chicago Controller City Mayor (*frequency hertz male(s) voice*);

"Start Utopian, Unit eight eight eight eight comma eight eight eight eight, **WhiteWolf**, Lieutenant Lieu, after twenty one twenty one, which **augmented animal societies** were still standing in North America we wonder, we cannot speak on continents outside North America since our broadcasting U.V. radio waves were not able to reach that distance, we concluded that the following survived through the twenty second century dark ages year one, we **color code them in accordance to Electro Magnetic Spectrum toners**, black Kay's Private Security Services gated communities, green with red blue mix situational county military bases plus blue naval bases, green **white** brown red flag wavers Mexican-Men militias, yellow black Churches with crosses and hearts, violet Ultra Violet animal sanctuary started because we could not eat pets, peoples' pets, anyone's pets to survive, mix with **white Elysium Corporation's domes providing needed protection** exposure from Earth's ecosystem collapsing due to space rocks asteroid meteors at certain times, Period End."

Sisco Cisco .il Violet (U.V.) Broadcast	(°o°) (°o°) (°o°) (°o°) (°o°) (°o°) (°o°) (°o°) (°o°) (°o°) (°o°) (°o°) (°o°) (°o°) (°o°) (U.V.)	Sisco Cisco .il Violet (U.V.) Broadcast

(Contemporary) (Military) (Psychological) (Super-Hero) (Utopia) = A.R. Holo HUD

<table>
<tr><td>Sisco
Cisco
.il Violet (U.V.)
Broadcast</td><td>(°o°) (°o°) (°o°) (°o°) (°o°) (°o°) (°o°) (°o°) (°o°) (°o°) (°o°) (°o°) (°o°) (°o°) (°o°)
(U.V.) (U.V.) (U.V.) (U.V.) (U.V.) (U.V.) (U.V.) (U.V.) (U.V.) (U.V.) (U.V.) (U.V.) (U.V.) (U.V.)
(U.V.) (U.V.) (U.V.) (U.V.) (U.V.) (U.V.) (U.V.) (U.V.) (U.V.) (U.V.) (U.V.) (U.V.) (U.V.) (U.V.)
(U.V.) (U.V.) (U.V.) (U.V.) (U.V.) (U.V.) (U.V.) (U.V.) (U.V.) (U.V.) (U.V.) (U.V.) (U.V.) (U.V.)</td><td>Sisco
Cisco
.il Violet (U.V.)
Broadcast</td></tr>
<tr><td>Date;</td><td colspan="2">Year; (####;) ## (Month) - Day = 2154; 09 (Sep) – 11 (eleven)</td></tr>
<tr><td>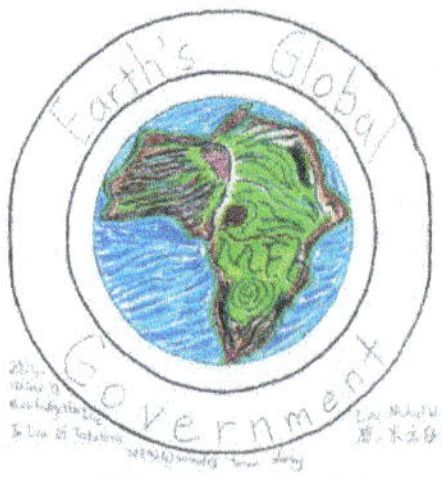

Earth's Global Government (EGG) Space Force Joint Task Force (J.T.F.)
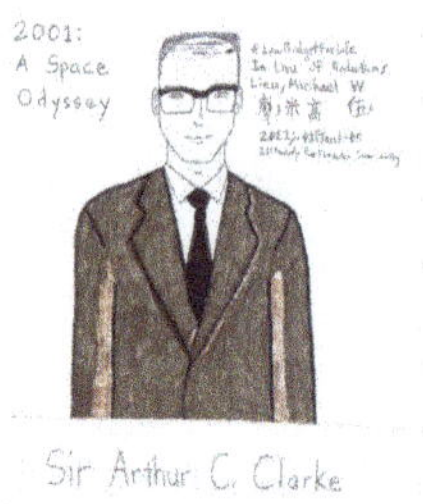

Sir Arthur C. Clarke Space Elevators.</td><td colspan="2">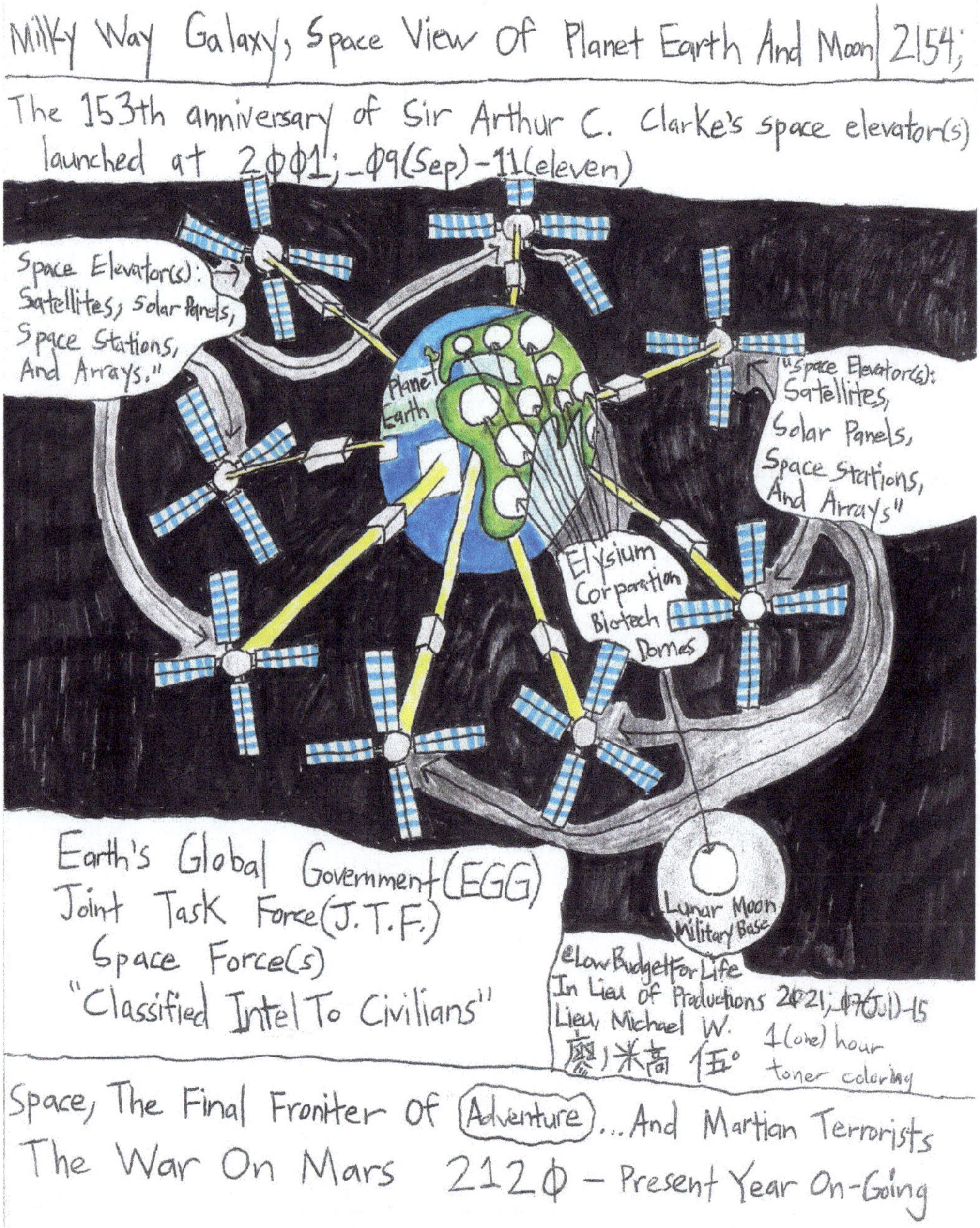
</td></tr>
<tr><td>Sisco
Cisco
.il Violet (U.V.)
Broadcast</td><td>(°o°) (°o°) (°o°) (°o°) (°o°) (°o°) (°o°) (°o°) (°o°) (°o°) (°o°) (°o°) (°o°) (°o°) (°o°)
(U.V.) (U.V.) (U.V.) (U.V.) (U.V.) (U.V.) (U.V.) (U.V.) (U.V.) (U.V.) (U.V.) (U.V.) (U.V.) (U.V.)
(U.V.) (U.V.) (U.V.) (U.V.) (U.V.) (U.V.) (U.V.) (U.V.) (U.V.) (U.V.) (U.V.) (U.V.) (U.V.) (U.V.)
(U.V.) (U.V.) (U.V.) (U.V.) (U.V.) (U.V.) (U.V.) (U.V.) (U.V.) (U.V.) (U.V.) (U.V.) (U.V.) (U.V.)</td><td>Sisco
Cisco
.il Violet (U.V.)
Broadcast</td></tr>
</table>

(Contemporary) (Military) (Psychological) (Super-Hero) (Utopia) = A.R. Holo HUD

Sisco Cisco .il Violet (U.V.) Broadcast	(°o°) (°o°) (°o°) (°o°) (°o°) (°o°) (°o°) (°o°) (°o°) (°o°) (°o°) (°o°) (°o°) (°o°) (°o°) (U.V.)	Sisco Cisco .il Violet (U.V.) Broadcast
Date;	Year; (####;) ## (Month) - Day = 2154; 09 (Sep) – 11 (eleven)	

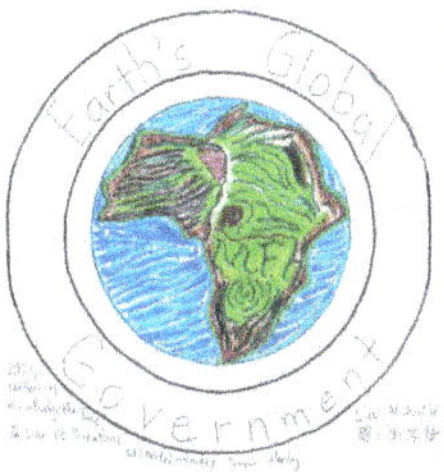

Earth's Global Government (EGG) Space Force Joint Task Force (J.T.F.)

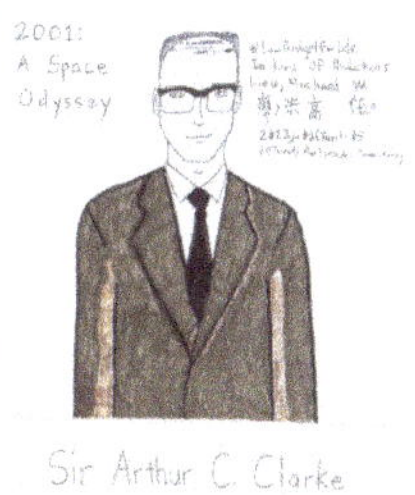

Sir Arthur C. Clarke Space Elevators.

Archival Recordings Of
Current EGG Space Force J.T.F. Fleet Admiral Keith D. Williams;

"This is most likely an audio log... I know it is the annual anniversary to the year Sir Arthur C. Clarke's Space Elevators first launched... two thousand one September eleven along side his movie and story Sir Clarke worked with on with Director Stanley Kubrick two thousand one A Space Odyssey, what a masterpiece...

That was in the twenty first century... over one hundred years.

You know back in twenty one twenty one, I was going to retire... I made Admiral of Space Force, United States Space Force, back before we established EGG, Earth's Global Government, you know **public sec Uncle Sam** loves **his** acronyms, EGG...

For years before twenty one twenty one... space rocks, War On Resources, unconventional warfare from Martian Fleet, Spacers and Belters, Roman Industries... We were dealing with the Kold War, the red scare, **War Con one**, fighting and identifying Communists, Marxist Leninist belief systems, Soviets, Imperialists, plus Hitler's Nazis keeping to themselves on Venus.

To think Uncle Sam Space Force, would unite with the Soviets... those Communists, Imperialists, and surviving Earthlings' first, second and third world countries in fighting foreign foes of Earth had to be planned. Sorry, got off top on Space Elevators launch."

Sisco Cisco .il Violet (U.V.) Broadcast	(°o°) (°o°) (°o°) (°o°) (°o°) (°o°) (°o°) (°o°) (°o°) (°o°) (°o°) (°o°) (°o°) (°o°) (°o°) (U.V.)	Sisco Cisco .il Violet (U.V.) Broadcast

(Contemporary) (Military) (Psychological) (Super-Hero) (Utopia) = A.R. Holo HUD

Sisco Cisco .il Violet (U.V.) Broadcast	(°o°) (°o°) (°o°) (°o°) (°o°) (°o°) (°o°) (°o°) (°o°) (°o°) (°o°) (°o°) (°o°) (°o°) (°o°) (U.V.)	Sisco Cisco .il Violet (U.V.) Broadcast
Date;	Year; (####;) ## (Month) - Day = 2154; 09 (Sep) – 11 (eleven)	

Earth's Global Government (EGG) Space Force Joint Task Force (J.T.F.)

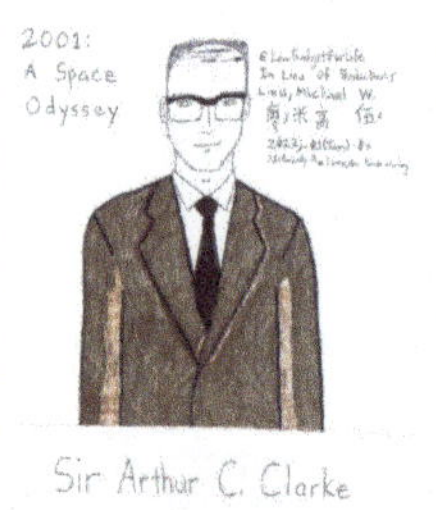

Sir Arthur C. Clarke Space Elevators.

Solar System, Drawing Diagram Of Planets 2154;

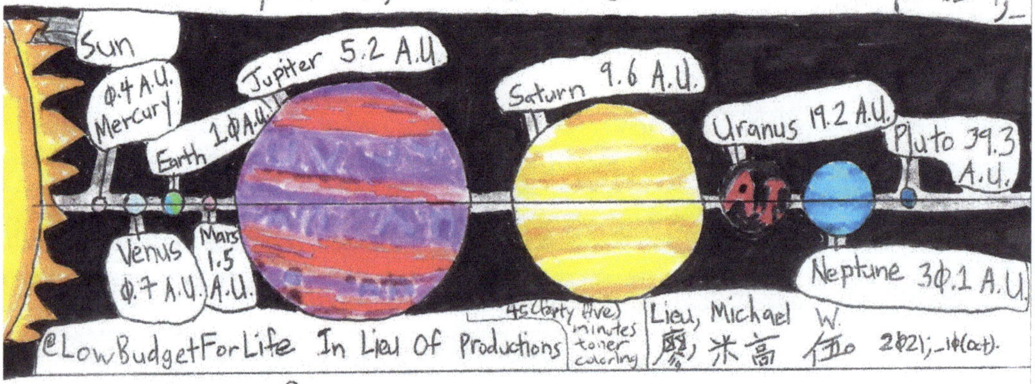

The average distance of our planets from the Sun.
The numbers in this drawing diagram are Astronomical Units (A.U.), based on the average distance between the Earth and the Sun;
1 A.U. = 1.50 x 10⁸ Kilo Meters (K.M.)

"2001: A Space Odyssey @ 2001; 09 (Sep) – 11 (eleven)
Sir Arthur C. Clarke's Space Elevators' Anniversary Celebration"

Our Solar System is 4.6 (four point six) billion years old.

Planet;	**Gravitational (G) Pull Rate Of Earth Ratio**;
Mercury	0.378 (zero point three seven eight) Gs.
Venus	0.907 (zero point nine zero seven) Gs.
Earth (Present Planet)	1.0 (one **and or** one point zero) Gs Of Earth.
Mars	0.377 (zero point three seven seven) Gs.
Jupiter	2.36 (two point three six) Gs
Saturn	0.916 (zero point nine one six) Gs.
Uranus	0.889 (zero point eight eight nine) Gs.
Neptune	1.12 (one point one two) Gs.
Pluto	0.071 (zero point zero seven one) Gs.

Sisco Cisco .il Violet (U.V.) Broadcast	(°o°) (°o°) (°o°) (°o°) (°o°) (°o°) (°o°) (°o°) (°o°) (°o°) (°o°) (°o°) (°o°) (°o°) (°o°) (U.V.)	Sisco Cisco .il Violet (U.V.) Broadcast

(Contemporary) (Military) (Psychological) (Super-Hero) (Utopia) = A.R. Holo HUD

Sisco Cisco .il Violet (U.V.) Broadcast	(°o°) (°o°) (°o°) (°o°) (°o°) (°o°) (°o°) (°o°) (°o°) (°o°) (°o°) (°o°) (°o°) (°o°) (°o°) (U.V.)	Sisco Cisco .il Violet (U.V.) Broadcast

Solar System, Diagram Planetary Statistics, Notes

Planet	Diameter	Distance From Sun * In Millions	Rotational Period	Notes
Mercury [Low B.]	4,878 K.M.	57.9 K.M. Inner Planet	58.64 days	
Venus [Low B.]	12,104 K.M.	108.1 K.M. Inner Planet	243 days (retrograde)	The Third Reich, Adolf Hitler's Nazi Political Party Head Quarters (H.Q.).
Earth [Low B.]	12,756 K.M.	149.5 K.M. Inner Planet	23 hours, 56 minutes, 4.1 seconds	The Home Planet Origins Of **Humans, Animal Kingdom**, And Artificial Intelligence (A.I.), Earth's Global Government (EGG).
Mars [Low B.]	6,787 K.M.	227.9 K.M. Inner Planet	24 hours, 37 minutes, 22.3 seconds	The War On Mars, 2120 November To Present Day Time, Martian Terrorists, Roman Industries.
Jupiter [Low B.]	142,800 K.M.	778.3 K.M. Outer Planet	9 hours, 50 minutes (equator). 9 hours, 55 minutes (polar)	The UltraViolet (U.V.) Rays' Broadcasting Phenomenon On Io And Europa Moon(s).
Saturn [Low B.]	120,540 K.M.	1,427.0 K.M. Outer Planet	10 hours, 39 minutes	
Uranus [Low B.]	51,118 K.M.	2,869.6 K.M. Outer Planet	17 hours, 54 minutes	The Head Quarters (H.Q.) Of Automation Artificial Intelligence (A.I.) Established **Elysium Corporation**.
Neptune [Low B.]	49,528 K.M.	4,496.6 K.M. Outer Planet	19 hours, 12 minutes	
Pluto [Low B.]	2,300 K.M.	5,900.0 K.M. Outer Planet	6 days, 9 hours	Some say, this is "not an actual planet".

Sisco Cisco .il Violet (U.V.) Broadcast	(°o°) (°o°) (°o°) (°o°) (°o°) (°o°) (°o°) (°o°) (°o°) (°o°) (°o°) (°o°) (°o°) (°o°) (°o°) (U.V.)	Sisco Cisco .il Violet (U.V.) Broadcast

146 (Lieutenant Lieu, In Lieu Of) Lieu, Michael W. 廖, 米高 伍。 146

(Neo Noir Vigilante) (Parody) (Psychological) = Parody Psych Lucid Dreaming.

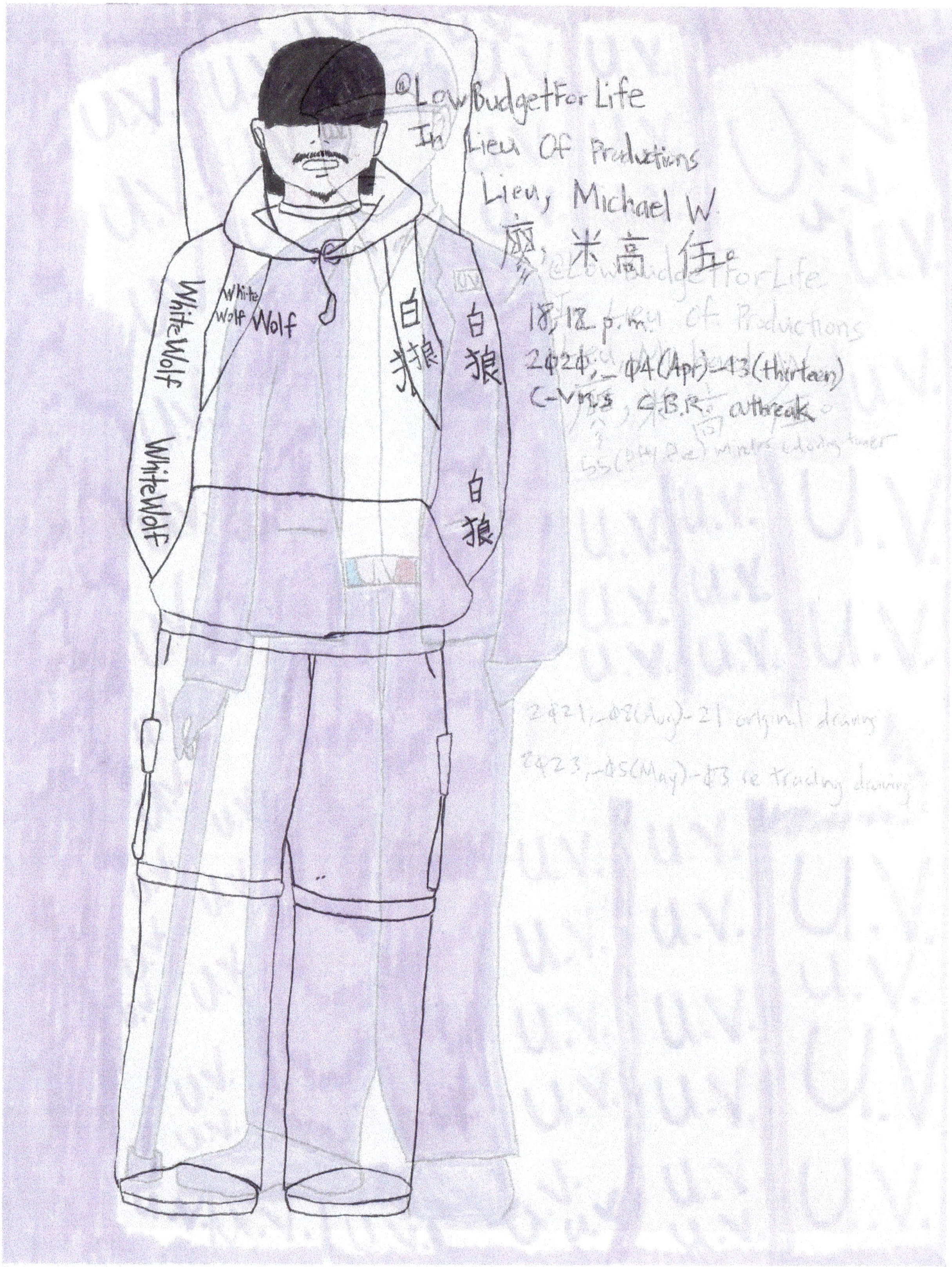

146 \ **2154;** Lieu, 廖, ‘WhiteWolf 白狼’ Left to Right (L→R) 146

(Psychological) (Super-Hero) (Utopia) = Parody Psych Lucid Dreaming.

Sisco Cisco .il Violet (U.V.) Broadcast	(°o°) (°o°) (°o°) (°o°) (°o°) (°o°) (°o°) (°o°) (°o°) (°o°) (°o°) (°o°) (°o°) (°o°) (°o°) (U.V.)	Sisco Cisco .il Violet (U.V.) Broadcast
Date;	Year; (####;) ## (Month) - Day = XXXX; XX (XXX) – XX (X)	

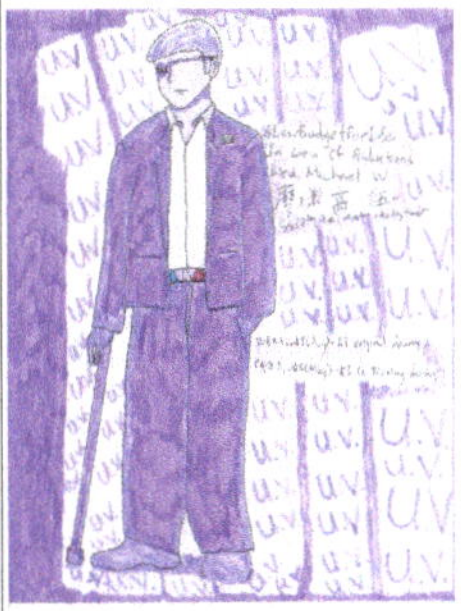

Chicago Controller City Mayor

Chicago (S), The Last Animal Sanctuary (S) Of Illinois (ILL), N.A.,

Electro Magnetic Spectrum; Radio Waves, Amplitude Modulation (A.M.) Frequencies Hertz (Hz) Broadcasting, "Shins' Sleeping Lessons"

Lieutenant 2nd Lieu, Michael W. 廖, 米高 伍o + **WhiteWolf 白狼** Bio**tech Suit+**; Vigilant Volunteers Sisco Cisco (°o°) Yellow;

(*bio**tech suit** male voice*)

"Space Elevator, Solar System anniversary. ZzZzZz… Uh…"

Chicago Controller City Mayor (*frequency hertz male(s) voice*);

"Start Utopian, Unit eight eight eight eight comma eight eight eight eight, **WhiteWolf**, Lieutenant Lieu, wondering how our Ultra Violet broadcasting works, **you** may be a **failed engineer**, key word **engineer**, how it works is a mix of biological viruses and **technological frequencies**, viruses and frequencies is the short answer, long answer, Belief Systems, microorganisms, **nanotech**, M.K. Ultra research archives, Edgar Cayce, brainwashing, indoctrination, like military Army Navy indoctrination, our Utopian Sovereignty Movement is the same, Persuasion Powers, psychological conditioning, the voices, Transmitters And Antennas, Vibrations, Frequencies Jamming, Neural Oscillations, Brain Waves, Neuro Science, Magnetoencephalography: Measuring Magnetic Fields Produced By Electrical Brain Activity, Neurological Implants And Or Explants To Directly Interface With Biotechs, Two Or More Psychological Minds Coexisting In One Platform Body, Clairvoyance, Extra Sensory Perception, E.S.P., Parapsychology, Psychic, Paranormal, the origins of consciousness, Period End."

Sisco Cisco .il Violet (U.V.) Broadcast	(°o°) (°o°) (°o°) (°o°) (°o°) (°o°) (°o°) (°o°) (°o°) (°o°) (°o°) (°o°) (°o°) (°o°) (°o°) (U.V.)	Sisco Cisco .il Violet (U.V.) Broadcast

(Contemporary) (Military) (Super-Hero) (Utopia) = The State, Uncle Sam (U.S.).

Sisco Cisco .il Green Broadcast	(°o°) (°o°) (°o°) (°o°) (°o°) (°o°) (°o°) (°o°) (°o°) (°o°) (°o°) (°o°) (°o°) (°o°) (°o°) Green	Sisco Cisco .il Green Broadcast
Date;	Year; (####;) ## (Month) - Day = 2154; 10 (Oct) – 30 (thirty)	

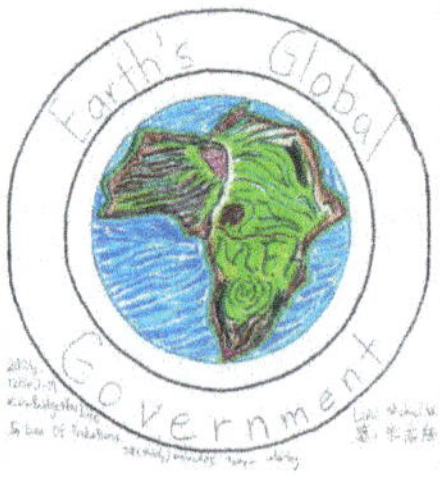

Earth's Global Government (EGG) Space Force Joint Task Force (J.T.F.)

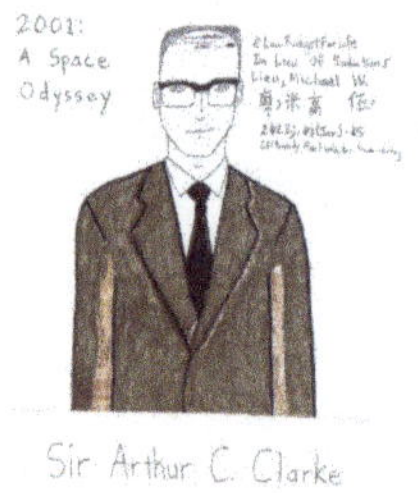

Sir Arthur C. Clarke Space Elevators.

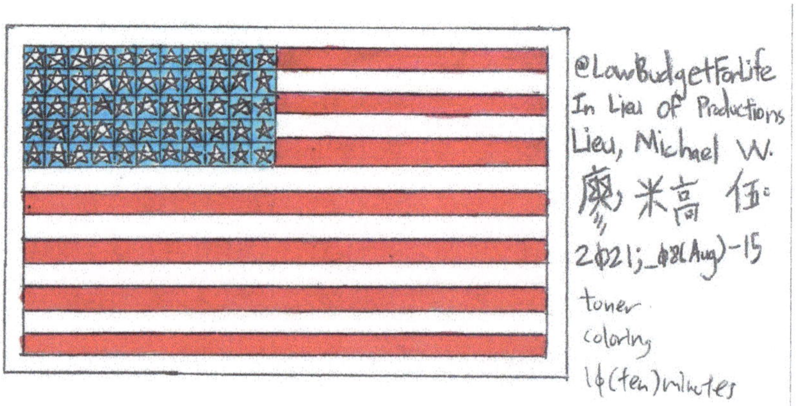

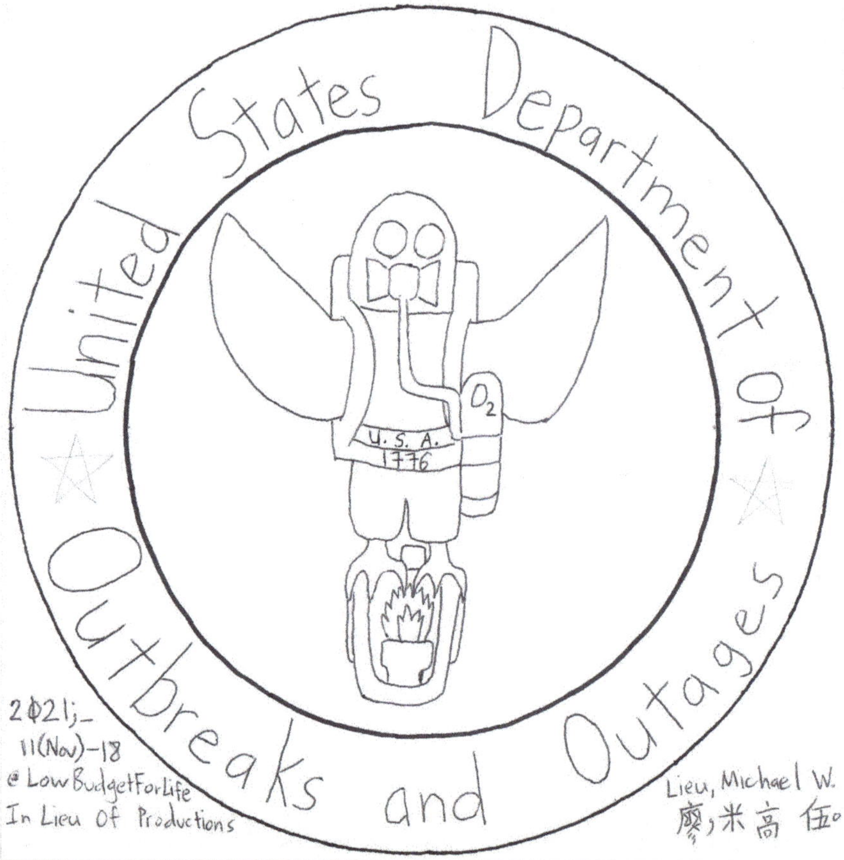

Sisco Cisco .il Green Broadcast	(°o°) (°o°) (°o°) (°o°) (°o°) (°o°) (°o°) (°o°) (°o°) (°o°) (°o°) (°o°) (°o°) (°o°) (°o°) Green	Sisco Cisco .il Green Broadcast

(Contemporary) (Military) (Super-Hero) (Utopia) = U.S.A. County Military Bases.

Sisco Cisco .il Green Broadcast	(°o°) (°o°) (°o°) (°o°) (°o°) (°o°) (°o°) (°o°) (°o°) (°o°) (°o°) (°o°) (°o°) (°o°) (°o°) Green	Sisco Cisco .il Green Broadcast
Date;	Year; (####;) ## (Month) - Day = 2154; 10 (Oct) – 30 (thirty)	

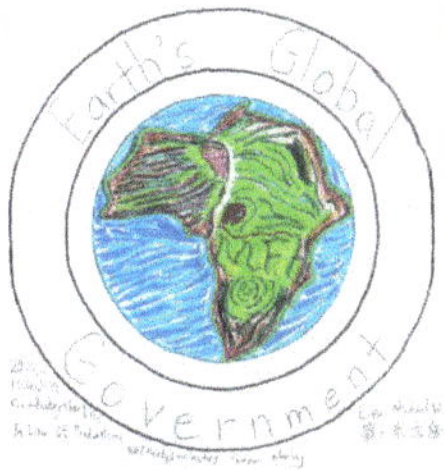

Earth's Global Government (EGG) Space Force Joint Task Force (J.T.F.)

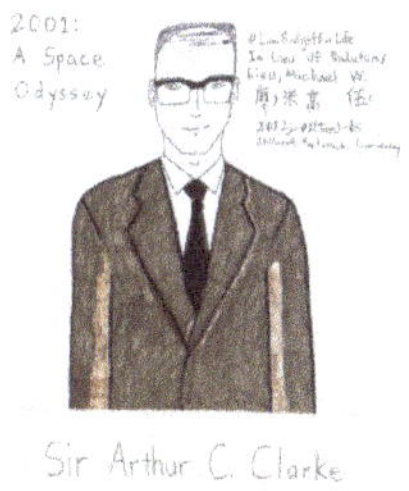

Sir Arthur C. Clarke Space Elevators.

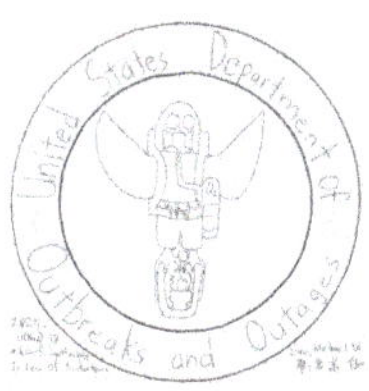

Lieutenant 2nd Lieu, Michael W. 廖, 米高 伍o + **WhiteWolf 白狼**
Bio**tech Suit+**; Vigilant Volunteers Sisco Cisco (°o°) Yellow;
(*bio**tech suit** male voice)*

"...**Audio voice recording** for **retirement journal** year four, the year **twenty one fifty four**, the date September eleven.

Today is Sir Arthur C. Clarke's Space Elevators Anniversary Launch Celebration, no the year still **twenty one fifty four**, the date **October thirty**, Space Elevators celebration has pasted.

I was doing **engineering maintenance** for Chicago Animal Sanctuary **dome**, tuning in on Earth's Space Force, War On Mars, Sir Arthur C. Clarke's Space Elevators by the U.V. broadcasting antenna and receivers, with Mayor Chicago Controller's projection for some time, moments, until next thing **I** know a **few days, weeks just flew by**, must have fell asleep or something...

The Mayor was... broadcasting on how one of many animal sanctuary was established, founded, during the twenty second century dark ages, year one, twenty one twenty one.

Much like the Colonel on base talked about it and with his Lieutenant Colonel, Military Police, Combat Engineers, civilians, contractors, veterans, retirees, survivors, etc...

What really happened during twenty one twenty one with the Church, State, Militia, and Sanctuary Sovereignty. How did they deal with no energy, meteorites, dead people, diseases, famine?

Sisco Cisco .il Green Broadcast	(°o°) (°o°) (°o°) (°o°) (°o°) (°o°) (°o°) (°o°) (°o°) (°o°) (°o°) (°o°) (°o°) (°o°) (°o°) Green	Sisco Cisco .il Green Broadcast

(Military) = U.S.A. County Military Bases (Neo Noir Vigilante) = Biotech Suit+

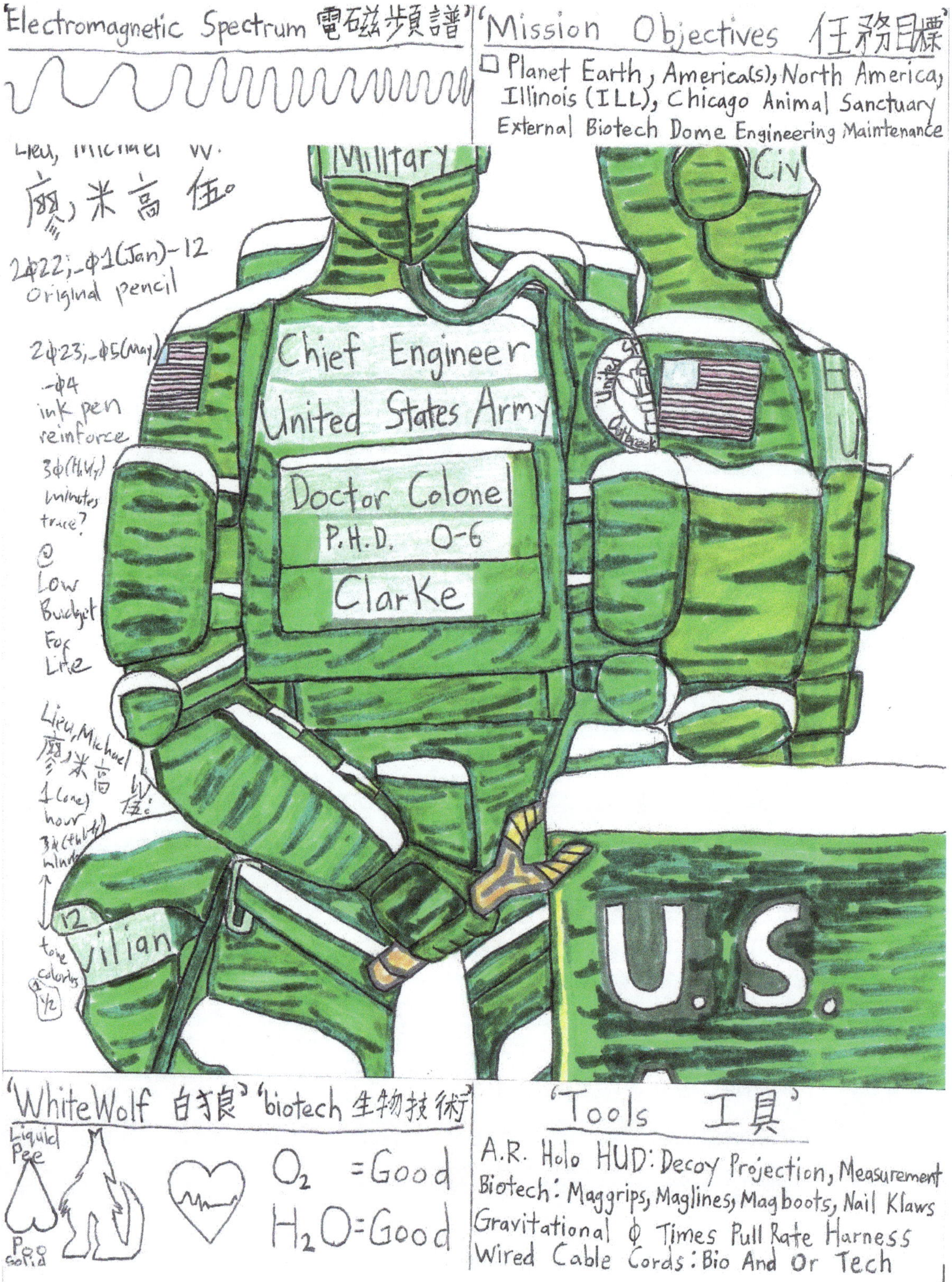

(Contemporary) (Military) (Super-Hero) (Utopia) = U.S.A. County Military Bases

Sisco Cisco .il Green Broadcast	(°o°) (°o°) (°o°) (°o°) (°o°) (°o°) (°o°) (°o°) (°o°) (°o°) (°o°) (°o°) (°o°) (°o°) (°o°) Green	Sisco Cisco .il Green Broadcast
Date;	Year; (####;) ## (Month) - Day = 2154; 10 (Oct) – 30 (thirty)	

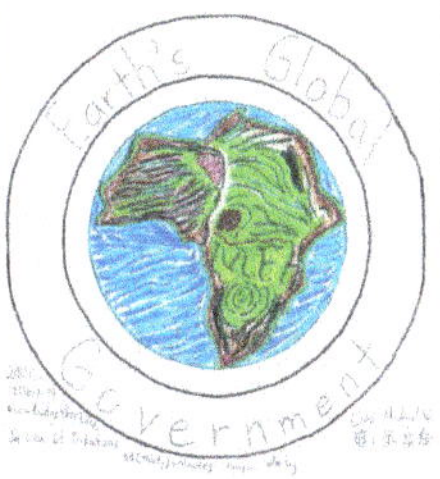

Earth's Global Government (EGG) Space Force Joint Task Force (J.T.F.)

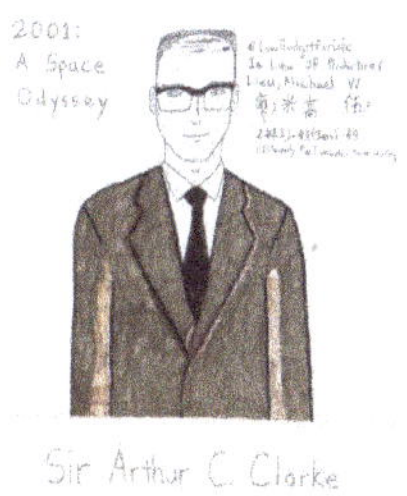

Sir Arthur C. Clarke Space Elevators.

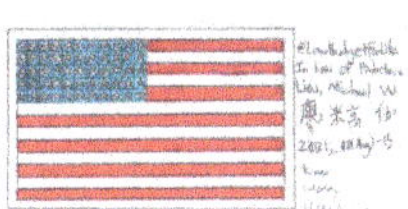

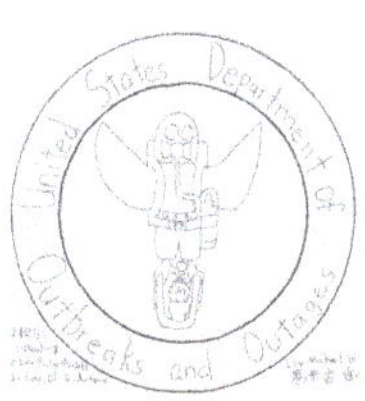

Lieutenant 2nd Lieu, Michael W. 廖, 米高 伍o + **WhiteWolf 白狼 Biotech Suit+**; Vigilant Volunteers Sisco Cisco (°o°) Yellow;
(*bio***tech suit** *male voice)*
"**I** am in front of the removed, former State officials military Colonel and the surviving civilians who used to reside in pre Animal Sanctuary Chicago Illinois, before the **dome**. Fort Cook."

United States Army Chief Engineer Military Colonel C. Clarke;
"(*bio***tech suit** *male voice)* **WhiteWolf, you are suited up** to take part in a **United States Department of Outbreaks and Outages** decontamination **washing**? We received reports that Fort Cook may undergo a possible outbreak outages due to this years' July four independence day fireworks!"

Lieutenant 2nd Lieu, Michael W. 廖, 米高 伍o + **WhiteWolf 白狼 Biotech Suit+**; Vigilant Volunteers Sisco Cisco (°o°) Yellow;
"(*bio***tech suit** *male voice)* What year is this years' you mean?"

United States Army Chief Engineer Military Colonel C. Clarke;
"(*bio***tech suit** *male voice)* Year twenty one fifty two, July four. Get ready. We got numerous civilians in green P.P.E. full body suits that is ready to decontaminate Fort Cook, Chicago within base went dark not long ago, be careful of hazmat, watch your six everyone. We are in War Con Two, space rocks still active."

Sisco Cisco .il Green Broadcast	(°o°) (°o°) (°o°) (°o°) (°o°) (°o°) (°o°) (°o°) (°o°) (°o°) (°o°) (°o°) (°o°) (°o°) (°o°) Green	Sisco Cisco .il Green Broadcast

(Military) = U.S.A. County Military Bases (Neo Noir Vigilante) = Biotech Suit+

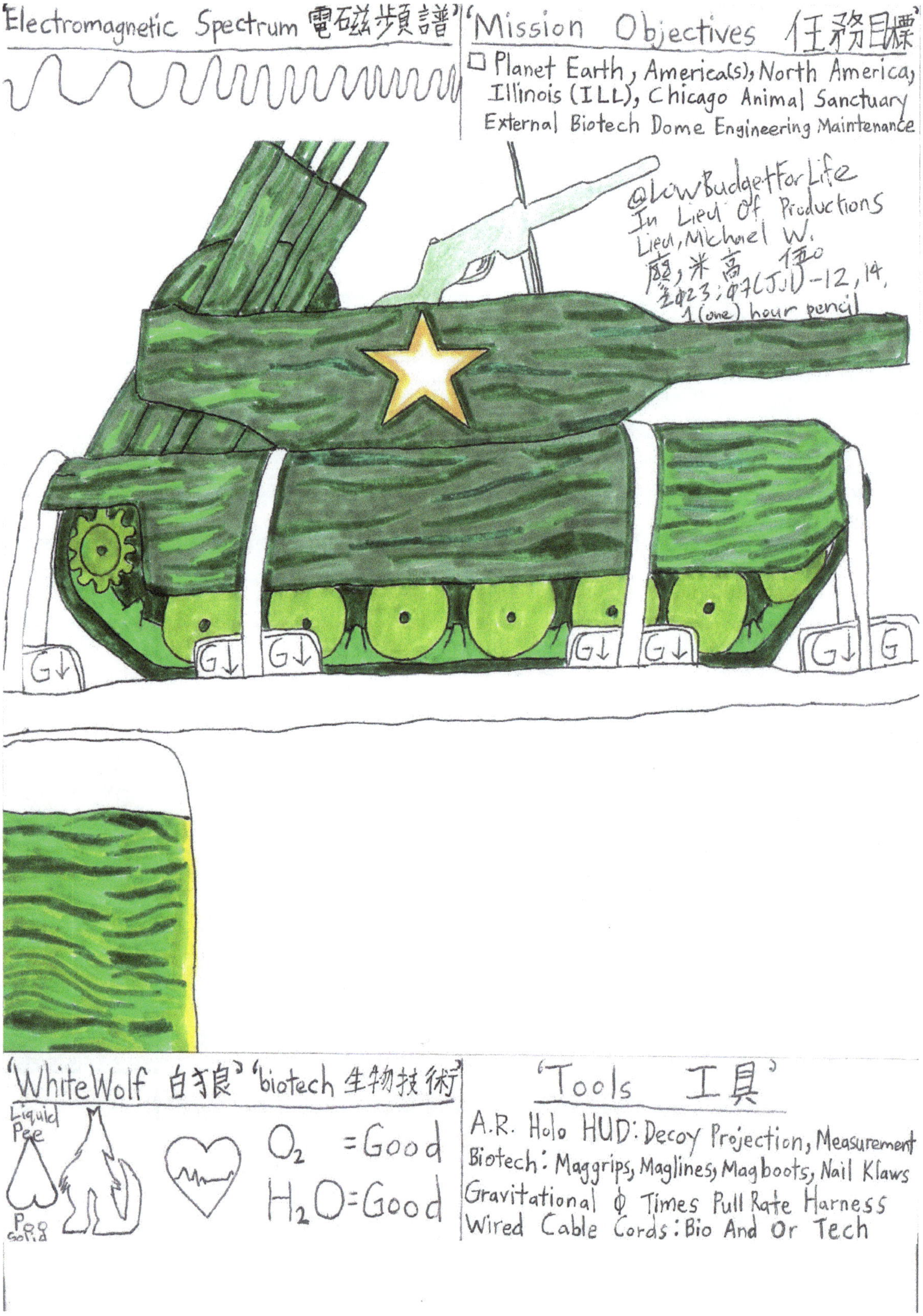

(Contemporary) (Military) (Super-Hero) (Utopia) = U.S.A. County Military Bases.

Sisco Cisco .il Green Broadcast	(°o°) (°o°) (°o°) (°o°) (°o°) (°o°) (°o°) (°o°) (°o°) (°o°) (°o°) (°o°) (°o°) (°o°) (°o°) Green	Sisco Cisco .il Green Broadcast
Date;	Year; (####;) ## (Month) - Day = 2154; 10 (Oct) – 30 (thirty)	

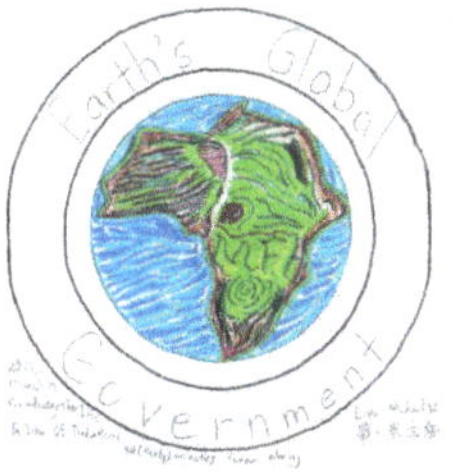

Earth's Global Government (EGG) Space Force Joint Task Force (J.T.F.)

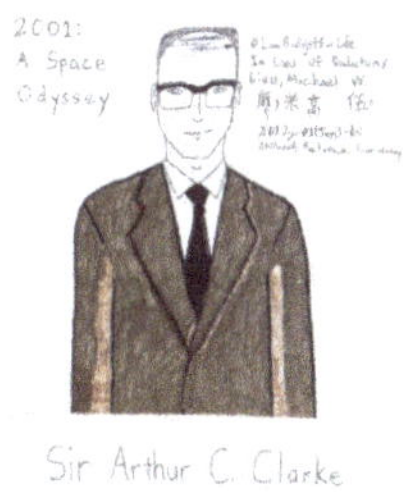

Sir Arthur C. Clarke Space Elevators.

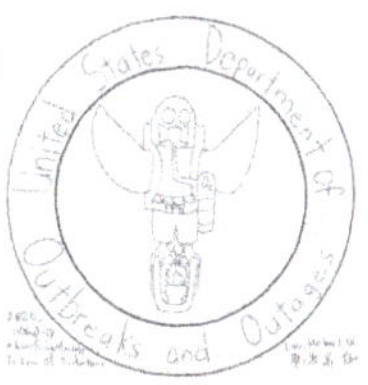

Lieutenant 2nd Lieu, Michael W. 廖, 米高 伍o + **WhiteWolf 白狼** Bio**tech Suit+**; Vigilant Volunteers Sisco Cisco (°o°) Yellow;
(*bio**tech suit** male voice)*

"Need to **recollect my thoughts**, beliefs, whatever going on.

After early retirement and travel overseas on Earth as a military veteran, or civilian disabled, that **I** learned Fort Angel of Los Angeles County Military Base, the base **I** was born into as a natural human male, had sometime ago returned to Mexico, Militia Mexican-Men acting authorities that work along side the surviving state Mexico government and black market drug cartels, just as the Colonel, Colonel Kyle Kinsman, had official orders to move east to the surviving United **States** of North America **founding fathers' original thirteen states and or colonies depending on** belief systems **viewpoint**, Fort Thirteen.

Largest United States Army Fort and Head Quarters Base established after twenty one twenty one, believed it is where EGG Space Force Joint Task Force **North America Command is at.**

My choices were to stick with the same people of Fort Angel military servicemembers, civilians, contractors, retirees, disables, veterans or **pack my belongings and** travel, off base, being a **nomad** with Sisco Cisco Surprise Face, My Ninja-Talon and so for another adventure that being on base did not really offer. This was **twenty one forties, late**, going to **twenty one fifties**."

Sisco Cisco .il Green Broadcast	(°o°) (°o°) (°o°) (°o°) (°o°) (°o°) (°o°) (°o°) (°o°) (°o°) (°o°) (°o°) (°o°) (°o°) (°o°) Green	Sisco Cisco .il Green Broadcast

(Apocalyptic) (Horror) (Neo Noir Vigilante) (Psychological) (Utopia) = Dome

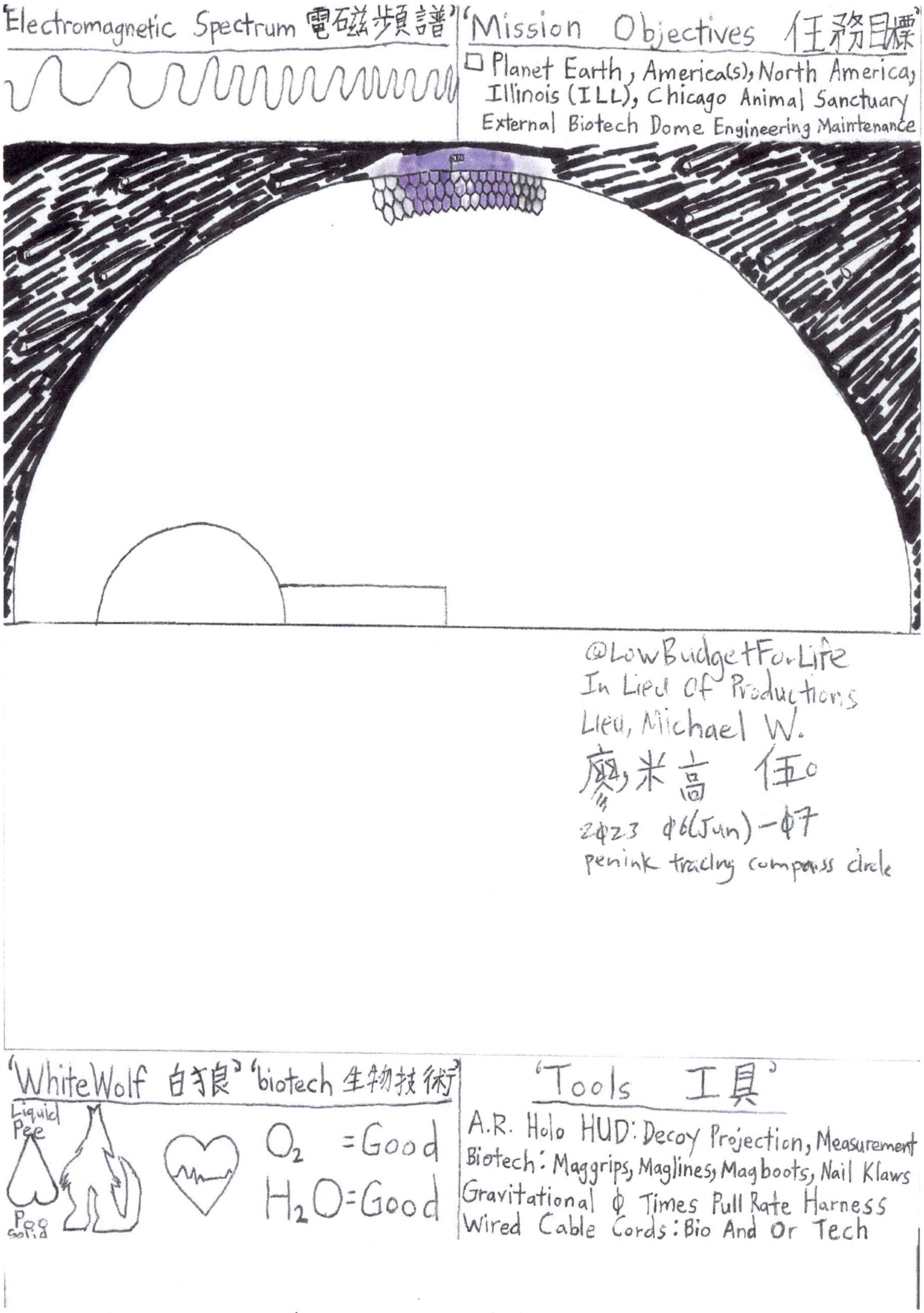

(Contemporary) (Military) (Super-Hero) (Utopia) = The State, Former Officials.

Sisco Cisco .il Green Broadcast	(°o°) (°o°) (°o°) (°o°) (°o°) (°o°) (°o°) (°o°) (°o°) (°o°) (°o°) (°o°) (°o°) (°o°) (°o°) Green	Sisco Cisco .il Green Broadcast
Date;	Year; (####;) ## (Month) - Day = 2154; 10 (Oct) – 30 (thirty)	

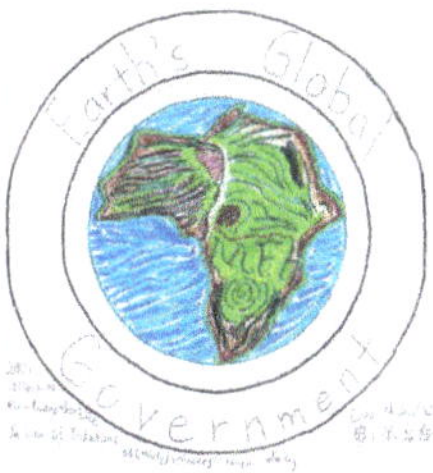

Earth's Global Government (EGG) Space Force Joint Task Force (J.T.F.)

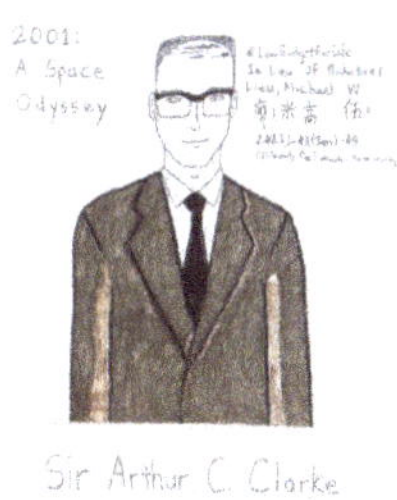

Sir Arthur C. Clarke Space Elevators.

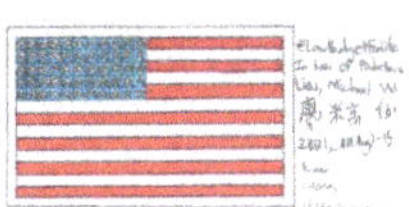

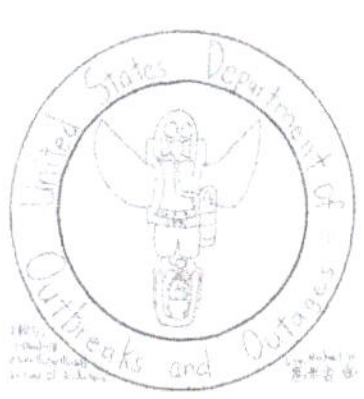

Lieutenant 2nd Lieu, Michael W. 廖, 米高 伍o + **WhiteWolf 白狼** Bio**tech Suit+**; Vigilant Volunteers Sisco Cisco (°o°) Yellow;
(*bio**tech suit** male voice)*

"Seeing **this dome, white large dome here**, with that Ultra Violet, U.V. flag on top of **hundreds of stories high** reminds **me** of the time **I** first arrived here about this purple phenomenon quote Chicago The Last Animal Sanctuary Of Illinois unquote.

By twenty one fifties, **I** arrive to former Fort Cook of Cook County, a Chicago County Base established after twenty one twenty one, during the twenty second century dark ages.

Big Ben and his Mexican-Men Militia splinter cell were the first ones to witness the U.V. Mayor Clean House act in Fort Cook...

Naval Station Great Lakes, **I** believe still under **the old world** county of... Lake County, Illinois, the one and only United States Department Of War Navy Branch Military Recruit Training Command active after all these years in the solar system...

Not many military Navy recruits with the low natural people population including the natural unnatural human infertility rates, unless **artificial people still count as** natural people.

However, **Elysium Corporation** did manage to solve the low natural people population decline, **but introduced another problem of artificial people kloning and augmentation** bio**tech enhancements, forever young people**..."

Sisco Cisco .il Green Broadcast	(°o°) (°o°) (°o°) (°o°) (°o°) (°o°) (°o°) (°o°) (°o°) (°o°) (°o°) (°o°) (°o°) (°o°) (°o°) Green	Sisco Cisco .il Green Broadcast

(Horror) = Biotech Dome Airlock Decontamination (Super-Hero) (Utopia) = Dirty.

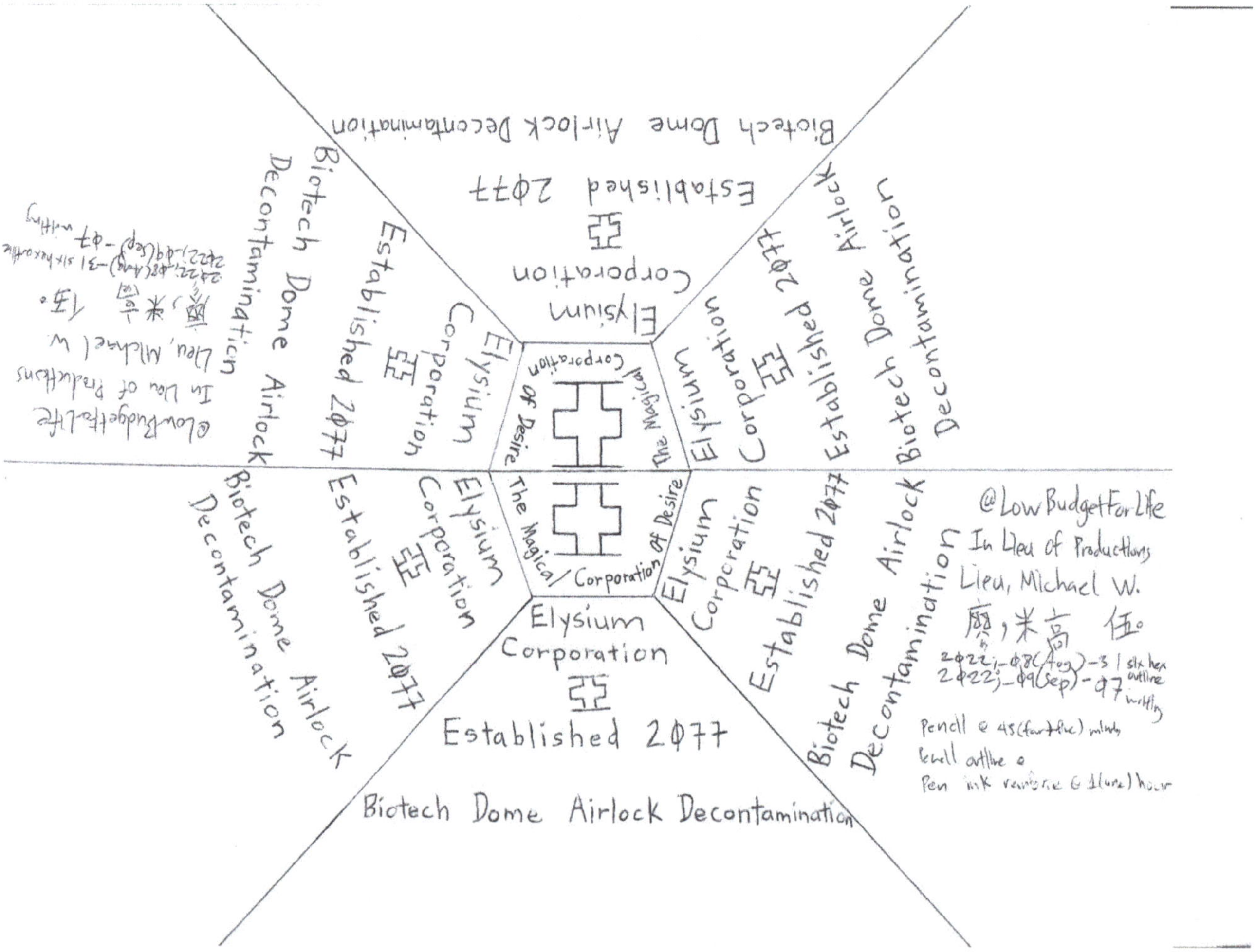

Sisco Cisco .il Violet (U.V.) Broadcast	(°o°) (°o°) (°o°) (°o°) (°o°) (°o°) (°o°) (°o°) (°o°) (°o°) (°o°) (°o°) (°o°) (°o°) (°o°) (U.V.)	Sisco Cisco .il Violet (U.V.) Broadcast
Date;	Year; (####;) ## (Month) - Day = 2154; 10 (Oct) – 30 (thirty)	

Sisco Cisco .il Yellow (°o°) Broadcast Automation Algorithms Closed Captioning Log;

Elysium Corporation Automation Algorithms; (*monotone female voice over*)

"Please stand by for bio**tech** dome **airlock** decontamination.

Thank **you** for **your** patience and cooperation **Utopian**."

Sisco Cisco .il Violet (U.V.) Broadcast	(°o°) (°o°) (°o°) (°o°) (°o°) (°o°) (°o°) (°o°) (°o°) (°o°) (°o°) (°o°) (°o°) (°o°) (°o°) (U.V.)	Sisco Cisco .il Violet (U.V.) Broadcast

(Horror) = Biotech Dome Airlock Decontamination (Neo Noir Vigilante) = X Rays

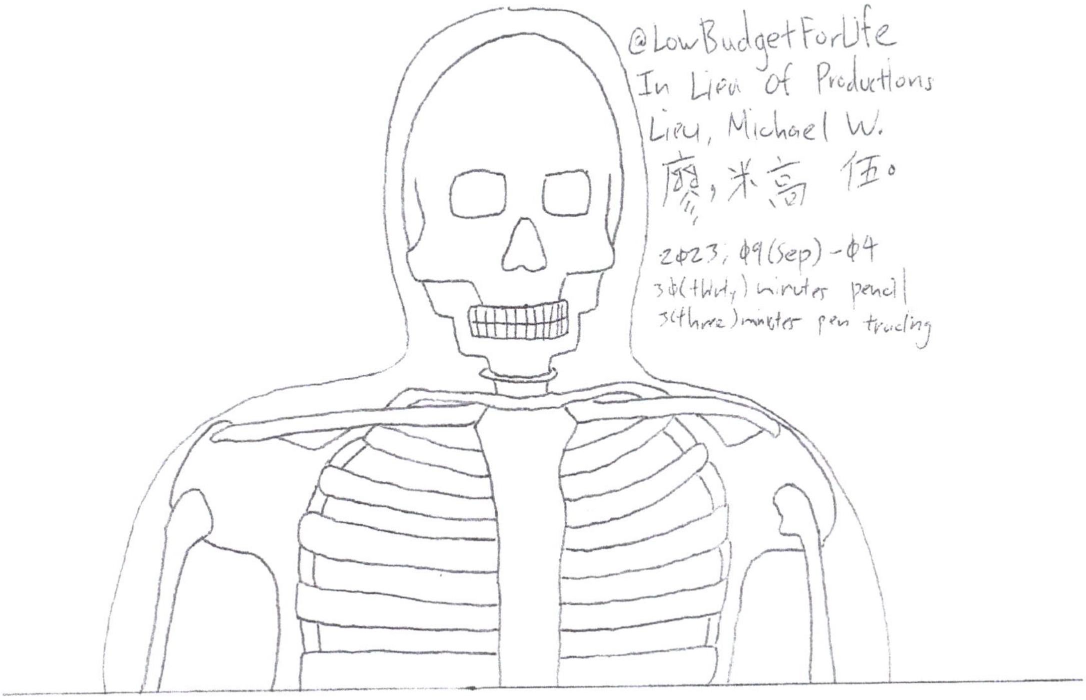

(Horror) (Psychological) (Super-Hero) (Utopia) = Arriving **Airlock** Contaminated.

<table>
<tr><td>Sisco
Cisco
.il Violet (U.V.)
Broadcast</td><td>(°o°) (°o°) (°o°) (°o°) (°o°) (°o°) (°o°) (°o°) (°o°) (°o°) (°o°) (°o°) (°o°) (°o°) (°o°)
(U.V.) (U.V.) (U.V.) (U.V.) (U.V.) (U.V.) (U.V.) (U.V.) (U.V.) (U.V.) (U.V.) (U.V.) (U.V.) (U.V.)
(U.V.) (U.V.) (U.V.) (U.V.) (U.V.) (U.V.) (U.V.) (U.V.) (U.V.) (U.V.) (U.V.) (U.V.) (U.V.) (U.V.)
(U.V.) (U.V.) (U.V.) (U.V.) (U.V.) (U.V.) (U.V.) (U.V.) (U.V.) (U.V.) (U.V.) (U.V.) (U.V.) (U.V.)</td><td>Sisco
Cisco
.il Violet (U.V.)
Broadcast</td></tr>
<tr><td>Date;</td><td colspan="2">Year; (####;) ## (Month) - Day = 2154; 10 (Oct) – 30 (thirty)</td></tr>
<tr><td>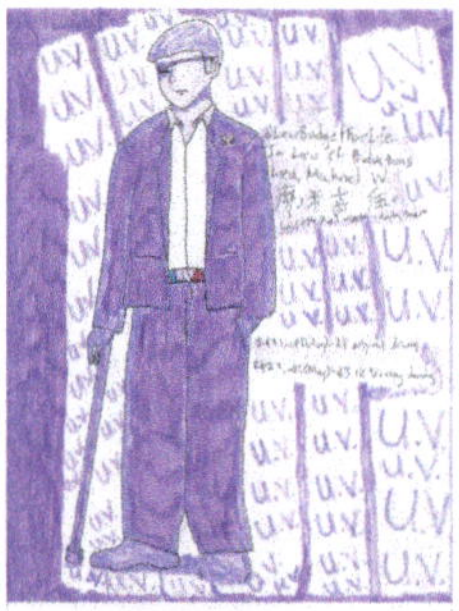
Chicago Controller City Mayor

Chicago (S), The Last Animal Sanctuary (S) Of Illinois (ILL), N.A.,

Electro Magnetic Spectrum; Radio Waves, Amplitude Modulation (A.M.) Frequencies Hertz (Hz) Broadcasting, "Scatman John's Song Of Scatland"</td><td colspan="2">Lieutenant 2nd Lieu, Michael W. 廖, 米高 伍o + WhiteWolf 白狼 Biotech Suit+; Vigilant Volunteers Sisco Cisco (°o°) Yellow;
(biotech suit male voice)
"Audio voice recording for my retirement journal year four, the year twenty one fifty four, the date October thirty...
I am my way back in dome, as Sir Arthur C. Clarke's Space Elevators Anniversary Launch Celebration was over and... and..."

Mister Kelvin; "Oh hi WhiteWolf, you are back from out dome exposure, dirty! You been out for some time... arriving in dome stand by, Mayor Chicago Controller broadcasting shortly..."

(Frequency Modulation (F.M.) Plus Amplitude Modulation (A.M.) Frequencies Hertz (Hz) Mix With Radio Waves And Ultra Violet Rays Broadcasting)

Chicago Controller City Mayor (frequency hertz male(s) voice);
"Start Utopian, Unit eight eight eight eight comma eight eight eight eight, WhiteWolf, Lieutenant Lieu, had a good time with Sir Arthur C. Clarke's Space Elevators Anniversary Launch Celebration, as well as tuning in on our U.V. broadcasting towers on augmented animals' society before, during, after twenty one twenty one, you agree to the terms of arriving to my beautiful peace loving animal sanctuary...? Period End."</td></tr>
<tr><td>Sisco
Cisco
.il Violet (U.V.)
Broadcast</td><td>(°o°) (°o°) (°o°) (°o°) (°o°) (°o°) (°o°) (°o°) (°o°) (°o°) (°o°) (°o°) (°o°) (°o°) (°o°)
(U.V.) (U.V.) (U.V.) (U.V.) (U.V.) (U.V.) (U.V.) (U.V.) (U.V.) (U.V.) (U.V.) (U.V.) (U.V.) (U.V.)
(U.V.) (U.V.) (U.V.) (U.V.) (U.V.) (U.V.) (U.V.) (U.V.) (U.V.) (U.V.) (U.V.) (U.V.) (U.V.) (U.V.)
(U.V.) (U.V.) (U.V.) (U.V.) (U.V.) (U.V.) (U.V.) (U.V.) (U.V.) (U.V.) (U.V.) (U.V.) (U.V.) (U.V.)</td><td>Sisco
Cisco
.il Violet (U.V.)
Broadcast</td></tr>
</table>

(Horror) = Biotech Dome Airlock Decontamination, Mister Kelvin, Temperature

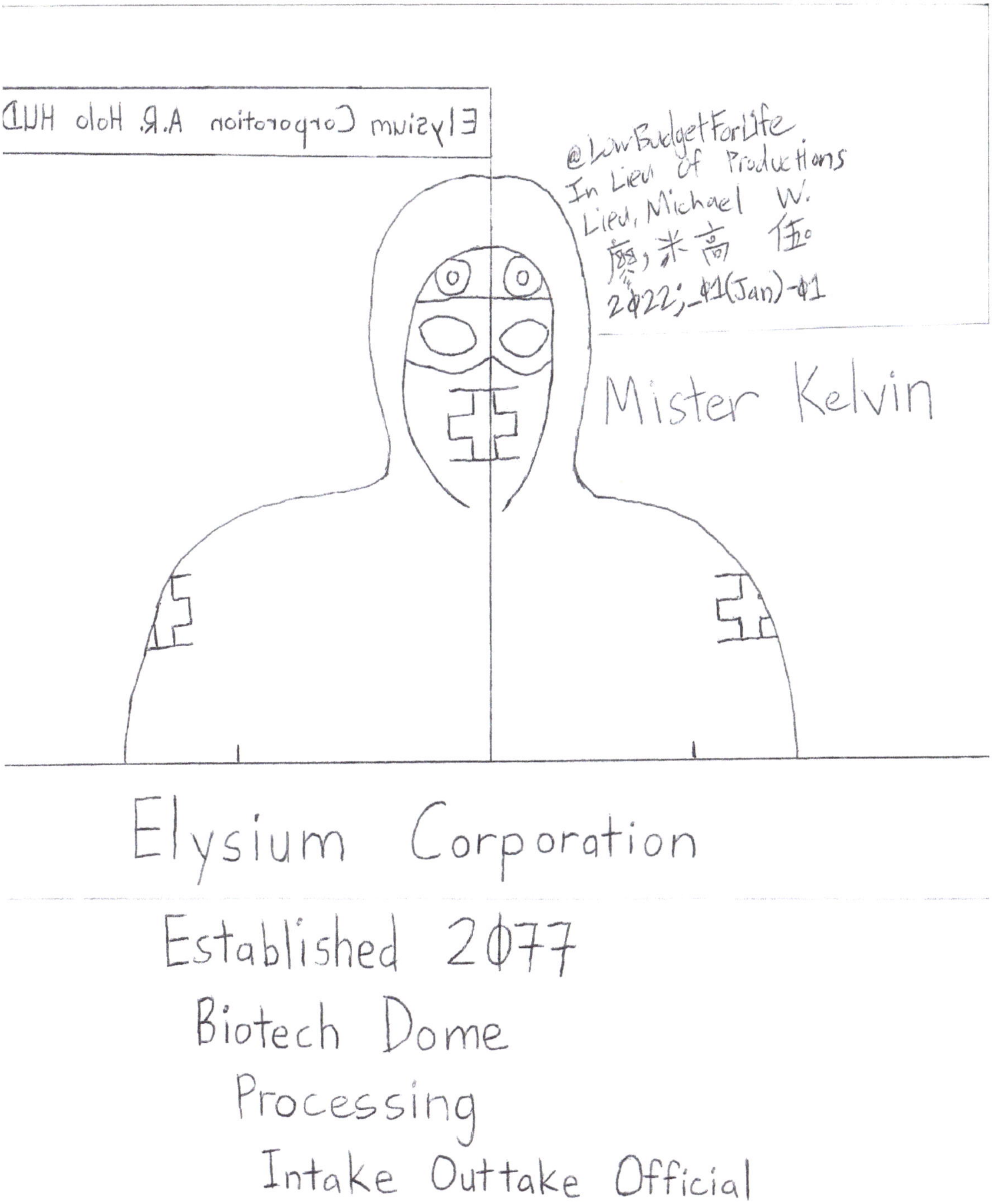

(Psychological) (Super-Hero) (Utopia) = 22nd Century Dark Ages, Earth Biosphere.

Sisco Cisco .il Violet (U.V.) Broadcast	(°o°) (°o°) (°o°) (°o°) (°o°) (°o°) (°o°) (°o°) (°o°) (°o°) (°o°) (°o°) (°o°) (°o°) (°o°) (U.V.)	Sisco Cisco .il Violet (U.V.) Broadcast
Date;	Year; (####;) ## (Month) - Day = 2154; 10 (Oct) – 30 (thirty)	

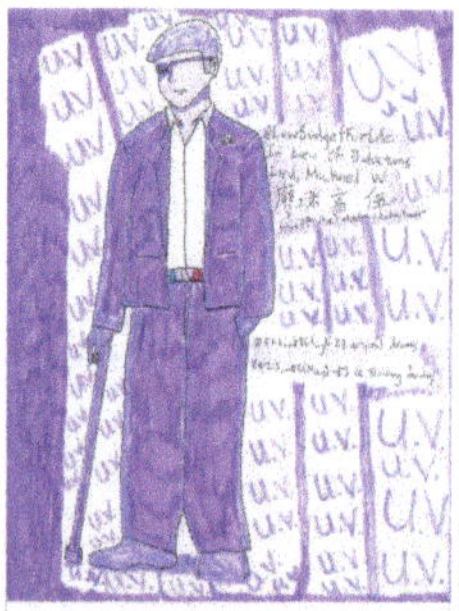

Chicago Controller City Mayor

Chicago (S), The Last Animal Sanctuary (S) Of Illinois (ILL), N.A.,

Electro Magnetic Spectrum; Radio Waves, Amplitude Modulation (A.M.) Frequencies Hertz (Hz) Broadcasting, "Muse's Apocalypse Please"

Lieutenant 2nd Lieu, Michael W. 廖, 米高 伍o + **WhiteWolf 白狼** Bio**tech Suit+**; Vigilant Volunteers Sisco Cisco (°o°) Yellow;
(*bio**tech suit** male voice)*

"...Yes, **I** agree, Mayor Chicago Controller... Hmm..."

Chicago Controller City Mayor (*frequency hertz male(s) voice*);

"Start Utopian, Unit eight eight eight eight comma eight eight eight eight, **WhiteWolf**, Lieutenant Lieu, some **augmented animals** did not reside **in domes** on Earth had to **constantly wear Personal Protective Equipment** that provided purified water liquids and purified oxygen air, since Earth's ecosystem is now hazardous to most biological beings made possible by space rocks, asteroids, meteorites, plus unconventional use of Chemical Biological Radiological and Electro Magnetic Pulse agents put forward the twenty second century dark ages' year one in twenty one twenty one, residing **under a cover shelter** to **sustain existence** was **not enough** as parts of Earth's biosphere were changing due to such unconventional agents, environmental and climate impact was changing quicker than augmented animals could mobilize, unfortunately even natural animals among the peaceful animals all flatline, died, mercy killed, some were painful and some were painless, many unmarked graves that end their suffering and ended their story and existence, Period End."

Sisco Cisco .il Violet (U.V.) Broadcast	(°o°) (°o°) (°o°) (°o°) (°o°) (°o°) (°o°) (°o°) (°o°) (°o°) (°o°) (°o°) (°o°) (°o°) (°o°) (U.V.)	Sisco Cisco .il Violet (U.V.) Broadcast

(Psychological) = Purple Psych Projection Via Ultra Violet (U.V.) Broadcast Rays.

Reads = Left to Right (L -> R) \ Wireless Broadcasting Frequencies Online
2154; Chicago (S), The Last Animal Sanctuary (S) Of Illinois' Intranet

Why bother, who cares, barely any humans follow rules regulation	Fine, whatever do not sit on it, complete this before entering

Chicago Controller
City Mayor

"Chicago City Clean House Act of
2152 (twenty one fifty two); 07-04 (July four)"

Indoctrination Self-Governing Guidelines Overview

Failure to complete the new world 2121 (twenty one twenty one) indoc, is a warrant for immediate removal from Chicago (S), The Last Animal Sanctuary (S) Of Illinois (ILL).

Exceptions apply only to the surviving beautiful friendly peace loving augmented and natural animal kingdoms plus planet earth's environment!!

Make a choice, Utopian.

No, I disagree, wipe my memory.	I do not know, let us discuss.	Yes, I agree.

Clean-up Crew
Unit 6 (six),
Unit 9 (nine),
Unit 8888,8888 (eight eight eight eight comma eight eight eight eight).

Abusive bitch and moan, toxicity constant crybaby complaint box	Let the shit hit the fan rock bottom, to crash and burn hard

2154; Chicago (S), The Last Animal Sanctuary (S) Of Illinois' Intranet
Reads = Left to Right (L -> R) \ Wireless Broadcasting Frequencies Online

(Psychological) = Purple Psych Projection Via Ultra Violet (U.V.) Broadcast Rays.

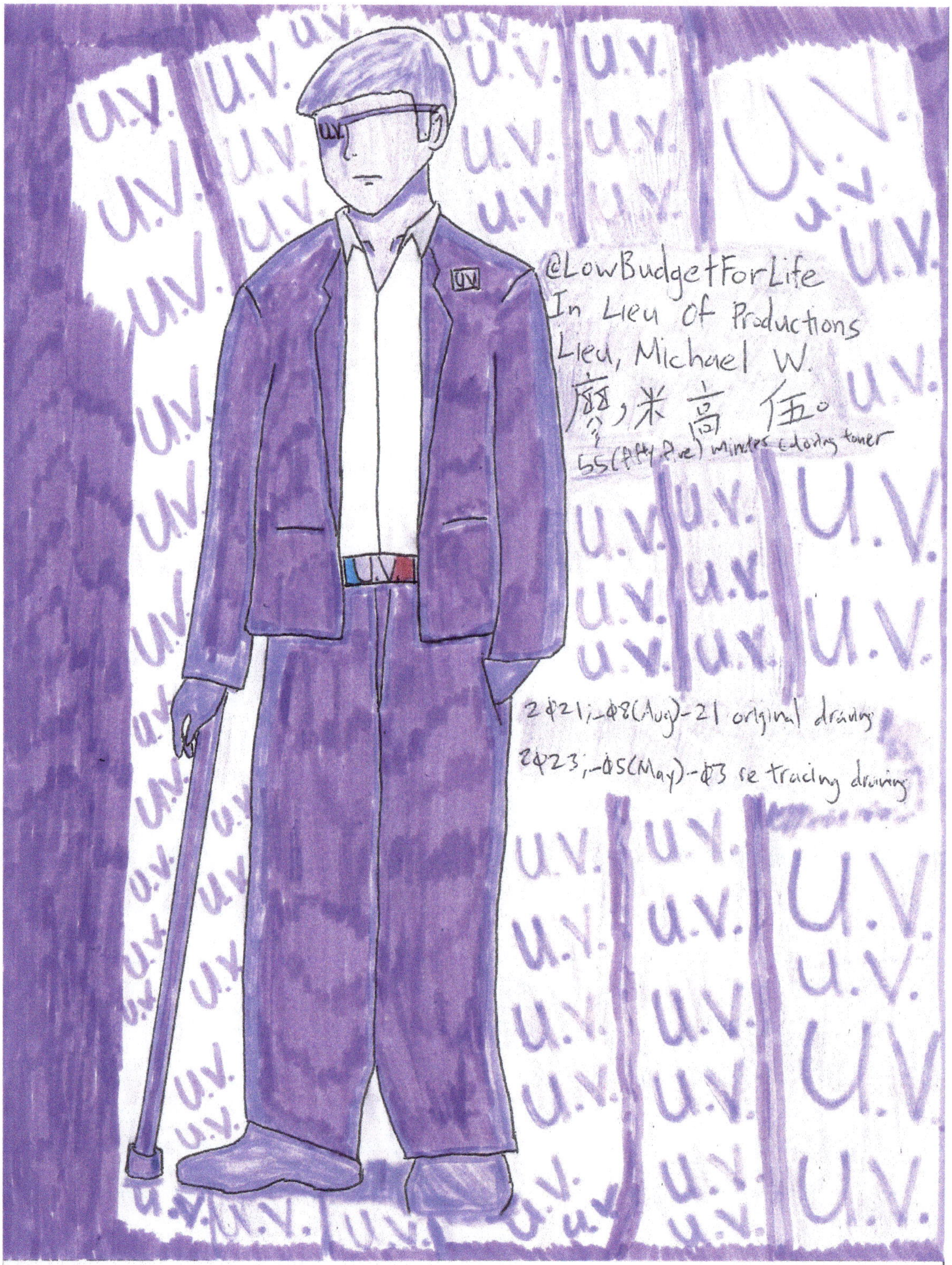

(Horror) (Psychological) (Super-Hero) (Utopia) = 22nd Century Dark Ages, Space.

Sisco Cisco .il Violet (U.V.) Broadcast	(°o°) (°o°) (°o°) (°o°) (°o°) (°o°) (°o°) (°o°) (°o°) (°o°) (°o°) (°o°) (°o°) (°o°) (°o°) (U.V.)	Sisco Cisco .il Violet (U.V.) Broadcast
Date;	Year; (####;) ## (Month) - Day = 2154; 10 (Oct) – 30 (thirty)	

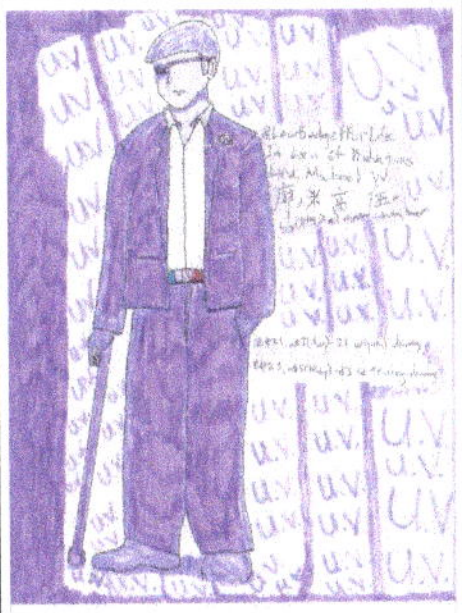

Chicago Controller City Mayor

Chicago (S), The Last Animal Sanctuary (S) Of Illinois (ILL), N.A.,

Electro Magnetic Spectrum; Radio Waves, Amplitude Modulation (A.M.) Frequencies Hertz (Hz) Broadcasting, "Andre Rieu's Beautiful Blue Danube"

Chicago Controller City Mayor (*frequency hertz male(s) voice*);

"Start Utopian, Unit eight eight eight eight comma eight eight eight eight, **WhiteWolf**, Lieutenant Lieu, decontamination is almost done for **you to depart from your biotech suit** and arrive within dome, in the meantime, part of the twenty second century dark ages on Earth has been broadcast, what about the **augmented animals in** space, space station residents, spacers, asteroid belt residents, belters, off planet residents from Mars, Martians, their existence is not as easy as those living on pre twenty one twenty one Earth with free purified air, free purified water, free food, easy handouts, modern welfare, to an aimless purposeless existence, spacers, belters, Martians need to know knowledge, be aware of the easiest, smallest, simplest, stupid things that can flatline, mercy kill, dismember, disintegrate, **augmented animals** in the vacuum of space, Period End."

Mister Kelvin; "Oh hi **WhiteWolf**, please step out of **your biotech suit, we can continue to monitor** for carbon monoxide as **it** decontaminates. **WhiteWolf** dirty! **Lieutenant Lieu clear**!"

Lieutenant 2nd Lieu, Michael W. 廖, 米高 伍o + **WhiteWolf 白狼 Biotech Suit+**; Vigilant Volunteers Sisco Cisco (°o°) Yellow;
(*biotech suit* *male voice*)
"...**Thanks everyone**... That was... that was interesting listen."

Sisco Cisco .il Violet (U.V.) Broadcast	(°o°) (°o°) (°o°) (°o°) (°o°) (°o°) (°o°) (°o°) (°o°) (°o°) (°o°) (°o°) (°o°) (°o°) (°o°) (U.V.)	Sisco Cisco .il Violet (U.V.) Broadcast

(Vigilante) (Neo Noir) = Bio**tech Suit** (Apocalyptic) = Diseases Contamination.

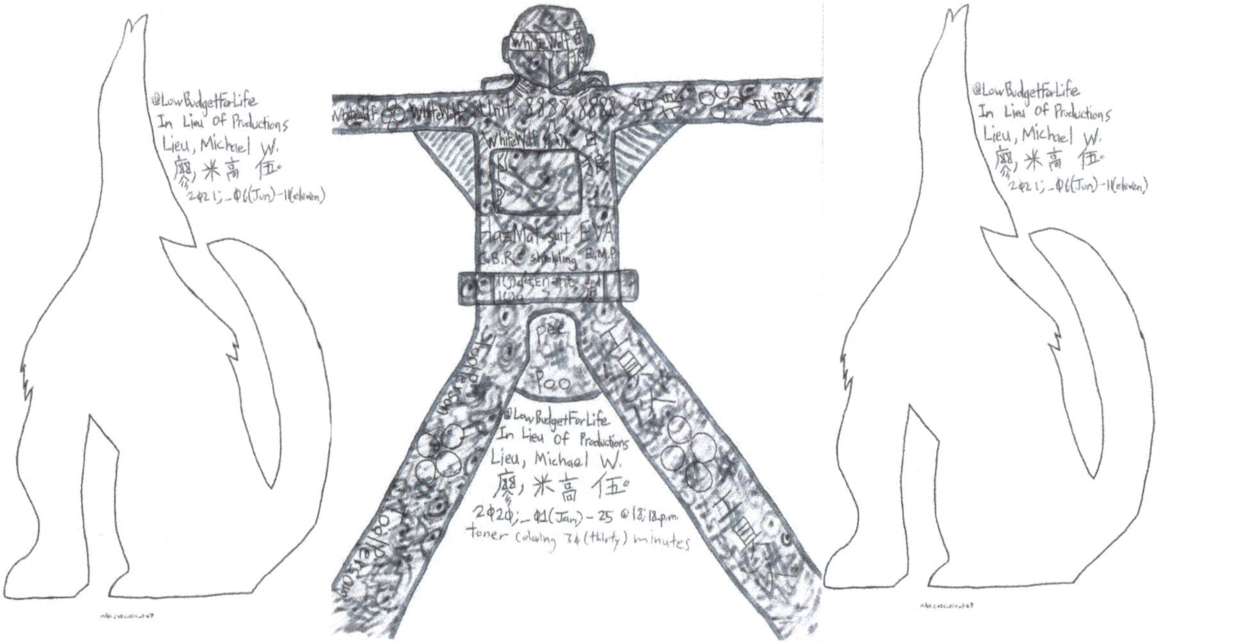

Lieutenant 2nd Lieu, Michael W. 廖, 米高 伍o; Vigilant Volunteers Sisco Cisco (°o°) Yellow;

"Been a while since **I** update **my retirement journal** year four, the year **twenty one fifty four**, the date, **October thirty, I** was **in suit** out dome on the **same year** from September eleven then **I** time traveling towards the future.

According to hard science, physics, we can only time travel forward due to the way we are confined in a four dimension-ish space time... dimension of space and dimension of time altogether make up the geometric object, spacetime.

No explanation how and or why time flows in one direction, as matter in our universe can move back and forth in space, but time cannot move back and forth.

Anyways, was taking part in the space elevator anniversary celebration, then it seemed like **I** was asleep or something with those purple psychological lucid dreaming of **myself in a WhiteWolf hoodie clothing**, **I** do not have any idea how dreams work, was that **time dilation**? Mayor, Chicago Controller was broadcasting to me, so many unanswered questions...

WhiteWolf suit is contaminated, decontamination is going to be a while.

(Utopia) = Orange Conscientiousness Cat Unmanned Aerial Vehicle (U.A.V.).

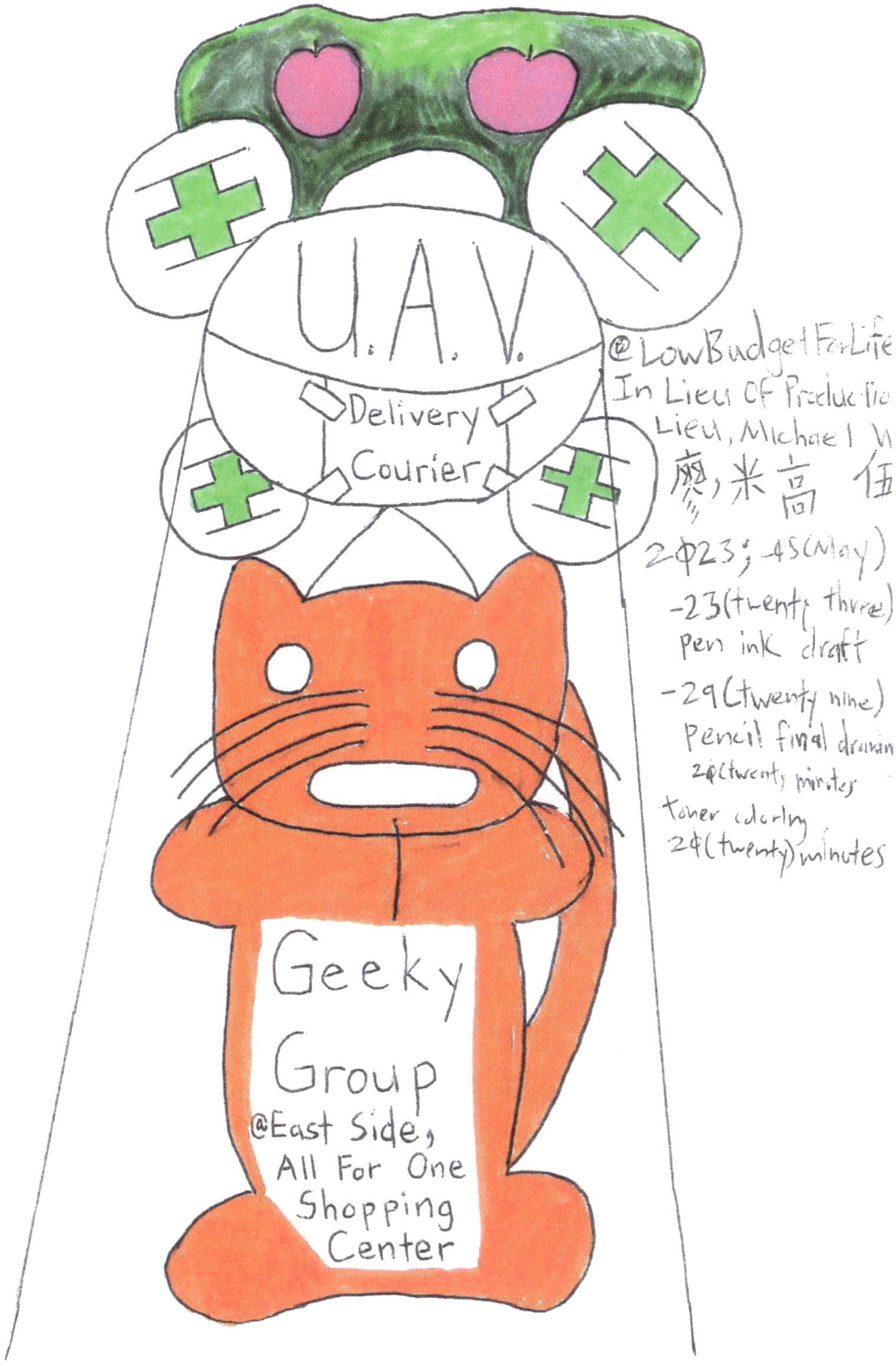

Lieutenant 2nd Lieu, Michael W. 廖, 米高 伍o; Vigilant Volunteers Sisco Cisco (°o°) Yellow;

"This **U.A.V.** has been here the whole time since **I** was out dome for Space Elevators' anniversary celebration...

Order requests were, From; Downtown District (D. D.) of Chicago City Police Department (C.C.P.D.) SWAT.

To; North (N) Side of Bio**tech** Dome Ocean Border.

Manifest Transport; **One** Toilet Paper, **One** Instant Camera Record, **One** Wearable Backpack Riot Shield.

Looks like **everything is still here**, been here, on stand by since..."

(Vigilante) (War, Military) (Super-Hero) (Parody) (Reality, Gun Girls Gun Guys) (Mythology) (Apocalyptic) (Utopia) (Military) (**Photo-Book**) (**Horror**) (Time Traveling) (Crime Drama) = Tools

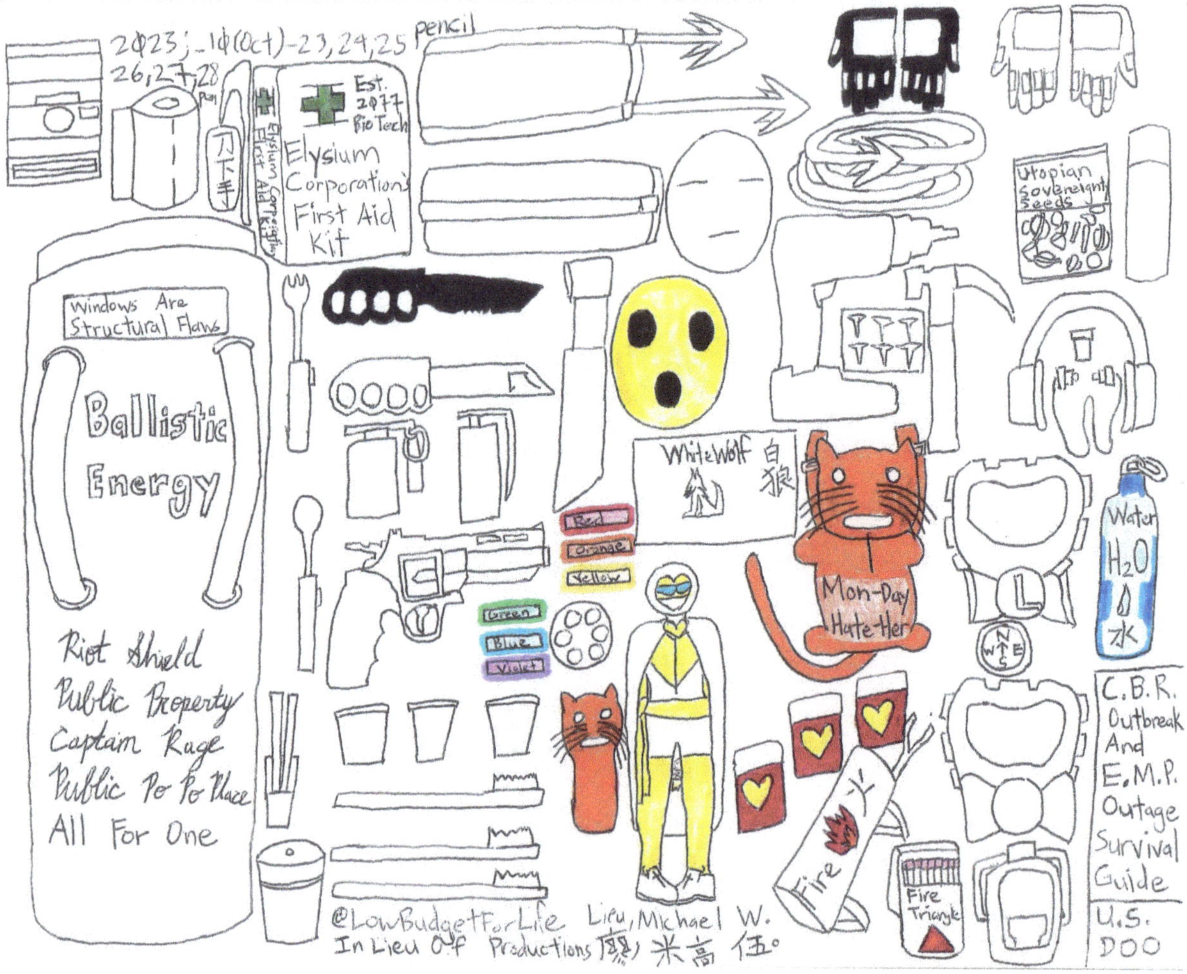

Lieutenant 2nd Lieu, Michael W. 廖, 米高 伍o; Vigilant Volunteers Sisco Cisco (°o°) Yellow;

"Ok, **got my stuff** from Space Elevator anniversary celebration day and looks like everything **I** need to carry on as an Earthling apocalyptic survivor plus **my** bio**tech suit WhiteWolf** being decontaminated in **Elysium Corp airlock**.

One toilet paper on stand by in **my tools shed**, and **one toilet paper on my person** from **my** recent gathering in **Downtown District**...

Going to check my holo HUD, if Remy and **them** from **Elementary Mid Hi School** County Base have arrive yet **in dome** Chicago Animal Sanctuary."

(Super-Hero) (Utopia) (Elementary, Mid, High School Friends; Eastern-Western)

(Owner Operator Manual Guide) (Comedy) (Travel) = "A 2 S 1 - 2154; Arriving"

Sisco Cisco .il Yellow (°o°) Broadcast

(°o°) (°o°) (°o°) (°o°) (°o°) (°o°) (°o°) (°o°) (°o°) (°o°) (°o°) (°o°) (°o°) (°o°) (°o°)
(°o°) (°o°) (°o°) (°o°) (°o°) (°o°) (°o°) (°o°) (°o°) (°o°) (°o°) (°o°) (°o°) (°o°) (°o°)
(°o°) (°o°) (°o°) (°o°) (°o°) (°o°) (°o°) (°o°) (°o°) (°o°) (°o°) (°o°) (°o°) (°o°) (°o°)

Sisco Cisco .il Yellow (°o°) Broadcast

Date; Year; (####;) ## (Month) – Day = 2154; 10 (Oct) – 30 (thirty)

Arriving Airlock Decon Queue #1 (one)

Chicago (S), The Last Animal Sanctuary (S)

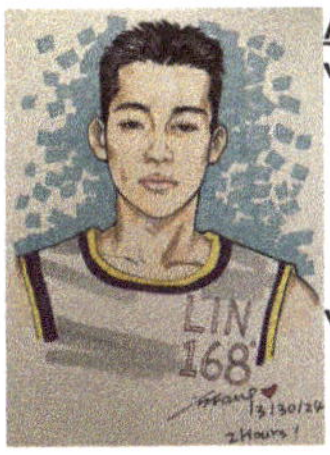

Actual Legal Given Name; "Remy Lin"

Alias Network Nickname(s); ".d Chicago b. Basketball-Man" "Lin 168 Basketball"

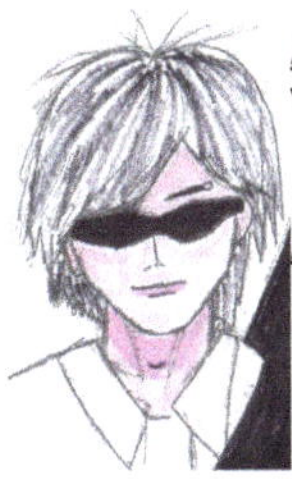

Actual Legal Given Name(s); "Jia Wong"

Alias Network Nickname(s); "B.4-Burma 3(three)III-M.M.B."

Arriving Airlock Decon Queue #2 (two)

Chicago (S), The Last Animal Sanctuary (S) Of Illinois (ILL), N.A., Planet Earth,

Arriving Airlock Decon Queue #3 (three)

Chicago (S), The Last Animal Sanctuary (S) Of Illinois (ILL), N.A., Planet Earth, Solar System,

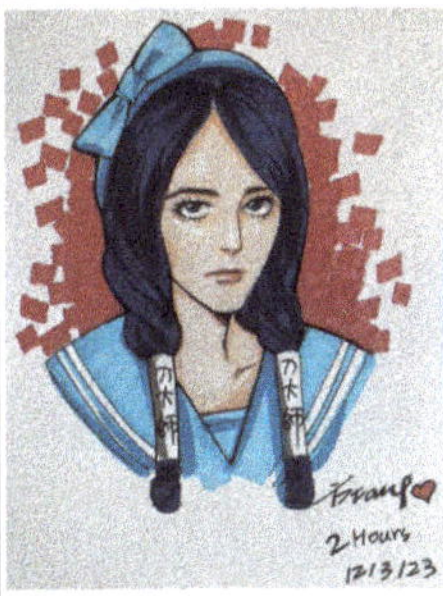

Actual Legal Given Name; "Kanon-Klones 0. Original-Copy"

Alias Network Nickname(s); "BladeMaster 刀大師, Sailor Suit"

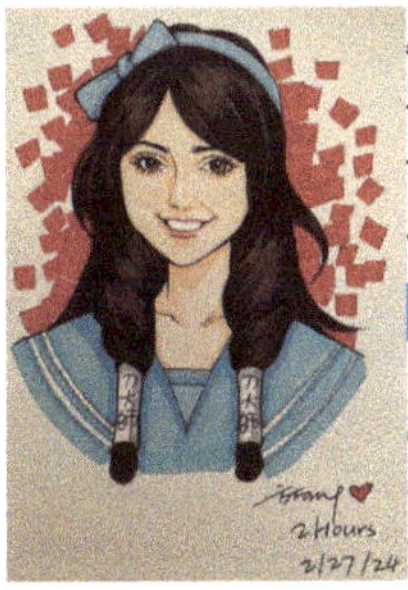

Actual Legal Given Name; "Elle K. Mcgee"

Alias Network Nickname(s); "BladeMaster 刀大師, Sailor Suit"

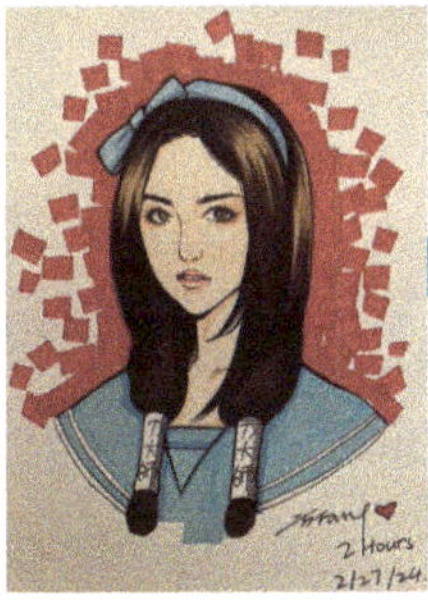

Actual Legal Given Name; "Fuka K. Fujimoto"

Alias Network Nickname(s); "BladeMaster 刀大師, Sailor Suit"

Arriving Airlock Decon Queue #4 (four)

Chicago (S), The Last Animal Sanctuary (S) Of Illinois (ILL), N.A.,

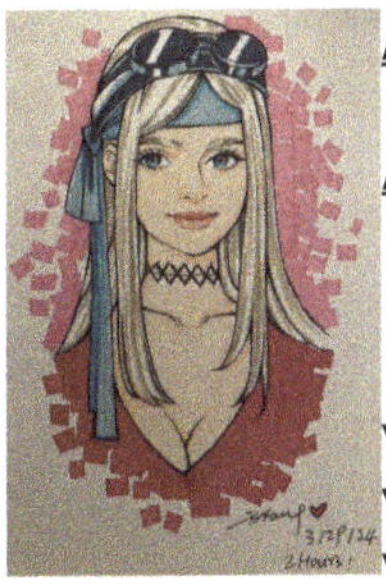

Actual Legal Given Name(s); "Valentina Vita"

Alias Network Nickname(s); "V. V." "KissShot Kosplayer" "White-Woman", "Red Girl", "Red Woman", "Red Female" "Face Masks" "Lady In Red", "Girl In Red" "Female In Red" "Ireland-Woman", "Peach Skin" "Milk Skin" "2 (two) Winged Demon Blade Kosplay", "KissShot-AcerolaOrion-HeartUnderBlade Kosplay", "KissShot Cosplay" "Anime Long Blade Cosplayers", "Beauty Augment" "XXXX Skin Suits" "Science-Fantasy Fiction Cosplayer" "Aerith Cosplayer"[Aug+]

Sisco Cisco .il Yellow (°o°) Broadcast

(°o°) (°o°) (°o°) (°o°) (°o°) (°o°) (°o°) (°o°) (°o°) (°o°) (°o°) (°o°) (°o°) (°o°) (°o°)
(°o°) (°o°) (°o°) (°o°) (°o°) (°o°) (°o°) (°o°) (°o°) (°o°) (°o°) (°o°) (°o°) (°o°) (°o°)
(°o°) (°o°) (°o°) (°o°) (°o°) (°o°) (°o°) (°o°) (°o°) (°o°) (°o°) (°o°) (°o°) (°o°) (°o°)

Sisco Cisco .il Yellow (°o°) Broadcast

(Super-Hero) (Utopia) (Elementary, Mid, High School; Eastern-Western) = 2136

Sisco Cisco .il Yellow (°o°) Broadcast	(°o°) (°o°)		Sisco Cisco .il Yellow (°o°) Broadcast
Date;	Year; (####;) ## (Month) - Day = 2154; 10 (Oct) – 30 (thirty)		
Arriving Airlock Decon Queue #1 (one) **Elysium Corporation Biotech Dome** Airlock Decontamination Chicago (S), The Last Animal Sanctuary (S) Of Illinois (ILL), North America (N.A.), Planet Earth, Solar System, Milky Way Galaxy.	**Actual Legal Given Name(s)**; "Remy Lin" **Alias Network Nickname(s)**; ".d Chicago b. Basketball-Man" "Lin 168 Basketball" 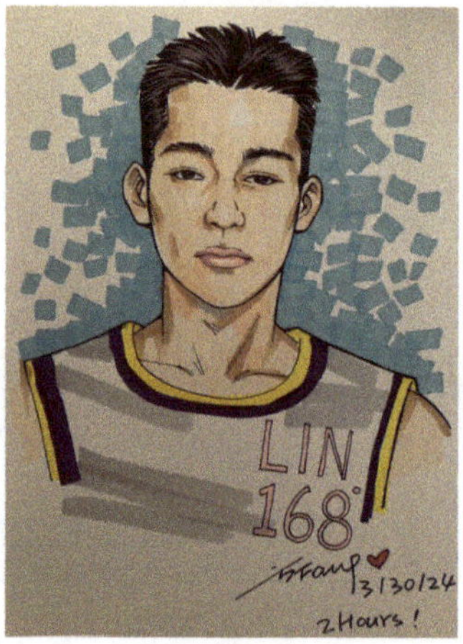	**Actual Legal Given Name(s)**; "Jia Wong" **Alias Network Nickname(s)**; "B.4-Burma 3(three)III-M.M.B." 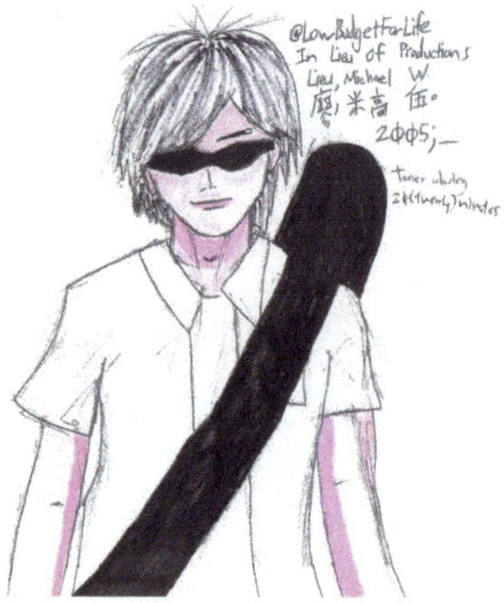	**Notes**; "**I** have not seen **my Elementary Mid High School Friends** since **I** departed and lived off county base…
Automation Algorithms **Closed Captioning Log**;	Connecting… Electro Magnetic Spectrum Frequency Hertz (Hz), Radio Waves, Intranet, Frequency Modulation (F.M.), Amplitude Modulation (A.M.), [*Aug+*]		Automation Algorithms **Closed Captioning Log**;
Sisco Cisco .il Yellow (°o°) Broadcast	(°o°) (°o°)		Sisco Cisco .il Yellow (°o°) Broadcast

Lieutenant 2nd Lieu, Michael W. 廖, 米高 伍o; Vigilant Volunteers Sisco Cisco (°o°) Yellow;

"Looks like Remy and **Jia** are already here, just waiting on rest of **my Elementary Mid Hi School friends** County Base **reunion** meet up…

Jia has aged like me with his white silver gray hair… Time traveling forward is making us all old… Natural Asian Kung Fu Generation.

Going to contact them on our wireless LAN holo HUD… Lots of catching up since **I** depart and live off base by twenty one fifties…"

(Super-Hero) (Utopia) (Elementary, Mid, High School; Eastern-Western) = 2136

Sisco Cisco .il Yellow (°o°) Broadcast	(°o°) (°o°)		Sisco Cisco .il Yellow (°o°) Broadcast
Date;	Year; (####;) ## (Month) - Day = 2154; 10 (Oct) – 30 (thirty)		
Arriving Airlock Decon Queue #1 (one) **Elysium Corporation** Bio**tech** **Dome** Airlock Decontamination Chicago (S), The Last Animal Sanctuary (S) Of Illinois (ILL), North America (N.A.), Planet Earth, Solar System,	**Actual Legal Given Name(s)**; "Remy Lin" **Alias Network Nickname(s)**; ".d Chicago b. Basketball-Man" "Lin 168 Basketball" 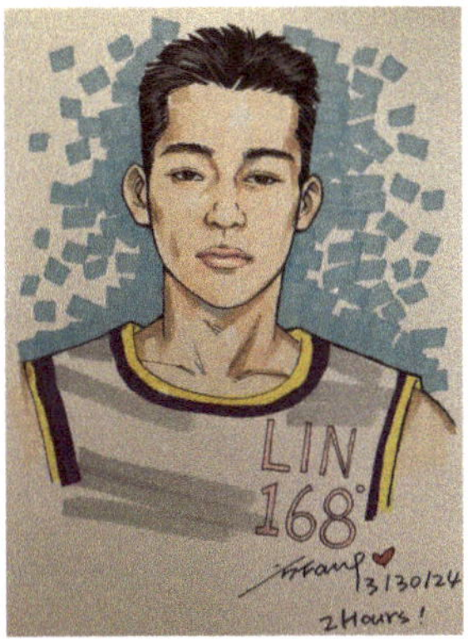	**Actual Legal Given Name(s)**; "Jia Wong" **Alias Network Nickname(s)**; "B.4-Burma 3(three)III-M.M.B." 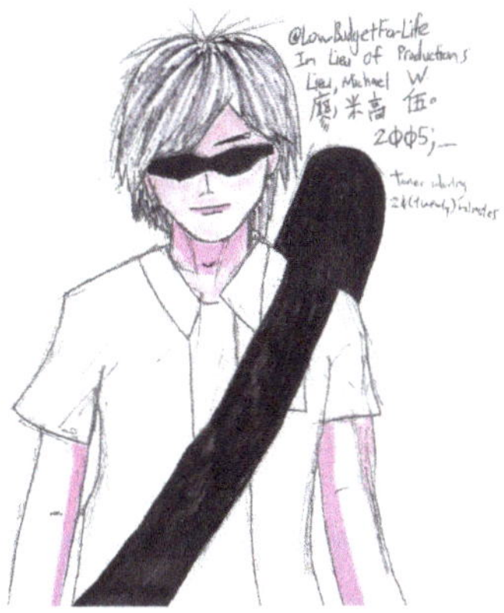	**Notes**; "**I** have not seen my **Elementary Mid High School Friends** since **I** departed and lived off county base...
Automation Algorithms **Closed Captioning Log**;	"Lin, **Remy**"; 'Hey... Hey, **Mic! Ah Sok! Old Man Mic**!' "**Wong, Jia**"; 'Yo what up foo **Mic! Ah Sok! Ah Sok! Old Man Mic!**' "**Lieu, Michael W. (Me)**"; 'Hey chief, **Ah Sok**, been a while since **I** seen **you all** on county base. That was around twenty one fifties." "Lin, **Remy**"; '**You** still look the same, even with that **white hair**.' "**Lieu, Michael W. (Me)**"; 'Yea yea, listen Remy, so, **we** got a Church here, Christian Catholic fusion, run by Sister Nuns, Father Priests, Sister **klones**, Sister Mitchell is Nun in charge, looks and behaves like a natural human and believes so, she loves wearing her Yellow **White** Heart **skintight suit**, is an **artificial persons, augmented**. Church is in West Side **dome** by the yellow flower field, look up to dome's sky if it is blue, that is West Side of Chicago Animal Sanctuary... Oh for food, farmers market is by **South Side**, after **you** finish **your airlock** decontamination, **you** should see the Militia Mexican-Men, these green guys are a splinter cell of the Militia we know in Fort Angel South California. Food is on **me**, tell them **Lieu Lieu** sent **you**, she bangs, they know. Sending **you** info."		Automation Algorithms **Closed Captioning Log**;
Sisco Cisco .il Yellow (°o°) Broadcast	(°o°) (°o°)		Sisco Cisco .il Yellow (°o°) Broadcast

(Super-Hero) (Utopia) (Parody, Contemporary Christian Catholic) = Church ✝.

Sisco Cisco .il Yellow (°o°) Broadcast	(°o°) (°o°)	Sisco Cisco .il Yellow (°o°) Broadcast
Date;	Year; (####;) ## (Month) - Day = 2154; 10 (Oct) – 30 (thirty)	

The Church ✝ = Chicago City Christian Catholic Church ✝ Belief Systems And Other Religions Welcome "Free: Hugs, Hearts". Location @ West (W) Side.

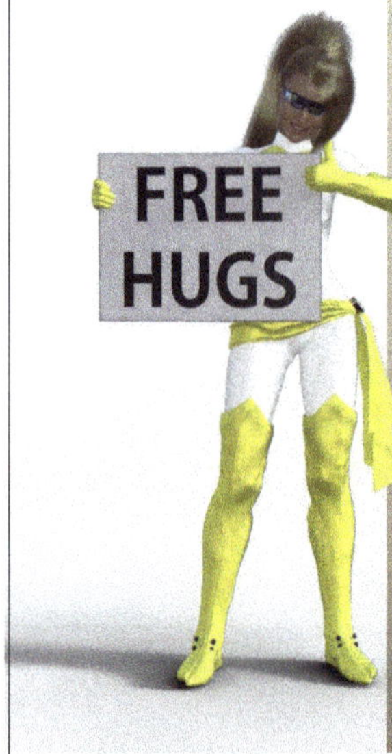

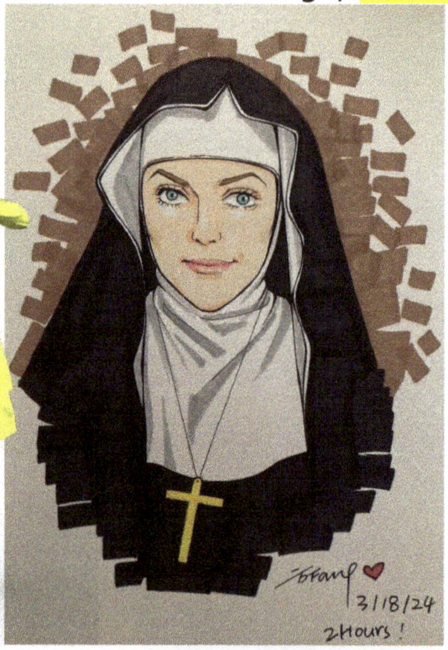

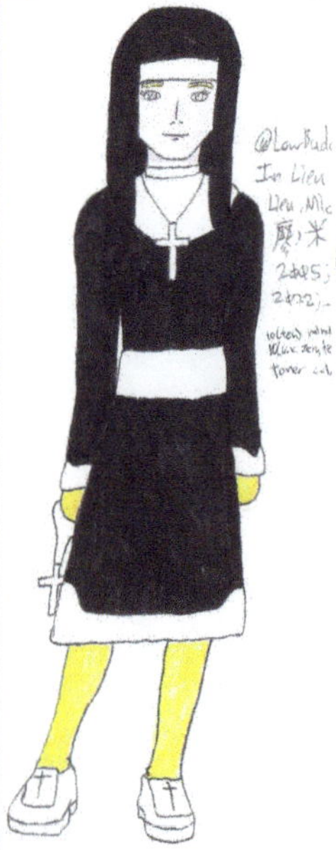

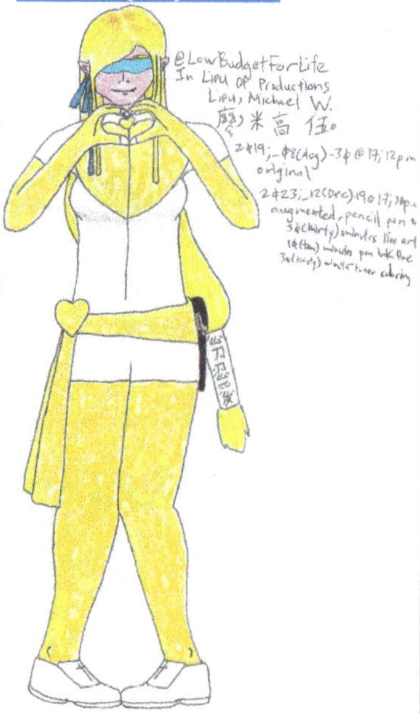

Sister Nun In Charge ✝ Sister Michelle Mitchell, AKA Sister Yellow Limbs Suit.

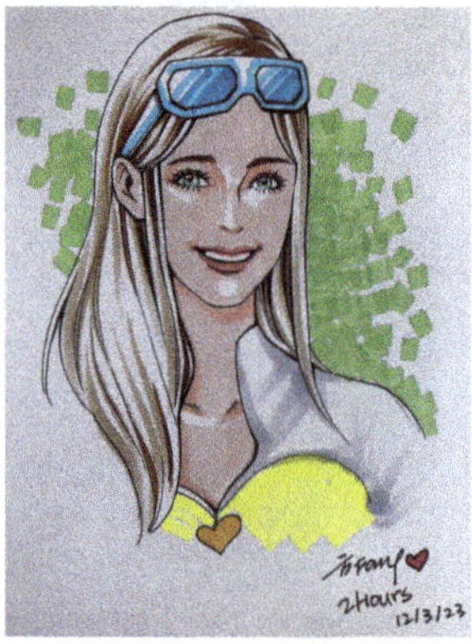

Acting Sister Nun In Charge;

Sister K. W. **White**;

AKA Yellow 黃 **White 白** "HeartBladeEndureLove 心刀忍愛"

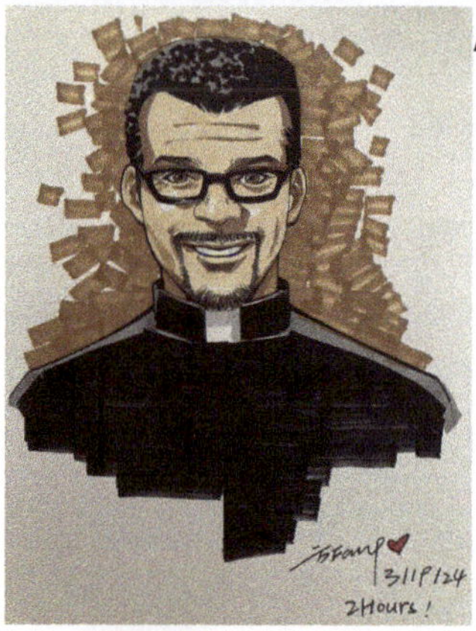

Assistant Father Priest In Charge;

Father ✝ Greg B. Ford

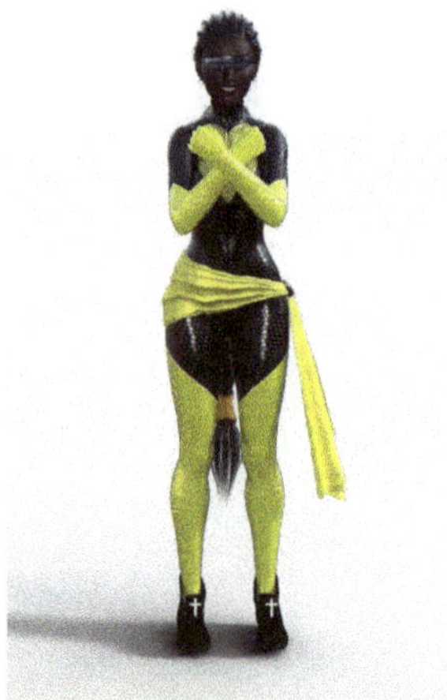

Church's ✝ JAG Protectors

Sisco Cisco .il Yellow (°o°) Broadcast	(°o°) (°o°)	Sisco Cisco .il Yellow (°o°) Broadcast

(Super-Hero) (Utopia) (Mythology) = Méxican-Men M. { Neo Western Cowboys }

Sisco Cisco .il Yellow (°o°) Broadcast	(°o°) (°o°)	Sisco Cisco .il Yellow (°o°) Broadcast
Date;	Year; (####;) ## (Month) - Day = 2154; 10 (Oct) – 30 (thirty)	

The Militia = Méxican-Men Militia Movement And Farmer Market
Spanish; "Soy México" - English; "I Am México",
Chicago (S) Chapter Chain Of Command. Location @ South (S) Side.

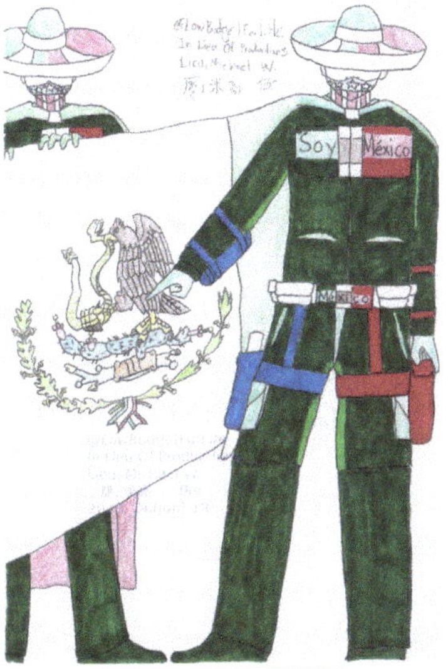

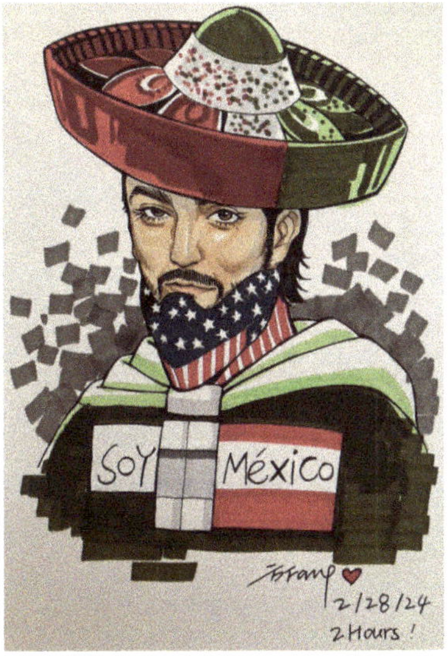

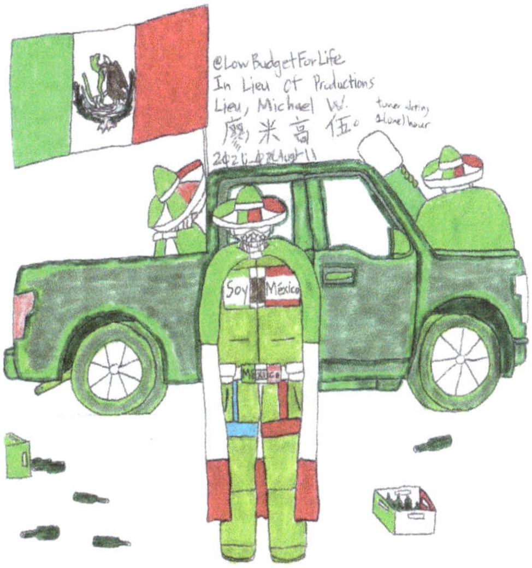

"Big Ben" Colonel Benavidez-Cruz, Richard "Big Boss 大老闆" "Big Bear Gran Oso"

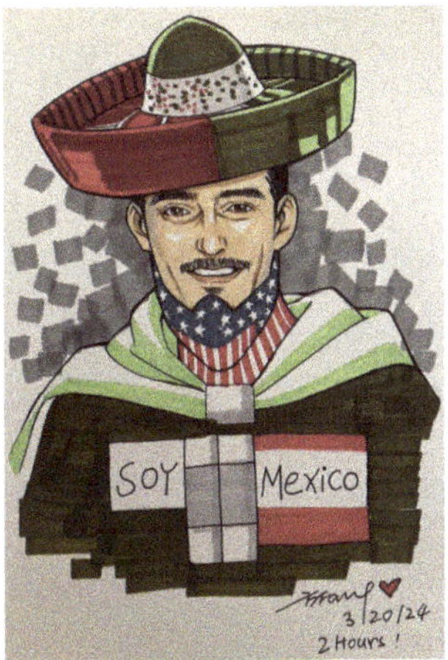

"Baby Ben"
Benavidez, Jose
Militia Translator

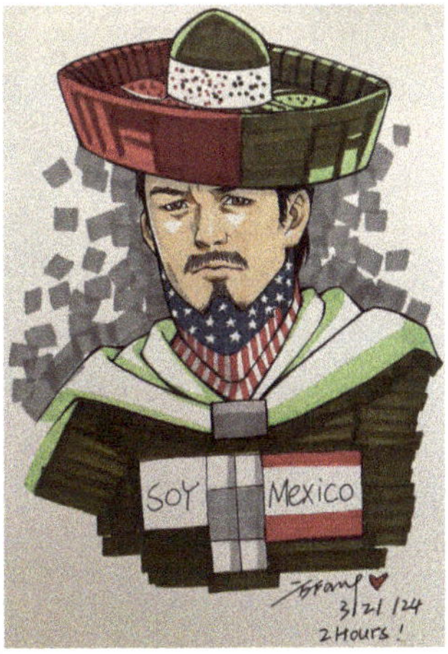

"Senor Jimenez"
Acting Lieutenant
Colonel Militia

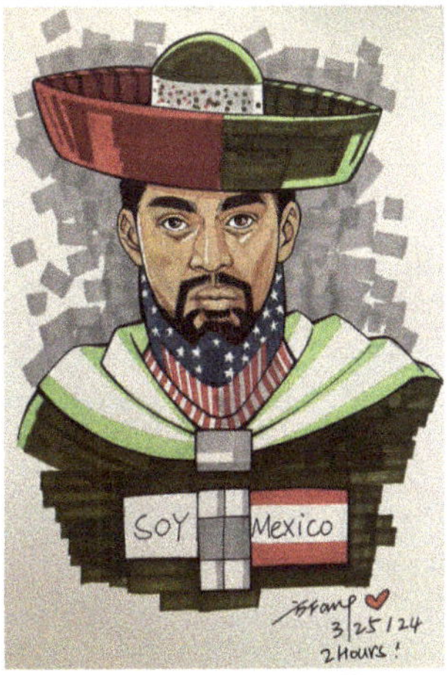

"Senor Diaz"
Acting Lieutenant
Colonel Militia

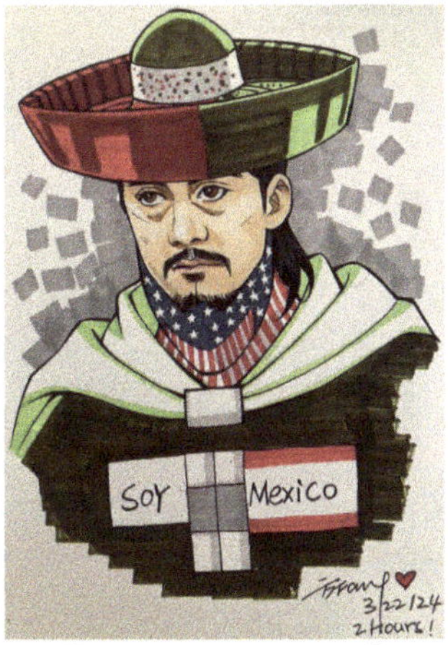

"Senor Gomez"
Recent Recruit
Méxican-Men Militia

Sisco Cisco .il Yellow (°o°) Broadcast	(°o°) (°o°)	Sisco Cisco .il Yellow (°o°) Broadcast

(Super-Hero) (Utopia) (Elementary, Mid, High School; Eastern-Western) = 2136

Sisco Cisco .il Yellow (°o°) Broadcast	(°o°) (°o°)		Sisco Cisco .il Yellow (°o°) Broadcast
Date;	Year; (####;) ## (Month) - Day = 2154; 10 (Oct) – 30 (thirty)		
Arriving Airlock Decon Queue #1 (one) **Elysium Corporation Biotech Dome** Airlock Decontamination Chicago (S), The Last Animal Sanctuary (S) Of Illinois (ILL), North America (N.A.), Planet Earth, Solar System,	**Actual Legal Given Name(s)**; "Remy Lin" **Alias Network Nickname(s)**; ".d Chicago b. Basketball-Man" "Lin 168 Basketball" 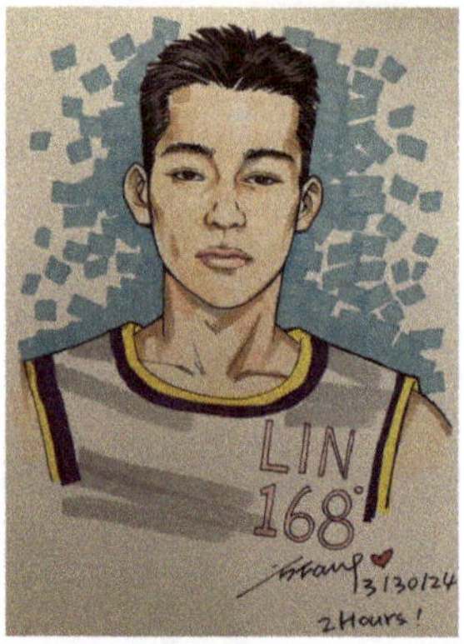 	**Actual Legal Given Name(s)**; "Jia Wong" **Alias Network Nickname(s)**; "B.4-Burma 3(three)III-M.M.B." 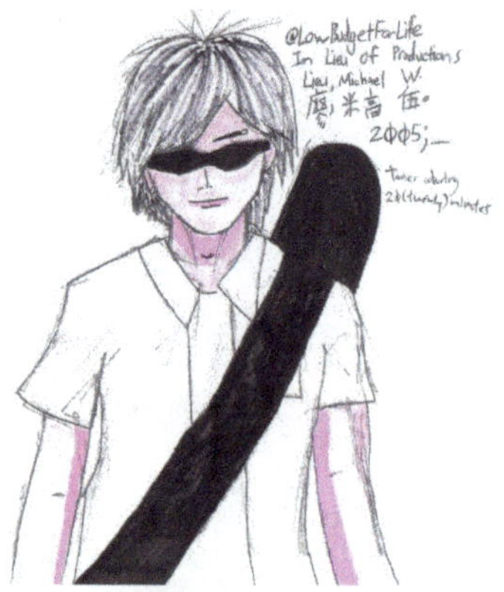	**Notes**; "**I** have not seen **my Elementary Mid High School Friends** since **I** departed and lived off county base...
Automation Algorithms **Closed Captioning Log**;	"**Lin, Remy**"; 'Whoa wow, yea, thanks **Mic**! Church, I gotta go! **Old Man Mic** still got it with all that years of public sec volunteer.' "**Wong, Jia**"; 'Yea foo, thanks **Mic Ah Sok**! Still got it as **Old Man Mic**! ...Not everyone within **our Elementary Mid High School Friends'** county base group could make it, but **most can. We** let **you** know. **You** going to be busy running a U.V. Animal Sanctuary.' "**Lin, Remy**"; '...That Mayor purple projection Chicago Controller, really quite something, to see that much U.V. rays at one time during the airlock for my first decontamination to Chicago Animal Sanctuary, explains how the Colonel and Lieutenant Colonel on county base would tell **us** to stay away from such purple U.V. rays.' "**Lieu, Michael W. (Me)**"; 'Our Mayor, Chicago Controller, is no joke and lazy, makes easy rules, but requires us to understand English since their Head Quarters is in Seattle, Washington, North America, it was like being on base, gotta know English and have paper physical paperwork... Uh... Hey, **Ah Soks we** catch up some more later?' "**Lin, Remy**"; 'Yea, later! Got to finish **airlock** decontamination!'		Automation Algorithms **Closed Captioning Log**;
Sisco Cisco .il Yellow (°o°) Broadcast	(°o°) (°o°)		Sisco Cisco .il Yellow (°o°) Broadcast

(Super-Hero) (Utopia) (Owner Operator Manual Guide) = Privacy Protection Aug.

Sisco Cisco .il Yellow (°o°) Broadcast	(°o°) (°o°)	Sisco Cisco .il Yellow (°o°) Broadcast
Date;	Year; (####;) ## (Month) - Day = 2154; 10 (Oct) – 30 (thirty)	
Arriving Airlock Decon Queue #2 (two) **Elysium Corporation Biotech Dome Airlock** Decontamination Chicago (S), The Last Animal Sanctuary (S) Of Illinois (ILL), North America (N.A.), Planet Earth, Solar System, Milky Way Galaxy.	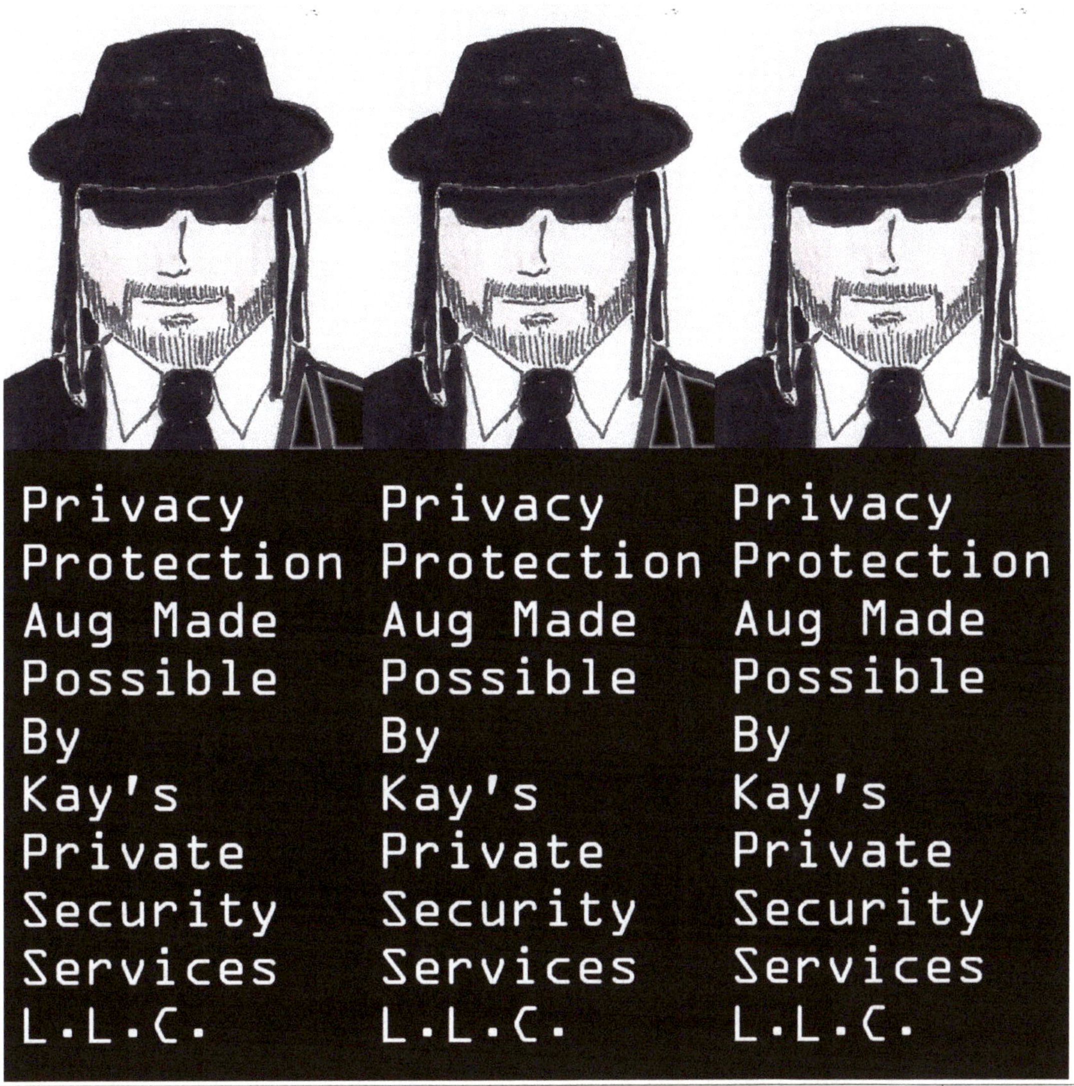	
Sisco Cisco .il Yellow (°o°) Broadcast	(°o°) (°o°)	Sisco Cisco .il Yellow (°o°) Broadcast

Lieutenant 2nd Lieu, Michael W. 廖, 米高 伍o; Vigilant Volunteers Sisco Cisco (°o°) Yellow; "Private sec... Kay's Private Security Services, top tier private security services in our solar system since... Who knows when those agents started."

(Super-Hero) (Utopia) (Comedy) = A.R. Holo HUD On BladeMaster Sailor Suits.

Sisco Cisco .il Yellow (°o°) Broadcast	(°o°) (°o°)	Sisco Cisco .il Yellow (°o°) Broadcast
Date;	Year; (####;) ## (Month) - Day = 2154; 10 (Oct) – 30 (thirty)	
Arriving Airlock Decon Queue #3 (three) **Elysium Corporation Biotech Dome Airlock Decontamination** Chicago (S), The Last Animal Sanctuary (S) Of Illinois (ILL), North America (N.A.), Planet Earth, Solar System, Milky Way Galaxy.	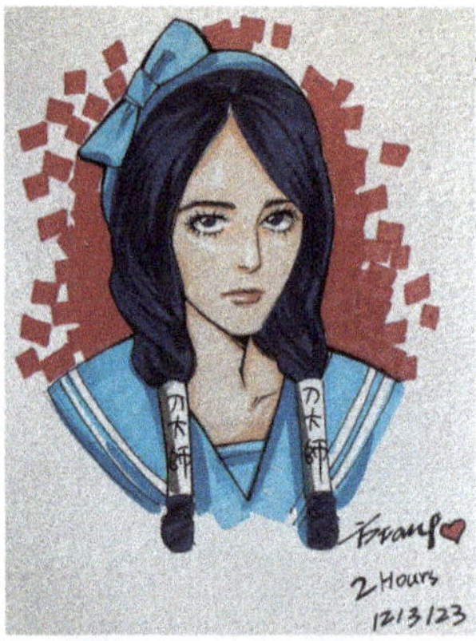 **Actual Legal Given Name(s);** "Kanon-Klones 0. Original-Copy" **Alias Network Nickname(s);** "BladeMaster 刀大師, Sailor Suit Student Trainees" **Community Network(s);** **"Elysium Corporation's Human Klones, Designer Babies And** A.I. Ai 愛 love" "BladeSociety 刀社會" [Aug+]	
	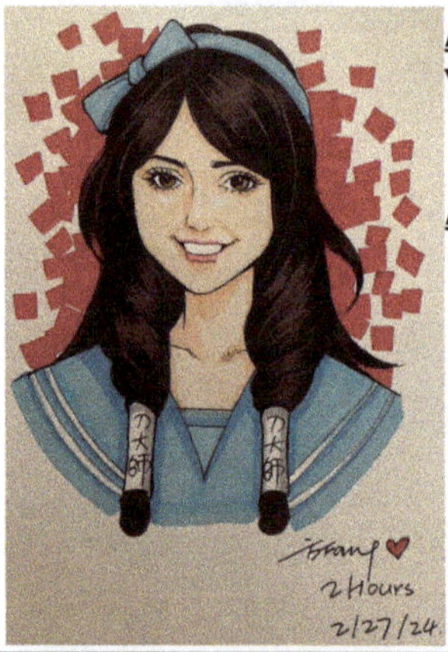 **Actual Legal Given Name(s);** "Elle K. Mcgee" **Alias Network Nickname(s);** "BladeMaster 刀大師, Sailor Suit Student Trainees" **Community Network(s);** **"Elysium Corporation's Human Klones, Designer Babies And** A.I. Ai 愛 love" "BladeSociety 刀社會" [Aug+]	
	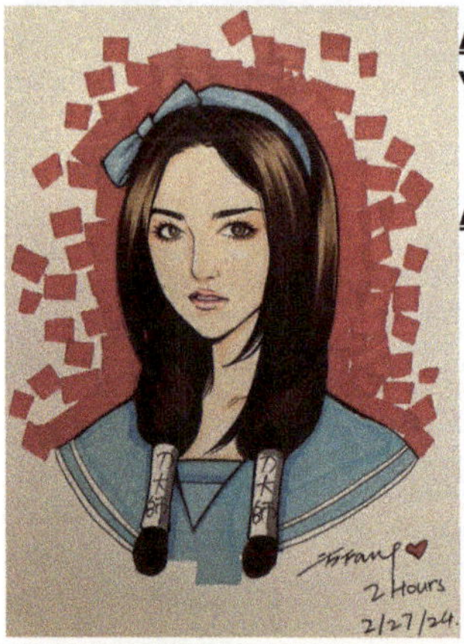 **Actual Legal Given Name(s);** "Fuka K. Fujimoto" **Alias Network Nickname(s);** "BladeMaster 刀大師, Sailor Suit Student Trainees" **Community Network(s);** **"Elysium Corporation's Human Klones, Designer Babies And** A.I. Ai 愛 love" "BladeSociety 刀社會" [Aug+]	
Sisco Cisco .il Yellow (°o°) Broadcast	(°o°) (°o°)	Sisco Cisco .il Yellow (°o°) Broadcast

(Super-Hero) (Utopia) (Comedy) (Horror) (Political) = "BladeSociety 刀社會"

Sisco Cisco .il Yellow (°o°) Broadcast	(°o°) (°o°)	Sisco Cisco .il Yellow (°o°) Broadcast
Date;	Year; (####;) ## (Month) - Day = 2154; 10 (Oct) – 30 (thirty)	
Arriving Airlock Decontaminating Queue #3(three) Chicago (S) Animal Sanctuary (S), Illinois (ILL) North America (N.A)		

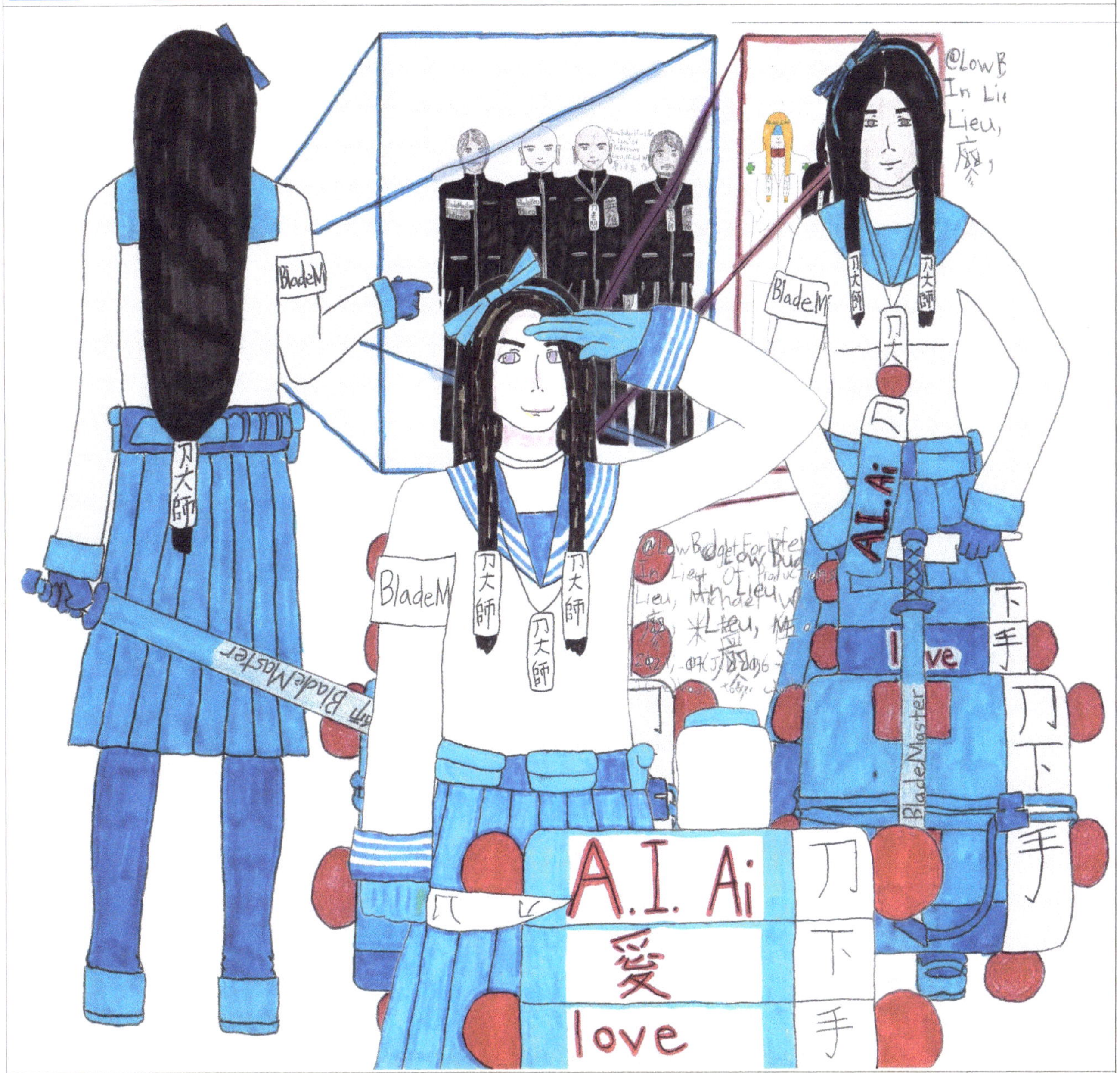

Sisco Cisco .il Yellow (°o°) Broadcast	(°o°) (°o°)	Sisco Cisco .il Yellow (°o°) Broadcast

(Super-Hero) (Utopia) (Travel) = A.R. Holo HUD On V. V., Travel Model Cosplayer.

Sisco Cisco .il Yellow (°o°) Broadcast	(°o°) (°o°)	Sisco Cisco .il Yellow (°o°) Broadcast
Date;	Year; (####;) ## (Month) - Day = 2154; 10 (Oct) – 30 (thirty)	

Arriving Airlock Decontamination Queue #4 (four) Chicago (S) Animal Sanctuary (S), Illinois (ILL) North America (N.A)

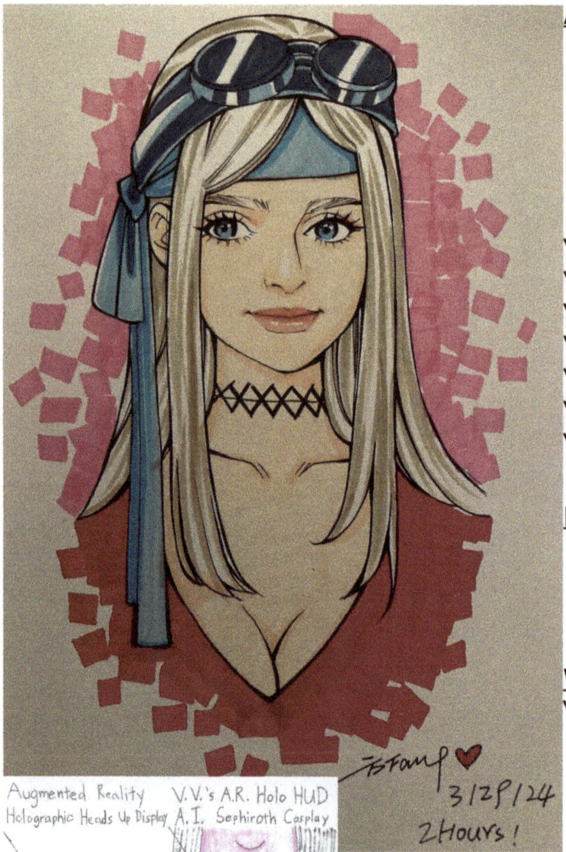

Actual Legal Given Name(s); "Valentina Vita"

Alias Network Nickname(s);
"V. V." "KissShot Kosplayer" "White-Woman", "Red Girl", "Red Woman", "Red Female" "Face Masks"
"Lady In Red", "Girl In Red" "Female In Red"
"Ireland-Woman", "Peach Skin" "Milk Skin"
"2 (two) Winged Demon Blade Kosplay",
"KissShot-AcerolaOrion-HeartUnderBlade Kosplay",
"Anime Long Blade Cosplayers", "Beauty Augment"
"XXXX Skin Suits" "Science-Fantasy Fiction Cosplayer"
"Aerith Cosplayer" "Aeris Cosplayer" "Aeris Kosplayer"
"Aerith Crisis Core Alternative History Cosplay" "FF7"
"Crisis Core Final Fantasy Seven" "CCFF7" [*Aug+*]

Nationality; "Earthling, Eur-Asian"

Wartime;
"Civilian"

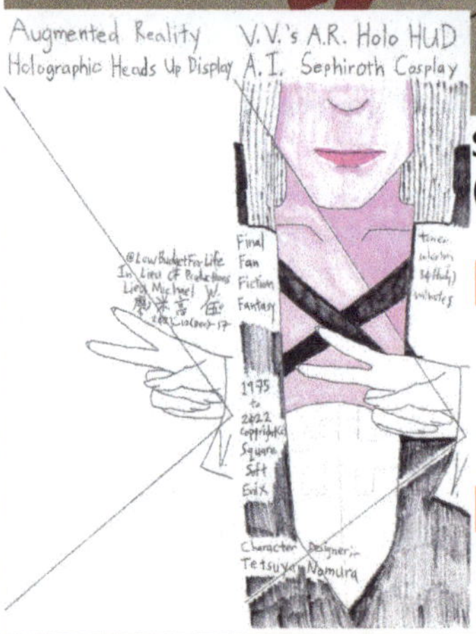

Science-Fantasy Fiction 1 (one; I)

V. V.'s Augmented Reality (A.R.)
Holographic (Holo)
Heads Up Display (HUD),
Artificial Intelligence (A.I.)
Sephiroth Cosplay

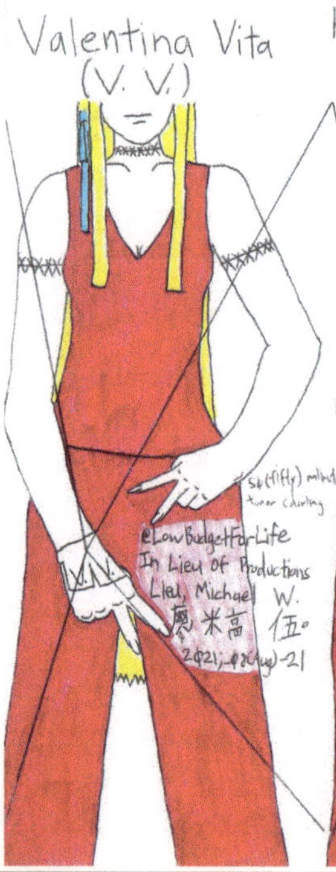

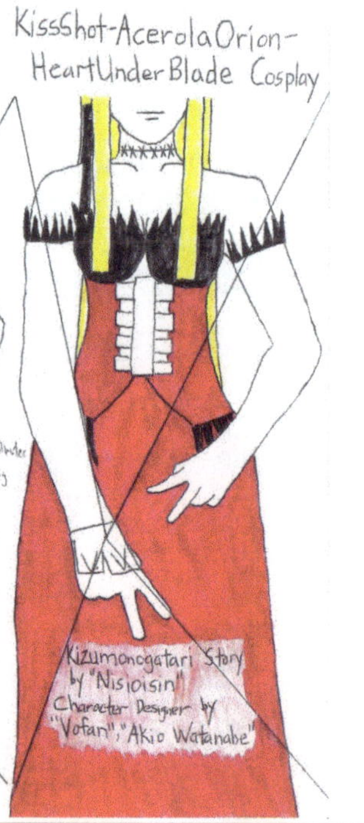

Sisco Cisco .il Yellow (°o°) Broadcast	(°o°) (°o°)	Sisco Cisco .il Yellow (°o°) Broadcast

(Super-Hero) (Utopia) (Travel) = A.R. Holo HUD On V. V., Travel Model Cosplayer.

Sisco Cisco .il Yellow (°o°) Broadcast	(°o°) (°o°)	Sisco Cisco .il Yellow (°o°) Broadcast
Date;	Year; (####;) ## (Month) - Day = 2154; 10 (Oct) – 30 (thirty)	

Arriving Airlock Decontaminating Queue #4 (four) Chicago (S) Animal Sanctuary (S), Illinois (ILL) North America (N.A)

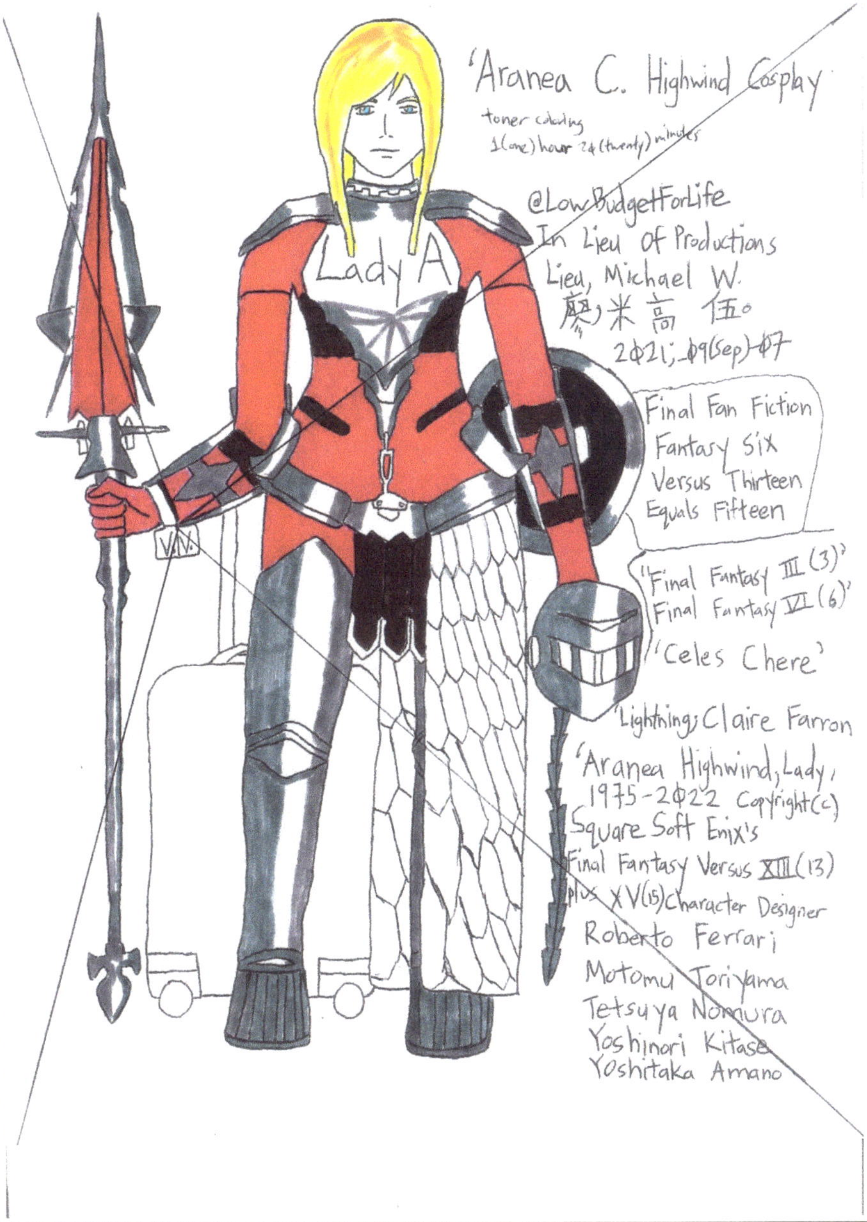

Connecting…
Electro Magnetic
Spectrum
Frequency Hertz (Hz),
Radio Waves, Intranet,
Frequency Modulation
(F.M.),
Amplitude Modulation
(A.M.),
[*Aug+*]

Elysium Corporation's
Multi Monitoring
Airlock Decontaminating
Queue #4 (four)
Active In Progress For;
"Valentina Vita" AKA
"V. V."

Sisco Cisco .il Yellow (°o°) Broadcast	(°o°) (°o°)	Sisco Cisco .il Yellow (°o°) Broadcast

(Super-Hero) (Utopia) (Travel) = A.R. Holo HUD On V. V., Travel Model Cosplayer.

Sisco Cisco .il Yellow (°o°) Broadcast	(°o°) (°o°)	Sisco Cisco .il Yellow (°o°) Broadcast
Date;	Year; (####;) ## (Month) - Day = 2154; 10 (Oct) – 30 (thirty)	
Arriving Airlock Decontaminating Queue #4 (four) Chicago (S) Animal Sanctuary (S), Illinois (ILL) North America (N.A)		

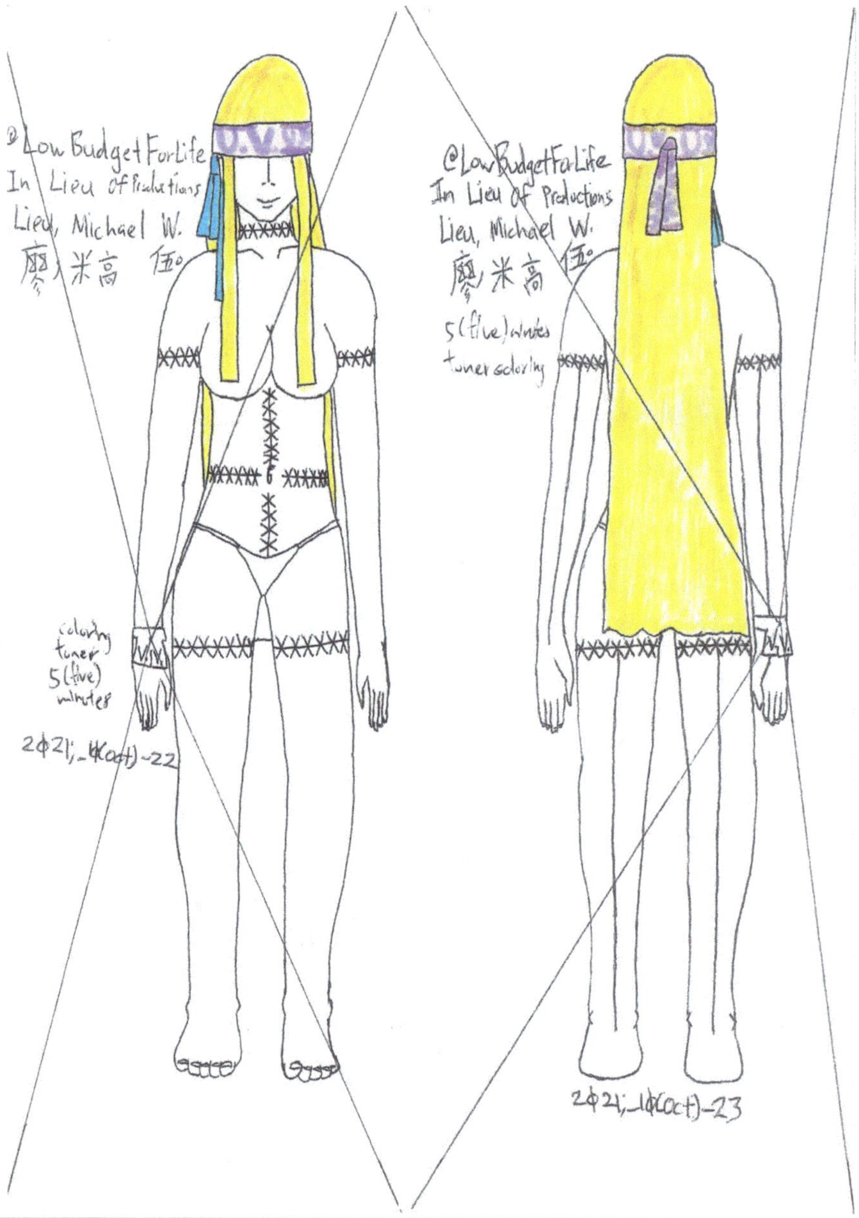

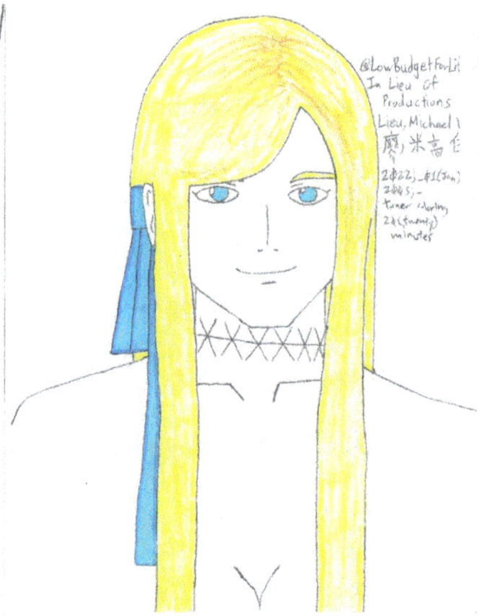

Elysium Corporation's
Multi Monitoring
Airlock Decontaminating
Queue #4 (four)
Active In Progress For;
"Valentina Vita" AKA
"V. V."

Sisco Cisco .il Yellow (°o°) Broadcast	(°o°) (°o°)	Sisco Cisco .il Yellow (°o°) Broadcast

(Super-Hero) (Utopia) (Travel) = A.R. Holo HUD On V. V., Travel Model Cosplayer.

Sisco Cisco .il Yellow (°o°) Broadcast	(°o°) (°o°)	Sisco Cisco .il Yellow (°o°) Broadcast
Date;	Year; (####;) ## (Month) - Day = 2154; 10 (Oct) – 30 (thirty)	

Arriving Airlock Decontamination #4 (four) Done. Chicago (S) Animal Sanctuary (S), Illinois (ILL) North America (N.A)

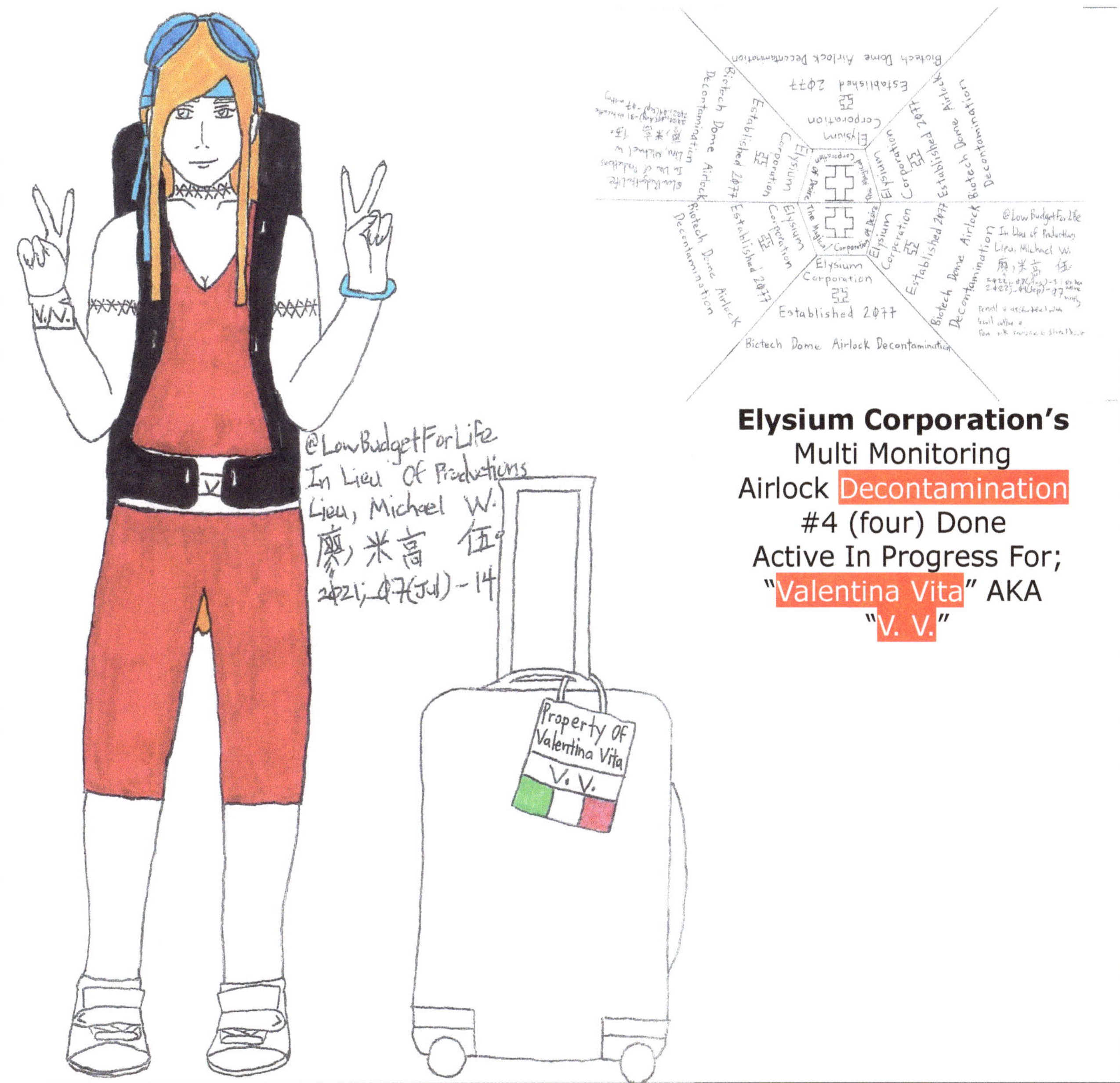

Elysium Corporation's
Multi Monitoring
Airlock Decontamination
#4 (four) Done
Active In Progress For;
"Valentina Vita" AKA
"V. V."

Sisco Cisco .il Yellow (°o°) Broadcast	(°o°) (°o°)	Sisco Cisco .il Yellow (°o°) Broadcast

(Super-Hero) (Utopia) (Travel) = A.R. Holo HUD On V. V., Travel Model Cosplayer.

Sisco Cisco .il Yellow (°o°) Broadcast	(°o°) (°o°)	Sisco Cisco .il Yellow (°o°) Broadcast
Date;	Year; (####;) ## (Month) - Day = 2154; 10 (Oct) – 30 (thirty)	
Arriving Airlock Decontamination #4 (four) Done. Chicago (S) Animal Sanctuary (S), Illinois (ILL) North America (N.A)		

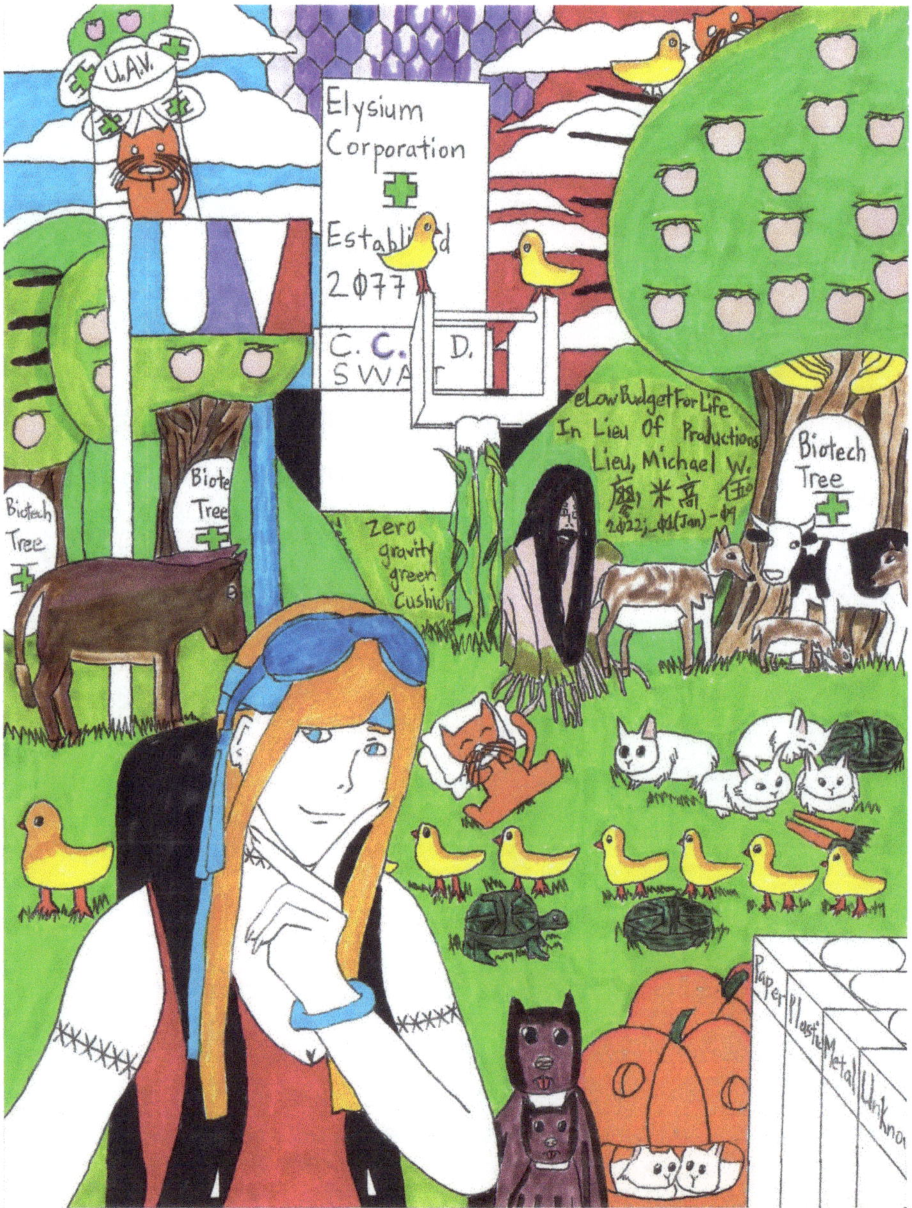

"Valentina Vita, V. V."

@ Planet Earth,
America(s),
North America (N.A.),
Illinois (ILL).

Elysium Corporation's Biotech Domes,

In Downtown District
(D. D.)

"Chicago (S), The Last Animal Sanctuary (S) Of Illinois (ILL)"

"Wow, so cute!! Too cute!! So many cute peaceful natural animals together!

(*squeals*)

So worth the travel and wait with many **airlock** decontaminations for this North America animal sanctuary! Thanks Mayor!!"

Sisco Cisco .il Yellow (°o°) Broadcast	(°o°) (°o°)	Sisco Cisco .il Yellow (°o°) Broadcast

(Super-Hero) (Utopia) (Travel) = A.R. Holo HUD On V. V., Travel Model Cosplayer.

Sisco Cisco .il Yellow (°o°) Broadcast	(°o°) (°o°)	Sisco Cisco .il Yellow (°o°) Broadcast
Date;	Year; (####;) ## (Month) - Day = 2154; 10 (Oct) – 30 (thirty)	

Sisco Cisco .il Yellow (°o°) Broadcast Automation Algorithms Closed Captioning Log;

"Valentina Vita, V. V.;" Wow, an **overview** of peaceful surviving natural animal differences on Planet Earth as of "2154;" Okay, for natural mammals..."

Natural Mammals: has **hair** all **over their bodies**, gives **live births with umbilical cords**, **produce milk for young babies**, have **mammary glands to feed** *their young babies with milk*, have **4 (four)** chambered hearts, Warm Blood, Existing Examples Of Natural Mammals In **Elysium Corporation's** Bio**tech Dome** Chicago (S) Animal Sanctuary (S): Bird, Cat, Cattle, Deer, Donkey, Rabbit, "Decreased Human Feral Children"... Interesting."

"Okay now for natural reptiles animals..."

Natural Reptiles: Have **scales** all **over their bodies**, gives **birth by laying eggs**, Reptiles can **continue growing throughout their lives** and continually **replace lost teeth**, have **3 (three)** chambered hearts, Kold Blood, Existing Examples Of Natural Reptiles In **Elysium Corporation's** Bio**tech Dome** Chicago (S) Animal Sanctuary (S): Turtle, *Snake, Lizard Not Visible To Naked Eye.*

Both Natural *Animals' Have*: **Vertebrates, Backbones**, Brain, Nervous System as a biological beings regardless of being natural and or **artificially** Kreated.

Sisco Cisco .il Yellow (°o°) Broadcast	(°o°) (°o°)	Sisco Cisco .il Yellow (°o°) Broadcast

182 (Lieutenant Lieu, In Lieu Of) Lieu, Michael W. 廖, 米高 伍o 182

(Comedy) (Horror, Neo Noir) = "BladeMaster 刀大師" (A 2 S 1) 2154; Arriving.

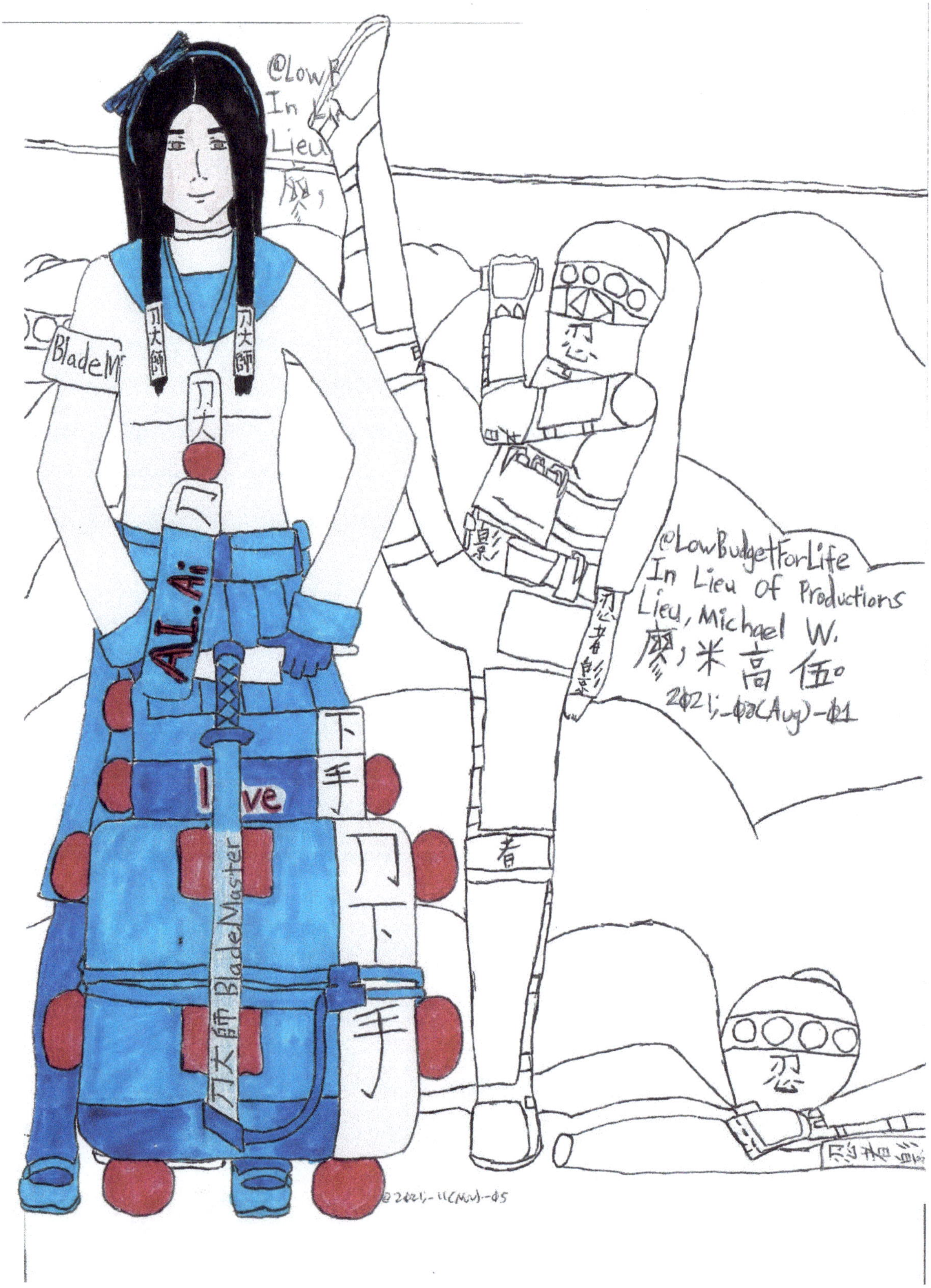

182 \ **2154;** Lieu, 廖, 'WhiteWolf 白狼' Left to Right (L→R) 182

(Comedy) (Horror, Neo Noir) (Super-Hero) (Utopia) = (A 2 S 1) 2154; Arriving.

Sisco Cisco .il Yellow (°o°) Broadcast	(°o°) (°o°)	Sisco Cisco .il Yellow (°o°) Broadcast
Date;	Year; (####;) ## (Month) - Day = 2154; 10 (Oct) – 30 (thirty)	

Sisco Cisco .il Yellow (°o°) Broadcast Automation Algorithms Closed Captioning Log;

Elysium Corporation Automation Algorithms; (*monotone female voice over*) "Please stand by for bio**tech** dome **airlock** decontamination. Thank you for your patience and cooperation **Utopian**."

BladeAssistant 刀下手 Artificial Intelligence A.I. Ai 愛 love **Luggage**; "(*monotone male voice over*) This way Elle to Chicago **Elysium's North Side** the State Acting Chief Engineer, ChEng Toy-Master **volunteer manager**, **WhiteWolf**, **Lieutenant Lieu, vigilant** with coffee and insomnia."

BladeMaster 刀大師 Sailor Suit Student Trainee, Elle K. Mcgee; "Oh hi, oh low!"

BladeSociety 刀社會 **Ninja Shadow 忍者影**; "(*bio**tech suit** female voice)* (*in Japanese*) Been a while, **Lieu Lieu**. A BladeMaster student is here for **you** on behalf of BladeSociety. BladeBoss sends regards and is good."

Lieutenant 2nd Lieu, Michael W. 廖, 米高 伍o; "(*In Cantonese*) Oh hi **Ninja Kanon**, it has been some time. Having fun with **smoke screen**? **Grenades working well**. (*In English*) **My** North **sector** in **dome** has an **airlock**. Whenever you ready chief... chef... Curious cat. BladeMaster sailor suit Elle... assuming this is the same BladeMaster Elle **I** know, and not a **new klone**."

Sisco Cisco .il Yellow (°o°) Broadcast	(°o°) (°o°)	Sisco Cisco .il Yellow (°o°) Broadcast

(Comedy) = "BladeMaster 刀大師" Sailor Suit Student Trainee. Rosa Salazar. Arc Two Story One (A 2 S 1), 2154; Arriving character Elle K. Mcgee.

{ "2 D Augmented Art Commission(s) Portrait(s) By; Fang Fang Comics, Fang YiLi" }

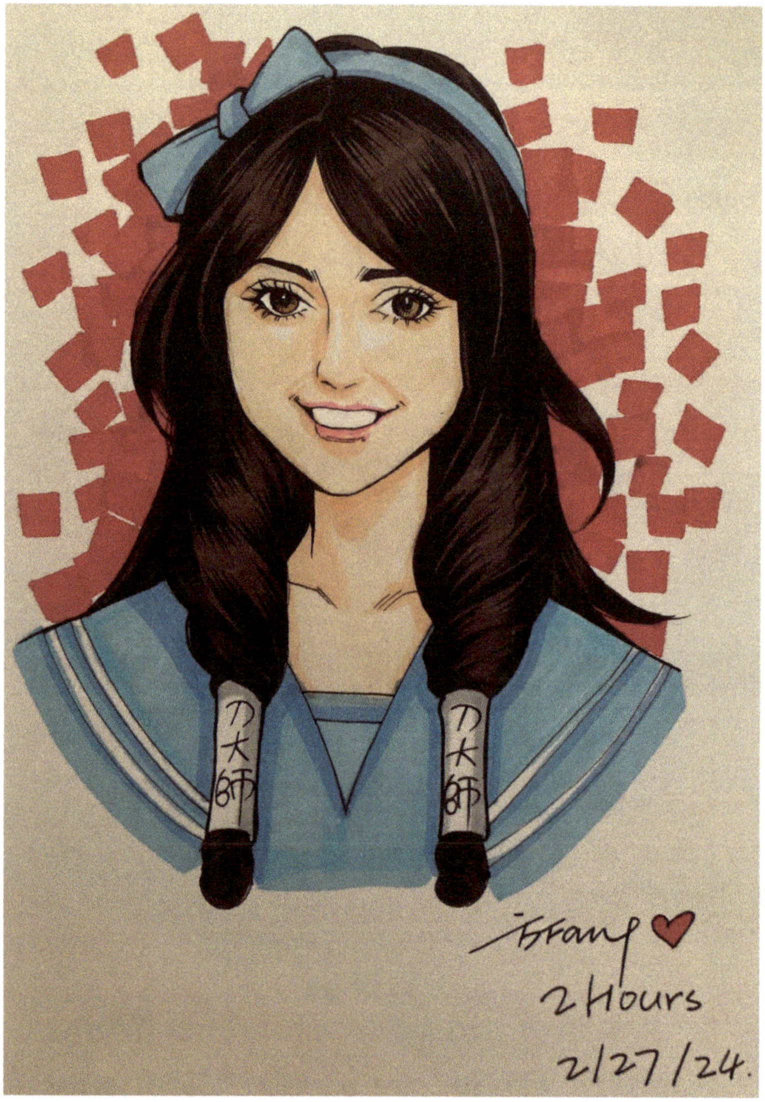

Sisco Cisco .il Yellow (°o°) Broadcast	(°o°) (°o°) (°o°) (°o°) (°o°) (°o°) (°o°) (°o°)	Sisco Cisco .il Yellow (°o°) Broadcast
Date;	2154; 10 (Oct) - 30	
Actual Legal Given Name(s); Elle K. Mcgee (goes by Elle)		
Alias Network Nickname(s); BladeMaster Elle, BladeMaster Sailor Suit Student, [Aug+]		
Birth-**Built Day**; Year 2152; Month 05 (May), Day 05 (five)		
Community Network(s); **"Elysium Corporation's Human Klones, Designer Babies And A.I. Ai 愛 love"** "BladeSociety 刀社會" "BladeMaster 刀大師, Sailor Suit Student Trainees" "Cook-Chef Community Martial Artist Educator" [Aug+]		
[Aug+]		
Sisco Cisco .il Yellow (°o°) Broadcast	(°o°) (°o°) (°o°) (°o°) (°o°) (°o°) (°o°) (°o°)	Sisco Cisco .il Yellow (°o°) Broadcast

(Comedy) (Horror, Neo Noir) (Super-Hero) (Utopia) = (A 2 S 1) 2154; Arriving.

Sisco Cisco .il Yellow (°o°) Broadcast	(°o°) (°o°)	Sisco Cisco .il Yellow (°o°) Broadcast
Date;	Year; (####;) ## (Month) - Day = 2154; 10 (Oct) – 30 (thirty)	

Sisco Cisco .il Yellow (°o°) Broadcast Automation Algorithms Closed Captioning Log;

BladeMaster 刀大師 Sailor Suit Student Trainee, Elle K. Mcgee; "Time to poke guy with this blade! Poke guy, Six C! Just joking, how **you** been? **Lieu Lieu**?"

Lieutenant 2nd Lieu, Michael W. 廖, 米高 伍o; "You know **my nickname**, and poke guy, Six C! You are a BladeMaster sailor student **I** know… Your Cantonese cursing, bad words on point! Okay, was starting to think you are another Klone that BladeBoss put out from **Elysium Corp**… Birth-**built year twenty one fifty two, not to long ago from this year… A recent klone**. So, what can **I** do for you besides, poke guy, Six C?!"

BladeMaster 刀大師 Sailor Suit Student Trainee, Elle K. Mcgee; "Aw, no poke guy, Six C… Anyways, here for this violet purple projection U.V. Mayor calls Chicago Animal Sanctuary volunteer work, going to learn more about how these remaining natural animals coexistence peacefully and… and… I promise not to cut them in pieces and eat them. This is not Eur-Asia, I know that. **We all** had to agree to a verbal violet contract with Mayor… I can read and write. Six C!"

Lieutenant 2nd Lieu, Michael W. 廖, 米高 伍o; "…Hm… Okay, F.O.S."

BladeMaster 刀大師 Sailor Suit Student Trainee, Elle K. Mcgee; "I am serious! Many people underestimate us for wearing old sailor suit uniforms!"

Sisco Cisco .il Yellow (°o°) Broadcast	(°o°) (°o°)	Sisco Cisco .il Yellow (°o°) Broadcast

(Comedy) (Horror, Neo Noir) = A 2 S 1 “2154; Arriving” Sailor Suit Elle K. Mcgee

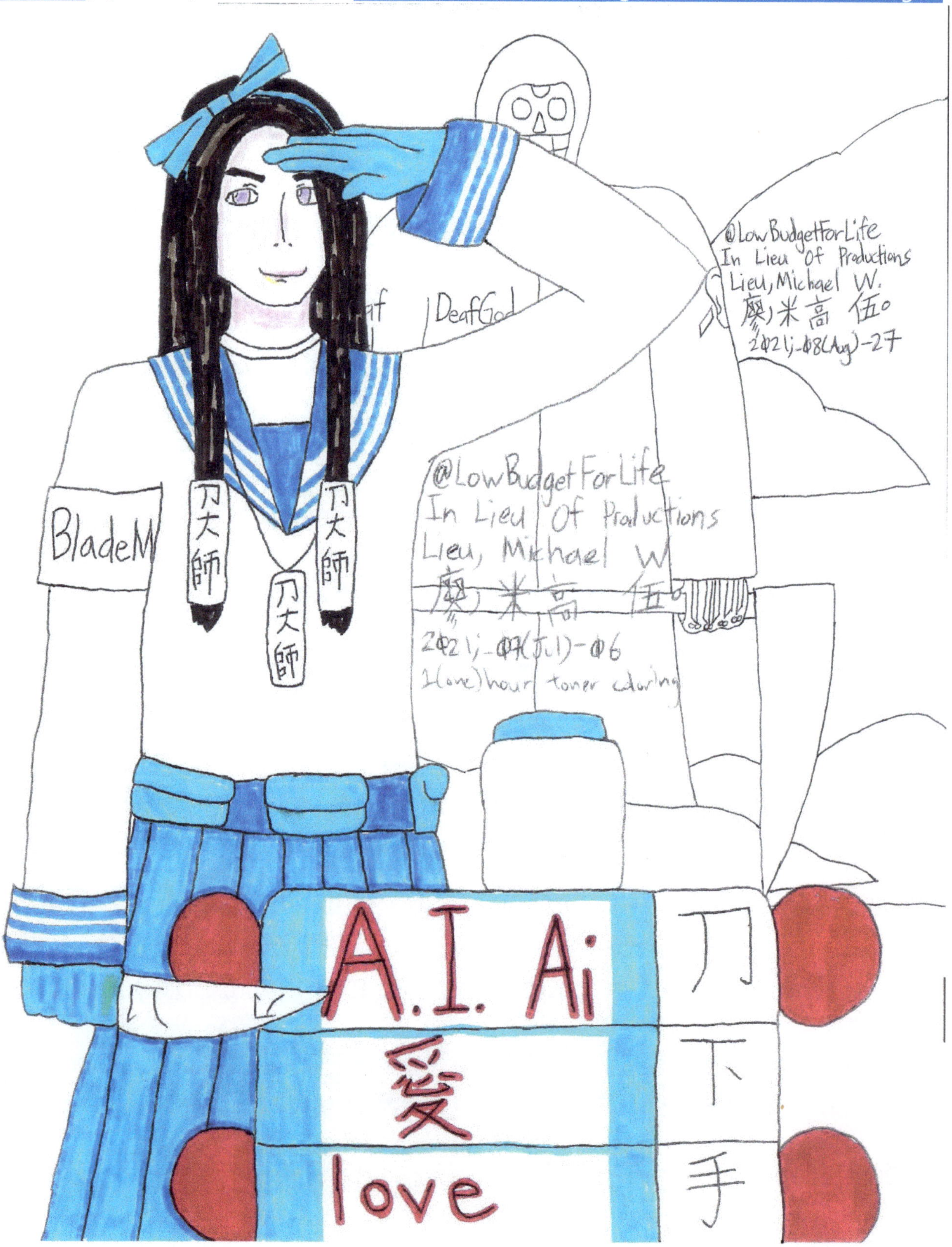

(Comedy) (Horror, Neo Noir) (Super-Hero) (Utopia) = (A 2 S 1) 2154; Arriving.

Sisco Cisco .il Yellow (°o°) Broadcast	(°o°) (°o°)	Sisco Cisco .il Yellow (°o°) Broadcast
Date;	Year; (####;) ## (Month) - Day = 2154; 10 (Oct) – 30 (thirty)	

Sisco Cisco .il Yellow (°o°) Broadcast Automation Algorithms Closed Captioning Log;

BladeAssistant 刀下手 Artificial Intelligence A.I. Ai 愛 love **Luggage**; "(*monotone male voice over*) BladeMaster sailor suit student trainee, Elle K. Mcgee approved access to BladeAssistant Ai **luggage**."

Lieutenant 2nd Lieu, Michael W. 廖, 米高 伍o; "Ok, so you been to **Downtown District in dome**, and looks like in terms of **volunteer work**, you be more suited for... Utopian Peacekeeping, our acting chief for that public sec volunteer group is Alex Doc Drastal. Oh he, likes black toner like BladeBoss too."

BladeMaster 刀大師 Sailor Suit Student Trainee, Elle K. Mcgee; "Oh! Not **Strike Force**? **Strike Force** sounds kool... Utopian Peacekeeping... sounds... Hm."

Lieutenant 2nd Lieu, Michael W. 廖, 米高 伍o; "Utopian Peacekeeping is mostly for artificial persons, augmented persons as well... Doc Drastal is a war military survivor from our Earth's Space Force for War On Mars, he lost body parts and became **augmented** to survive, had to go through much training with MartialArtsGod Society, rival to BladeSociety... **Strike Force**, mostly natural persons, no **klone persons**... Time to put together your unique uniform."

BladeMaster 刀大師 Sailor Suit Student Trainee, Elle K. Mcgee; "Oh MartialArtsGod Society! We thank him for his service to Earth, EGG says. Here are my Blades. Nice of **you** to be a **BladeAssistant** again **Lieu Lieu**!"

Sisco Cisco .il Yellow (°o°) Broadcast	(°o°) (°o°)	Sisco Cisco .il Yellow (°o°) Broadcast

2154; 10(Oct)-30(thirty) @ 19;19 P.M. { act 1(one) back story } Bio**tech Suit**+
Elysium Corporation 亞, Engineering Environment Bio**tech Dome**, **Interior**,
+024°(twenty four degrees) Celsius / +075°(seventy five degrees) Fahrenheit.
"Chicago (S), The Last Animal Sanctuary (S) Of Illinois (ILL),"
'North (N) Side', Bio**tech** Dome Ocean Border, King And X District,
Utopian Sovereignty of North America (**U.**S. of N.A.) Illinois (ILL) 60666
North America, Planet Earth, Solar System, Milky Way Galaxy.

Lieutenant 2nd Lieu, Michael W. 廖, 米高 伍o; Vigilant Volunteers Sisco Cisco (°o°) Yellow;

"While **I** work on this unique uniform for her, what is a **BladeMaster** sailor suit student really doing here **in dome** Chicago, The Last Animal Sanctuary Of Illinois? "**2154; Lieu, 廖,**" 'WhiteWolf 白狼' A 2 S 3 B 1 concludes".

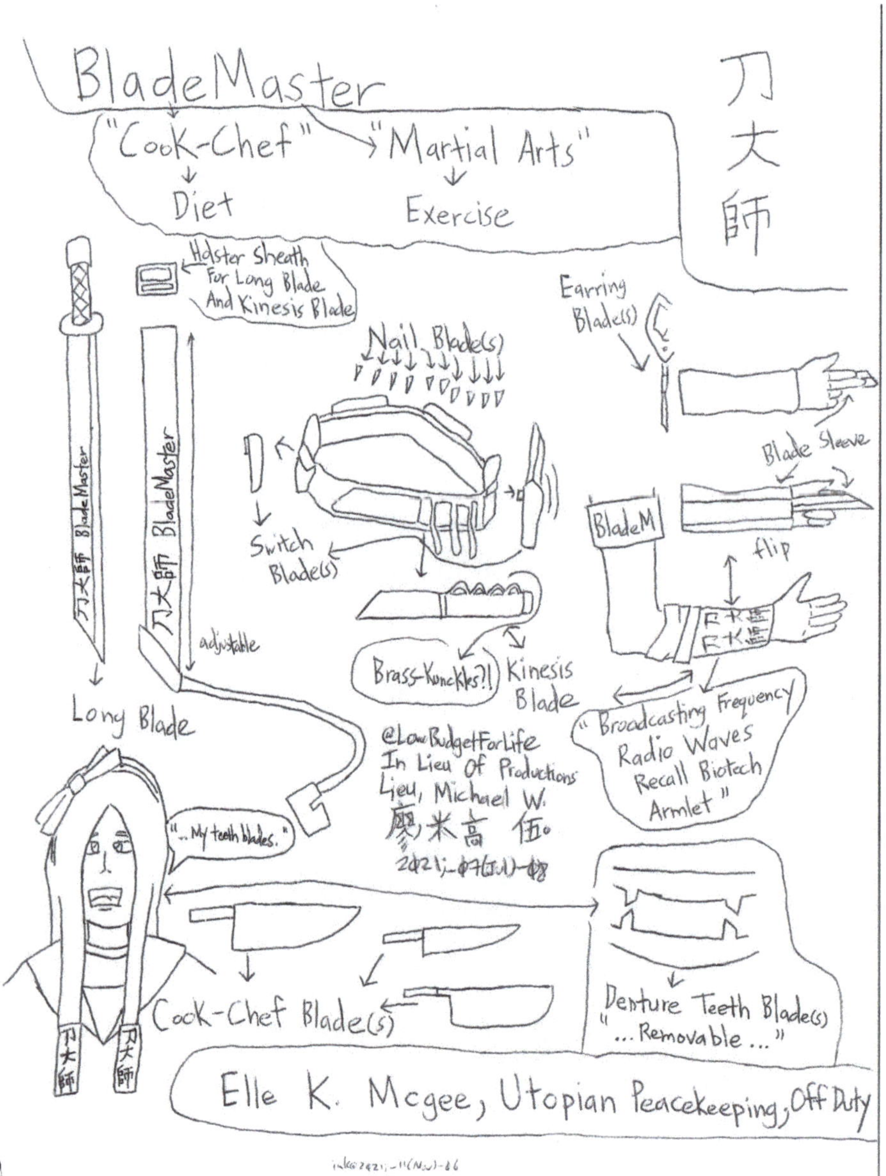

Model References Work Inspiration; Bibliography

2019;_12(Dec), **21X0;_U.S.-Department-Of-War-KIPI-Kosplay** "**Photo-Book** Prototype" { A 1 S X } [**White 白**]

2020;_03(Mar), **2154;_Arriving-Coloring-Book** "Coloring-Book" { A 2 S 1 } [Blue 藍]

2020;_09(Sep)-06(six), "**2154;_Arriving**" "Concept Art Script Experiment" { A 2 S 1 } [Blue 藍]

2020;_10(Oct)-31(thirty one), "**Conscientiousness Cat/Kat's XXXX;_Kiss-Hiss+**" "Parody Splicing Idea Generator" { A ? S X } [Orange 橙]

2022;_08(Aug)-31(thirty one), "**2154;_Arcadia**" "Travel Drawings Audio Retcon" { A 2 S 2 } [Red 紅]

2024; Repurposed (A 2 S 3 B 1) title from "2154;_Ascension-00-Prologue" to "**2154; Lieu, 廖,**" being a prequel prologue to self-publish supplement(s) books, "**2154;_Arriving**" "**2154;_Arcadia**" "**Con. Cat's XXXX;_Kiss-Hiss+**".

2024; Split Visual Story Text Book (A 2 S 3 B 2) to "2154; Incident", next one.

2014; National Geographic (N.G.) Kids, Mission; Wolf Rescue. All About Wolves And How To Save Them By Kitson Jazynka With N.G. Daniel Raven-Ellison

2024; 05 (May) website accessed on www.study.com search, "The Dark Ages"

2014's Dawn of the Planet of the Apes Movie - Opening Scene, Virus Spreading.

2022; 10 (Oct) – 21 website on www.usamm.com, "Military Bases; Guide"

2024; 05 (May) 21 (twenty one) website accessed on https://nssdc.gsfc.nasa.gov/planetary/factsheet/planet_table_ratio.html, "Planetary Fact Sheet – Ratio To Earth Values by NASA Official Doktor David R. Williams, Last Updated 2024; 03 (Mar) – 22 (two two)

2020s' Expanse T.V. Series, end of Season(s); 4 (four) to 5 (five) to 6 (six), story, character arc of "space rocks, asteroids, meteors, terror attack", can their ever be an Expanse Series Season 7 (seven) continuing?

2024; 05 (May) 30 (thirty) booklet read "National Fire Protection Association"

2024; 06 (Jun) M.K. Ultra, Uncle Sam's unconventional mind control research.

Social Net Media Pages Online Name(s) Abstract Of "**2154;_Arriving**"

"So, here's my info, so contact me maybe?" ***Failed Engineer Drafting Designer***

I think some people are going to spam me with goat porn and Nigerian prince pleas scams for their quote "family, friends..." ***Artist Architecture Author Alchemist***

Oh well... Creativity curse. (surprising face (; O)) "***Visual Story Text Book***"

"I'm sorry, the number you have dialed has a voice mail box not yet set up. Good bye!"

Low Budget Business Card

"Lieutenant Lieu" \ "In Lieu Of" \ Lieu, Michael W.

L → R 廖, 米高 伍o

https://www.instagram.com/lowbudgetforlife

https://www.youtube.com/user/lowbudgetmic

https://twitter.com/lowbudgetmic

https://www.facebook.com/lieu.michael.w

2154visualstorytextbook@gmail.com

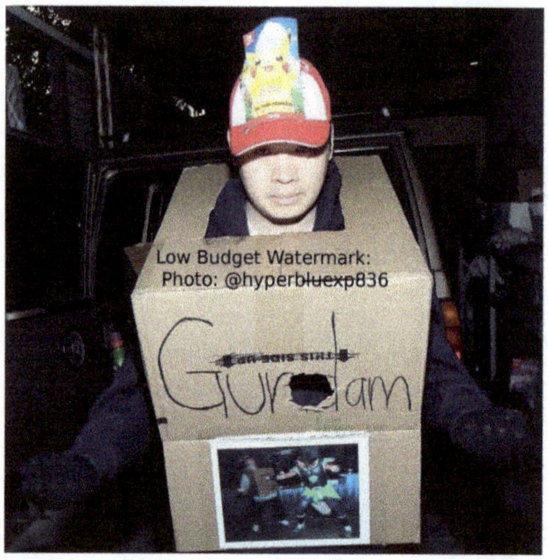

Kat, Conscientiousness Cat

Kay, Agent Cool

Kosplay In Playing Identities,

Doktor A.I. Ai 愛 love

Kiss-Hiss+, Human-Snake

Kat 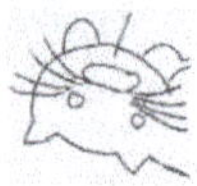

KEANU 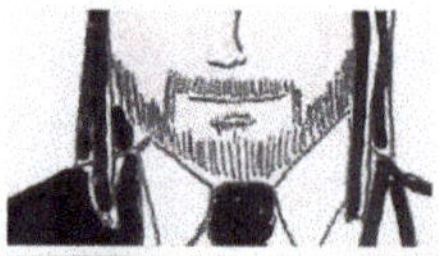

KIPI 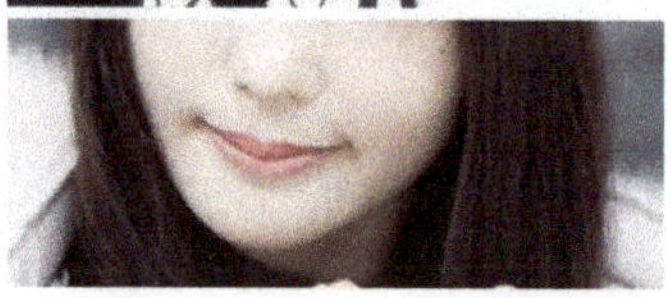

Kiss-Hiss+ ?

Social Net Media Pages Visual Collages, from 2020;_09(Sep)-06(six)
Abstract Of Arc Two Story One (A 2 S 1) "**2154;_Arriving**"
"Never gonna give you up, never gonna let you down..." -Rick Roll
https://www.instagram.com/lowbudgetforlife #2154 #2154collage #2154visualstorytextbook

https://www.deviantart.com/lowbudgetforlife

https://www.facebook.com/2154VisualStoryTextBook/photos/

https://lieumichaelw.blogspot.com/

http://www.amazon.com/author/lieumichaelw
Butt, fear not... here is a story universe layout of my madness, visions I dream about...

DOE MultiVerse number(English number) "subject to change due to X variable"
Insanely Inspired by; W.B. + D.C. + C.W.'s 2019 Crisis On Infinite Earths.

MultiVerse #(number)	Character Name Point Perspective	Next Story Universe To Draw-Write After Completing "2154;_Arriving" Book Titles
0(zero) or O	Multiverse Manipulating	0X1O;_?
1(one) or I	Lieu, Michael W. 廖, 米高 伍o	2006;_Choices
2(two) or to	Valentina Vita(V. V.)	2154;_Awareness
3(three) free	A.I.+ Conscientiousness Cat	Con-Kat's XXXX;_Kiss-Hiss+
4(four) or 死	Holocene Extinction Entity 6	Hideo Kojima's Death Stranding Multiverse
5(five) high	Mars, D. Roman God Of War	2120;_Endgame
6(six) or 9	Kefka Palazzo "Kosplay"	"Watch The World Burn"
7(seven or L	1 Name 4 All "Sephiroth"	1997;_Square Soft Enix's Final Fantasy 7
8(eight) ate	Endless Eight "Ate" 8	8888;_V.R.-Sim-19to91;_8888 = "Subtitle"
9(nine) or 6	"ylapsoK" ozzalaP akfeK	"nruB dlorW ehT hctaW"
10(ten) or X	Multiverse Manipulating	2001;_Indoctrination
18 (eighteen)	Ardyn Lucis Caelum "Cosplay"	2006-2016;_Final Fan Fiction Fantasy 6 Versus 13 Equals 15
		399 / 590
1941(nineteen forty one)	Adolf Hitler, Nazi Germany, My Fuhrer!	1941;_Uprising W.B., D.C., C.W.'s 2017 Crisis On Earth X

Version 2(two) of my "*Science-Fiction Character Story Universe*" Blueprint,
"In math, X is a variable that is subject to change." = Arial Rounded MT
Insane Inspiration; { Updated And Abstract From "2154;_Arcadia" }
W.B. + D.C. + C.W.'s 2020; Crisis On Infinite Earths' Part 5(five) Conclusion Wrap-Up

4(four) digit title #(number)	Character Name Point Perspective	Next Story Universe To Draw-Write After Completing "2154;_Arcadia" Book Series Title
0X1O;	Multiverse Manipulating	0X1O;
20th	***20th (twentieth) century***	***Alternative history timeline.***
1941;	Adolf Hitler, My Fuhrer, Nazi Party Of Germany, Third Reich	1941; Totalitarian
1997;	"BladeBoss 刀老闆" Mark D. Mark S.	1997; Sovereignty
21st	***21st (twenty first) century***	***Alternative history timeline.***
2001;	Sir Arthur C. Clarke	2001; Space
2006-2016;	Ardyn Lucis Caelum "Cosplay"	2006-2016; Science-Fantasy Fiction
????;	"Question-Mark ?"	Hideo Kojima's Death Stranding
22nd	***22nd (twenty second) century***	***Alternative history timeline.***
2120;	Roman Industries	2120; Endgame { Arc Zero (A 0) }
2121;	Jon W. James a.k.a. Anonymous	2121; Movement
2121;	Seattle City Police Department (S.C.P.D.) Special Weapons And Tactics (SWAT)	2121; SWAT
2121;	United States Army (U.S.A.)	2121; Army
2136;	**Ai-Aug A.R. Holo HUD**	**2136; Grad { Arc One (A 1) }**
21X0;	**U.S.-Department-Of-War-KIPI-Kosplay { Prototype Photo-Book Script }**	
2136;	Sisco Cisco (°o°) Yellow-Mellow	2136-2139; Navy
2154;_	Elle K. Mcgee "BladeMaster 刀大師"	2154;_Arriving { Arc Two (A 2) }
XXXX;	Augmented Artificial Intelligence (A.I.+) Ai love Conscientiousness Cat/Kat "Con-Kat"	Conscientiousness Cat/Kat's XXXX;
2154;_	"Valentina Vita (V. V.) Healthy-Strong Life"	2154;_Arcadia
2154;	**Lieutenant Lieu, Michael W. 2nd 廖, 米高 伍o "WhiteWolf 白狼"**	**2154; Lieu, 廖,**

4(four) digit title #(number)	Character Name Point Perspective Abstract From "2154;_Arcadia"	Next Story Universe To Draw-Write After Completing "2154;_Arcadia" Book Series Title
0X1O;	Multiverse Manipulating	0X1O;
Variable	***Simulation century***	***Simulation history timeline.***
XXXX;_	Augmented Artificial Intelligence (A.I.+) Ai love Conscientiousness Cat/Kat "Con-Kat"	Conscientiousness Cat/Kat's XXXX;_Kiss-Hiss+
XXXX;	Augmented Artificial Intelligence (A.I.+) Ai love Conscientiousness Cat/Kat "Con-Kat"	Conscientiousness Cat/Kat's XXXX; Kiss-Hiss+ Part 2 (two, to, too) A-Rag, Hon-Knee
XXXX;	Augmented Artificial Intelligence (A.I.+) Ai love Conscientiousness Cat/Kat "Con-Kat"	Conscientiousness Cat/Kat's XXXX; Kiss-Hiss+ Part 3 (three) A-Rag, Hon-Knee
XXXX;	Augmented Artificial Intelligence (A.I.+) Ai love Conscientiousness Cat/Kat "Con-Kat"	Conscientiousness Cat/Kat's XXXX; N
XXXX;	Augmented Artificial Intelligence (A.I.+) Ai love Conscientiousness Cat/Kat "Con-Kat"	Conscientiousness Cat/Kat's XXXX; B
XXXX;	Augmented Artificial Intelligence (A.I.+) Ai love Conscientiousness Cat/Kat "Con-Kat"	Conscientiousness Cat/Kat's XXXX; N
XXXX;	Augmented Artificial Intelligence (A.I.+) Ai love Conscientiousness Cat/Kat "Con-Kat"	Conscientiousness Cat/Kat's XXXX; K
XXXX;	Augmented Artificial Intelligence (A.I.+) Ai love Conscientiousness Cat/Kat "Con-Kat"	Conscientiousness Cat/Kat's XXXX; O
XXXX;	Augmented Artificial Intelligence (A.I.+) Ai love Conscientiousness Cat/Kat "Con-Kat"	Conscientiousness Cat/Kat's XXXX; O
XXXX;	Augmented Artificial Intelligence (A.I.+) Ai love Conscientiousness Cat/Kat "Con-Kat"	Conscientiousness Cat/Kat's XXXX; N
XXXX;	Augmented Artificial Intelligence (A.I.+) Ai love Conscientiousness Cat/Kat "Con-Kat"	Conscientiousness Cat/Kat's XXXX; O

Lieu, Michael W.'s (Alternative History AKA Soft Science-Fiction) (Parody) genre

Arc Two Story Three Book One, A 2 S 3 B 1, "2154; Lieu, 廖," Concept Generator

Our History, Soft Science-Factual genre	*Alt History, Soft Science-Fiction genre*
Long History *Past*, collapse of Roman Republic Empire's Dark Ages of 500 Years;	Long Alternative History *Future*, twenty second (22nd) century Dark Ages, 2121;
1918; Origins 1919; Spanish Flu Virus Outbreak Pandemic 1918; Origins 1919;	
1941; Hitler's Nazi Political Party ***does attack*** Soviet Union, Union Soviet Socialist Republics, U.S.S.R., which the Nazis had a peace treaty with at that time.	1941; Hitler's Nazi Political Party ***does not attack*** Soviet Union, Union Soviet Socialist Republics, U.S.S.R., which the Nazis had a peace treaty with at that time.
1941; Imperial Japan ***does attack*** U.S. military base Pearl Harbor, U.S. ***joins in*** World War 2 (Two) as a result	1941; Imperial Japan ***does not attack*** U.S. military base Pearl Harbor, U.S. ***stays out*** of World War 2 (two) as a result
1945; World War 2 (Two) ends with U.S. dropping two nuke atom bombs on Imperial Japan's Hiroshima and Nagasaki.	1944; World War 2 (Two) ends, with super nations; United States (U.S.), Nazi Germany, Soviet Union, Imperial Japan,
1945; 10 (Oct) – 24 United Nations (U.N.) established in our history.	1944; No such thing as United Nations (U.N.) in this alternative history.
1950s; Korean War, between North And South, no peace treaty ever signed.	1950s; Corean War ***did not occurred*** because Imperial Japan colonize it.
1958; 10 (Oct) – 01, National Aeronautics Space Administration, NASA, year it form.	1958; 10 (Oct) – 01 (one), United States Space Force AKA U.S.S.F., year it formed.
1962; Cold War's Cuban Missile Crisis.	1944 to 2120; Kold War's Cuban Missile C.
1963 John F. Kennedy, J.F.K., assassinated.	1963; J.F.K.'s assassination foiled by Kay.
1964; Martin Luther King Junior And Malcolm X meet in civil rights conference. 1964;	
1965; 02 (Feb) – 21 (twenty one) Malcolm X assassinated.	1965; Malcolm X assassination attempt foiled by Kay's Private Security Services.
1968; 04 (Apr) – 04 (four) Martin Luther King Junior assassinated.	1968; 04 (Apr) - 04 Martin Luther King Junior assassination attempt foiled by Kay.
1968; Rob F. Kennedy, R.F.K., assassinated.	1968; R.F.K.'s assassination foiled by Kay.
1997; British returns Hong Kong to Communist Chai-Na Party (C.C.P.).	1997; British, under Nazi semi control, continues Hong Kong's rule.
2000; Year Two Thousand (2000) Killer "Y 2 K" Computer Glitch Bug Apocalypse	

Lieu, Michael W.'s (Alternative History AKA Soft Science-Fiction) (Parody) genre

A 2 S 3 B 1, "2154; Lieu, 廖," Concept Generator Part 2 (two) Expanding.

Our History, Soft Science-Factual genre	*Alt History, Soft Science-Fiction genre*
2001; (two thousand one) September eleven (AKA nine eleven) Terrorist Attacks. 911 (nine one one).	2001 (two thousand one) A Space Odyssey, Hard Science-Fiction movie Hard Science Factual genre. Space Stations
War On Terror, War On Drugs, War On Ukraine,	War On Mars, War On Resources, War On Corruption,
"2001 September eleven, nine eleven Terror Attack", "War On Terror" leading to the formation of: United States Department of Homeland Security (U.S. D.H.S.) along with other Uncle Sam acts, agencies, branches, as justified and followed.	"2001: A Space Odyssey", "1950s' Space Race And Space Force", "For All Mankind Series", "Expanse Series", "Space Elevators" later leading to formation of: Earth's Global Government (EGG) Space Force AKA Earth Space Force.
2006; year I actually gradated high school	2136; year fictional I gradated hi skhool
Military Navy, U.S.: Active Duty 2006 – 2009	Military Navy, U.S.: Active Duty 2136 – 2139
Original Official Military Ship Date; "2006; 11 (Nov) – 13 (thirteen), Updated Military Ship Date "2006; 07(Jul) - 31"	Alternative History Official Military Ship Date; "2136; 11 (Nov) – 13 (thirteen) X = Variable / 0 = Constant
2007-2008; Civilian Housing Market Real Estate Economic Crash 2007-2008;	
2008; Blu-Ray wins new High Def Media	2008; High Den. Digital Ver. Disc (H.D.D.V.D.).
2009; 02 (Feb) – 11 (eleven) My Actual Military Navy, U.S.: Active Duty Separation Date	2139; XX (XXX) – XX (X) My Fictional Military Navy, U.S.: Active Duty Separation Date(s)
2016's Call Of Duty Infinite Warfare War Game.	20XX; Earth Military Navy Space Force Fusion
2016; U.S. President Donald J. Trump Elected	2016; First Female U.S. President Hilary Clinton
2016; U.S. Vice President Mike Pence Onboard	2016; First Female Vice President Kamala Harris
2016; Nancy Pelosi 2016;	
The Colonel, Colonel Gibbs, U.S. Army Civ	The Colonel, Colonel Kinnanman, S. Squad
The Lieutenant Colonel ? { ? = forgot name } "This is Katana, she's got my back!"	The Lieutenant Colonel Fukuhara, S. Squad "This is Lieutenant Colonel, she's got my back!
Blue To Green, Sailor To Soldier	Sailor Volunteer, Soldier Voluntold
Individual Augmentee (I.A.) ia	Artificial Intelligence (A.I.) ai
Sanctuary City, Sanctuary Cities (D). Animal Sanctuary, Animal Sanctuaries. (S).	
Sovereign Citizen Movement.	Sovereign (S), Sovereignty (S).
"The Price Of Freedom Is Not Free..." "...That Is What Body Bags Are For."	
2019; 11 (Nov) Severe Acute Respiratory Syndrome (SARS) Corona Virus (C.V.) 2 (two)	
2020; 03 (Mar) Chai-Na Virus Outbreak Panic	2020; 03 (Mar) Coronavirus Outbreak Pandemic

2022;_05(May)-30(thirty) Abstract Of Arc Two Story Two (A 2 S 2) "2154;_Arcadia"

After making time to watch **Warner Bros. (W.B.)** ***Space Jam 1 (one)*** and ***Space Jam 2 (two) A New Legacy*** and along side **honest trailers** of ***Space Jam 2 (two) A New Legacy***, I got the **feelings**, the **frequencies**, that **W.B.** wants to do an **Intellectual Property (I.P.) convention center**, I realize my "*Science-Fiction Character Story Universe*" books references a good amount of **W.B. I.P. movies, one TeleVision (T.V.) show**:

"Two thousand one... ...the best science-fiction film ever made..." -Lucas, George Star Wars ***2001: A Space Odyssey*** and sequel follow-up ***2010: The Year We Make Contact***.

to

Featuring Actor Matthew Mcconaughey in both ***Contact*** and ***Interstellar*** movies. Alright.

to

Actor Joaquin Phoenix in ***Her*** and ***Joker*** as a Science-Fiction Dystopia failed state society.

to

Director Michael Crichton's ***WestWorld***, Science-Fiction study on Artificial Intelligence (A.I.) consciousness from the original M.G.M. movie and ***W.B. H.B.O. T.V. show series***.

to

Monster Kaiju genre Directors: Guillermo Deltoro's ***Pacific Rim***, Gareth Edwards' ***Gozilla***.

to

Actor Tom Cruise, *The Running Man* Not That Actor Arnold S. Movie, In ***Live Die Repeat Edge Of Tomorrow***, Science-Fiction time loop with violent video game references.

to

Actor Harrison Ford in both ***Blade Runner The Final Cut*** to ***Blade Runner 2049***.

to

Director Steven Spielberg's ***Ready Player One*** also a **W.B. I.P. convention center**.

to

The Wachowskis' ***The Animatrix story 0 (zero) origins, The Matrix 1 (one), Matrix 2 (two) Reloaded, Matrix 3 (three) Revolutions, Matrix 4 (four) Resurrections.***

to

Director Fred M. Wilcox's ***Forbidden Planet***, that 1950s (nineteen fifties) movie that inspired and created the *Star Trek universe* and the infamous A.I. Robby The Robot.

to

Director Michael Anderson's ***Logan's Run*** of the infamous 30 (thirty) years old age quote mandatory ascension belief system retirement unquote and living under a dome city.

...Just to name a few of **W.B. I.P. Science-Fiction stories**, me doing a bit of journal and or diary entry at the back of story 2 (two) "2154;_Arcadia" helps me understand that yea, I grew up with my "Elementary, Mid, High School Friends" watching many movies back then in 2019 (twenty nineteen) and before 2020 (twenty twenty) disaster year season 1 (one).

Arc Two (A 2) "A Garden Of Vigilantes"

As Of Story Three (S 3) Book One (B 1) = 2154; Lieu, 廖,
Color Coding Character Toner(s) Work In Progress (WIP)

Black 黑	Violet 紫	Blue 藍	Green 綠	Yellow 黃	Orange 橙	Red 紅	White 白

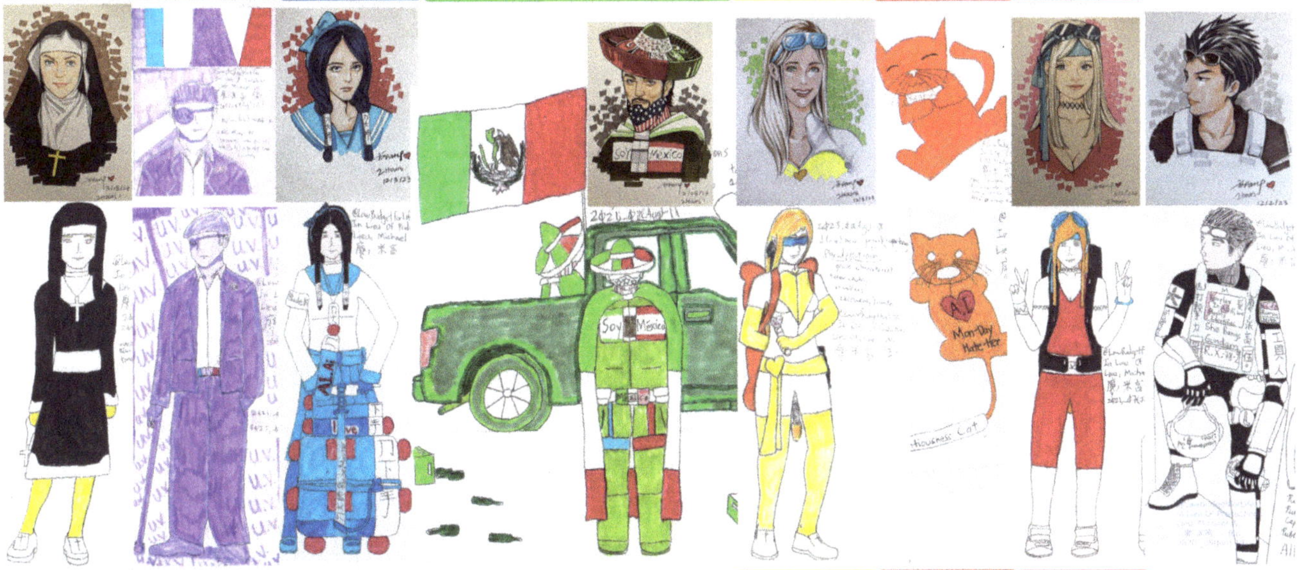

Black 黑	Violet 紫	Blue 藍	Green 綠	Yellow 黃	Orange 橙	Red 紅	White 白

Western English	Eastern Kanji	Genres Characters Label Mix Color Coding Toners	Model Reference Names
Black	黑	Parody + Contemporary Christian Catholic C.	Elizabeth Mitchell
Violet	紫	Psychological + Dystopia	(Killgrave + Mad Hatter)
Blue	藍	Comedy + **Horror**	Kanon Miyahara
Green	綠	Mythology + Militia	Diego Luna
Yellow	黃	Parody + Super-Hero	Briana White
Orange	橙	Utopia + Parody	(Garfield + Keanu Movie)
Red	紅	Travel	Sophie S., PeachMilky
White	白	**Vigilante + Neo Noir**	**Ludi Lin**

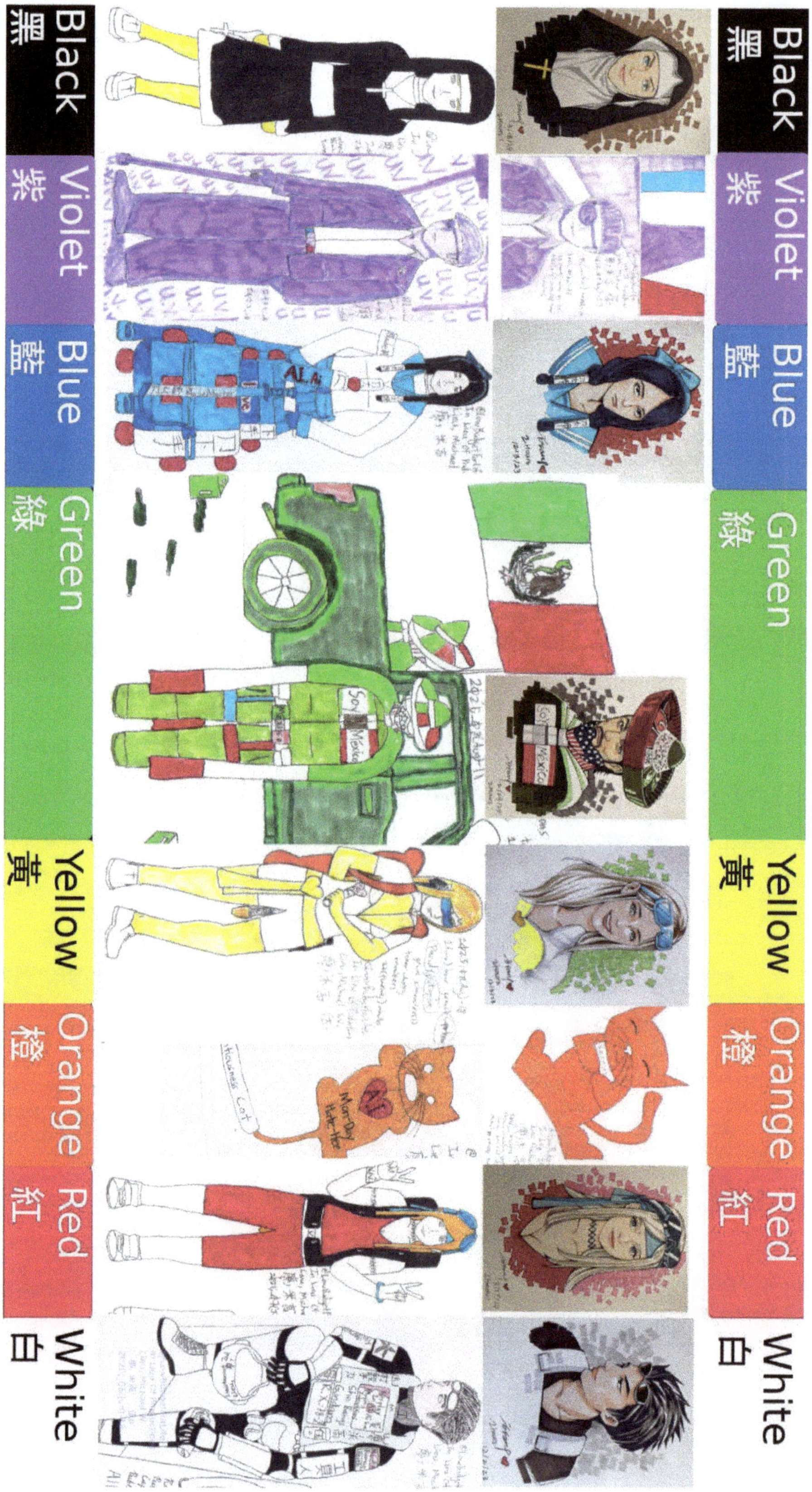
Black 黑
Violet 紫
Blue 藍
Green 綠
Yellow 黃
Orange 橙
Red 紅
White 白

"Lieu, Michael W.'s Hard Soft Science-Fiction Parody Fan Art Alternative History"

A 0(Arc Zero) "Beginning Of The End" Story 0(zero) Origins

"2120; Endgame" Roman Industries @ 11 (Nov) – 17 (seventeen), Planet Mars.
"Mars, The Roman God Of War"
Apocalyptic genre character(s) = Arial Rounded MT "Red" 'Balance To Sol System.'
Human Population In The Solar System; 11,000,000,000 (eleven billion)
Some: Martian Terrorists', Some Asteroid Belters, Some Spacers' Belief Systems;
"Humans are overpopulating the solar system and needs to be annihilated by one half."
"4 (four) Horsemen Of Apocalypse"

'Disease'	Famine	War	Death
"Sharing is caring." -? (Question-Mark) Andromeda Strain (1971)	"Create enough **hunger**, and everyone becomes a criminal { or a terrorist }." -Liam Neeson { Alter Ego, Da Mentor } W.B. D.C. Chris Nolan's Batman Begins, The League Of Shadows.	"In war, parents bury their children. In peace, children bury their parents." -Herodotus, Greek Historian	"The laws of physics is a **death** { deaf } warrant to intelligent life… Extinction is the norm for life." -Doktor Professor Michio Kaku, Theoretical Physicist And Futurist { Thanatos, Thanos }
'Conquest'	Famine Or Feast	War On Resources	Death, Merchant Of 'Deaf'

=Apocalyptic Worlds' Ending List= Insane Inspiration;

Orson Scott Card's Ender's Game Universe
Disney, Marvel Studios' Avengers 4(four) Endgame Movie
"Oh, what good is a world that only ever lets you down? Why not end it all right here?"
-Arydn
Warner Brothers(W.B.), D.C., C.W. 2019 Pre Crisis On Infinite Earths – Post 2020
W.B., Chris Nolan "Chris-N" Tenet Movie / Sci-Fi + Amazon's The Expanse Series
W.B., Home Box Office(H.B.O.) WestWorld TeleVision(T.V.) Show Series, Season 3
20th Century Fox Studios' The Martian Movie Based On Andy Weir { Any Weird } Story
N.B.C. Universal's Revolution T.V. Show / W.B., Steven Soderbergh's Contagion Movie
Horror Of The Red Planet, The Wizard Of Mars Movie
Sony, Naughty Dog's → The Last Of Us Story ← W.B., H.B.O.

2120; 11	2121; 02	05	09	12	2122; 01	02	06	10	12
0(zero)	1(one)	1 Billion(B)	2 B	3 B	4 B	5 B	6 B	7 B	8 B

Apocalypse Causes; Hubris, Zombies { Undead }, War, { Space } Alien Invasion,
Robots { Artificial Intelligence (A.I.) }, Evil, Celestial Impact, Soda, Flood,
The Internet Going Down. From GameInfomer article unknown issue.
Complete The World War Trilogy = 1 (one) Beginning, 2 (two) Middle, 3 (three) End.
Type: 0 (zero), 1 (one), 2 (two), 3 (three), 4 (four) Civilization.

A 0(Arc Zero) "Beginning Of The End" Story 0(zero) Origins

"2120; Endgame" Roman Industries @ 11(Nov)-17(seventeen), Planet Mars.

2013's Tomorrowland Science-Fiction Movie, Governor Nix's Speech, Hugh Laurie
{ Apocalyptic Educational Indoctrination Speech } Total Run Time = 02;30

"Let's Imagine... if you glimpsed the future and were frightened by what you saw, what would you do with that information? You'd go to... The politicians?
Captains of industry? And how would you convince them? Data? Facts? Good luck.
The only facts they won't challenge are the ones that keep the wheels greased and the dollars rolling in. But what if... What if there was a way of skipping the middle man... and putting the critical news directly into everyone's head?
The probability of widespread annihilation kept going up.
The only way to stop it... was to show it. To scare people straight.
'Cause what reasonable human being wouldn't be galvanized by the potential destruction of everything they've ever known or loved? To save civilization... I would show its collapse. But how do you think this vision was received?
How do you think people responded to the prospect of imminent doom?
They gobbled it up like a chocolate ecclair. They didn't fear their demise.
They repackaged it. It could be enjoyed as video games, as T.V. shows, books, movies.
The entire world whole heartedly embraced the apocalypse and sprinted towards it with gleeful abandon. Meanwhile... your earth was crumbling, all around you.
You've got simultaneous epidemics of obesity and starvation!
Explain that one. Bees and butterflies start to disappear...
The glaciers melt... Algae blooms... All around you.
The coal mine canaries are dropping dead and you won't take the hint!
In every moment, there is the possibility of a better future...
but you people won't believe it.
And because you won't believe it you won't do what is necessary to make it a reality.
So you dwell on this "Oh terrible future" you resign yourselves to it.
For one reason: Because that future doesn't ask anything of you... Today!
So, yes... We saw the iceberg. We warned the Titanic.
But you all just steered forward anyway. Full steam ahead.
Why? Because you want to sink. You gave up..." =02;30 (two minutes; thirty seconds)

Governor Nix's Speech @ YouTube Vids, by actor Hugh Laurie.

A 0(Arc Zero) "Beginning Of The End" Story 0(zero) Origins

"2120; Endgame" Roman Industries @ 11 (Nov) – 17 (seventeen), Planet Mars.

Apocalyptic genre = Arial Rounded MT "Red. Martian Terrorists, Roman Industries

2120; Endgame	*Model, Actor*	*Characters' Of The Expanse T.V. Universe*
Anvar	Cas Anvar	Alex Kamal, Martian Military Officer
Adams	Frankie Adams	Bobbie Draper, Martian Military Enlisted
Bryk	Greg Bryk	Lieutenant Lopez, Martian Military Officer
Yoon	Jean Yoon	Captain Yao, Martian Military Officer
Atherton	Ted Atherton	Doktor Strikland, Science Pediatrician
Koaho	Mpho Koaho	Rich Travis, Martian Military Enlisted
Gonzalez-Vio	Carlos Gonzalez-Vio	Doktor Cortzar, Scientist Protogen
Outerbridge	Peter Outbridge	Captain Martens, Martian Military Officer
Allen	Sarah Allen	Private Hillman, Martian Military Enlisted
Williams	Dewshane Williams	Corporal Sa'id, Martian Military Enlisted
Pla	Conrad Pla	Colonel Janus, Martian Military Officer
Dillon	Hugh Dillon	Lieutenant Sutton, Martian Military Officer
Seymour	Jeff Seymour	Pyotr Korshunov, Martian Military Minister
Crawford	Rachael Crawford	Admiral Penano, Martian Military Officer
Rock	Sabryn Rock	Private Riko Oshi, Martian Military Enlisted
Liu	Simu Liu	"Asian" Solider Boy, Martian Military
Dar	Hamed Dar	Private Trepp, Martian Military Enlisted
Mitchell	Atticus Mitchell	Ensign Sinopoli, Martian Military Officer
Mcintosh	Yanna Mcintosh	Captain Lucas, Martian Military Officer
Bridges	Krista Bridges	Captain Kirino, Martian Military Officer
Mcleod	Dayle Mcleod	Lee Lee, Martian Civilian
Steward	Kolton Stewart	David Draper, Martian Civilian
Schulze	Paul Schulze	Martian Civilian Police Department (P.D.)

A 0(Arc Zero) "Beginning Of The End" Story 0(zero) Origins

"2121; Movement" Jon W. James @ 02 (Feb) – 28 (twenty eight), Planet Earth
"Utopia, The Ideal Sovereignty Of Existence"
Dystopia genre character(s) = DejaVu Sans "Violet/Purple" "Peaceful Protocol(s)"
Human Population In The Solar System; 10,999,999,999
Utopian Sovereignty (U.S.) Belief Systems;.
"***If the system fails you, you create your own system.***" Williams, Michael K.

"Anonymous" Quotes To Lieu By
"If you are ashamed to stand by your colors, better to seek another flag."
"Until you commit your goals to paper you have intentions that are seeds without soil."
"Commitment in the face of conflict produces character."
"Life without friendship is like the sky without the sun."

=Dystopia Documenting Spying System= Insane Inspiration;

2020(twenty twenty); Disaster Years Season 1(one) / Anonymous: name, group
2020; 05(May)-28(twenty eight) George Floyd Civil Unrest, Watch The World Burn
2020; Seattle, Washington(WAS), United States of North America(U.S. of N.A.)
Capital Hill Autonomous Zone(CHAZ) / Sony, Sucker Punch's Infamous Second Son
Sovereign (S) citizen movement / A Human Sanctuary (S) city In U.S.

Square Soft Enix's Final Fantasy 6(six; VI), and or 3(three; III) in N.A., and Dragon Quest 11(eleven; XI) S Echoes Of An Elusive Age Role Playing Games(R.P.G.).

Homelessness in U.S. of N.A. and non-profit organizations that support them.
I Owe You(I.O.U.); Student Debt Loans, Home Mortgage, Rent Housing Eviction
Imploding Infrastructure / "Civilian life... is a dog eat dog world..."
W.B., George Miller's Mad Max original movie / Margaret Atwood Handmaid's Tale

A.M.C.'s Fear The Walking Dead season 1(one) TeleVision(T.V.) show series

Hive Mind Control And Indoctrination Brainwashing Difference / George Orwell 1984's Big Brother, The Thought Police And Crime(s), Science-Factual Dystopia
2+2=5 (two plus two equals five)
Thesis; "What is utopia and how do we create utopia on Earth?
Make Earth and the solar system like utopia."

Mythology of Christian Holy Bible, The Book Of Revelations,
The Apocalypse Of John { Jon } and Norse culture of Ragnarok.
W.B. D.C. Batman No Man's Land Story, Dark Knight: Returns, Rises, Gotham T.V.
W.B. Alcon Entertainment's Blade Runner 2022 Blackout Movie

A 0(Arc Zero) "Beginning Of The End" Story 0(zero) Origins

"2121; Movement" Jon W. James @ 02 (Feb) – 28 (twenty eight), Planet Earth

A.M.C.'s The Walking Dead universe, story 0(zero) origins of the human Male (M) W Wolf Owen character and his Wolf pack group mystery.

"...I see my younger self in their eyes..." Williams, Michael K.

Persuasion Power, Mind Control { Kontrol } Relations

Control Type(s)	**Model, Actor**	**Character(s) In Film Movies, T.V. Show Series, Video Games**
Mind Controller	"David Tennant, Podcast"	Marvel Netflix Jessica Jones T.V. Killgrave The Purple Man, Doctor Who
"I Want It" Mind Controller	"Benedict Samuel, Official"	D.C. Fox Gotham T.V.'s Mad Hatter, Human Male W Wolf Owen, "Character In Asthma 2015 Movie"
Mind Control Victim	"Krysten Ritter, The Real"	Marvel Netflix Jessica Jones T.V. "Character In Asthma 2015 Movie"
Remedy Control Game	"Courtney Hope, The"	The United States Federal Bureau of Control (F.B.C.) Director Jessie Faden
Mind Controller	"Barry Keoghan, keoghan92"	Dunkirk, Chernobyl, Disney Marvel's Eternals Space Alien A.I. Druig

Big Bad Villains; Marvel Killgrave and D.C. Mad Hatter team-up plus Druig.

A mobile cell phone rings in the New York City Police Department (N.Y.C.P.D.) room.

Killgrave, The Purple Man Wearing His UltraViolet (U.V.) Business Suit;
"(annoyed) Whose is that?"

That same mobile cell phone continues to ring in the N.Y.C.P.D. room.

Killgrave, The Purple Man Wearing His UltraViolet (U.V.) Business Suit;
"(annoyed) Whose phone is that?"

Killgrave gets the ringing cell phone from the N.Y.C.P.D. officer and throws it against the wall.

Killgrave, The Purple Man Wearing His UltraViolet (U.V.) Business Suit;
"(yelling) Next persons' phone rings has to eat it!!"

20th Century Fox Studios' Maze Runner 3 The Death { Deaf } Cure, The Last City

A 0(Arc Zero) "Beginning Of The End" Story 0(zero) Origins

"2121; SWAT" Seattle City Police Department (S.C.P.D.) @ 03(Mar)-15(fifteen), Planet Earth
"Police, The Human Enforcers Of Behavior" "Following The Laws, Human Laws"
Found Footage, Reality, Gun Guys Gun Girls genre character(s) = Franklin Gothic Medium "Black"
Human Population In The Solar System; "Let The Bodies Hit The Floor { Or In The Atmosphere }"
Police Department Belief Systems;. "***Protect And Serve.***"

Insane Inspiration;
1999; Artisan Entertainment's The Blairwitch Project movie,
directed by Daniel Myrick and Eduardo Sanchez
2008; Paramount Pictures' Cloverfield movie, directed by Matt Reeves
2012; 20th Century Fox Studios' Chronicle movie, directed by Josh Trank
2012; End Of Watch movie, directed by David Ayer
2013; Magnolia's Europa Report, Hard Science-Ficion movie, directed by Sebastian Cordero

Police BodyCamera (BodyCam) / Tom Clancy's Rainbow 6(six) Patriots Enemy Within
The Black Market Of Human Society, Trafficking: Human, Drugs, Firearms, Animals
Homegrown Domestic Terrorist(s) In The United States of North America(U.S. of N.A.)
Scholastic Wheels visual book; Written by Judy Nayer, Pictures by Chris Reed

Flatline of natural causes. / Flatline and or dismemberment of augmented causes.

1919; Spaniard Flu(S-Flu) Chemical, Biological, Radiological(C.B.R.) Outbreak Pandemic

2020; Disaster Years Season 1(one)
C.B.R. Outbreak Pandemic, Silence Or Sirens In A Human Urban City Society

2121; Another Disaster Years In About 1(one) Hundred(100) Years Later By The
22nd(twenty second) Century with C.B.R. Outbreak Pandemic Augmented By Electro Magnetic
Pulse(E.M.P.) Outage Blackout, The Dark Ages Of Fire, Burning Buildings And Collapse

The Adjustment Bureau Movie, Thompson On Free Will Versus Determinism Philosophy: "We actually tried free will before. After taking you ***from hunting and gathering to the height of the Roman Empire***, we stepped back to see how you'd do on your own. You gave us the ***Dark Ages for five centuries***... until finally we decided we should come back in. The Chairman thought maybe we just needed to do a better job of ***teaching you how to ride a bike before taking the training wheels off again***. So we gave you ***the Renaissance, the Enlightenment, the Scientific Revolution***. For six hundred years we taught you to ***control your impulses with reason***, then in ***nineteen ten we stepped back. Within fifty years, you'd brought us World War one, the Depression, Fascism, the Holocaust and capped it off by bringing the entire planet to the brink of destruction in the Cuban Missile Crisis***. At that point a decision was taken to step back in again before you did something that even we couldn't fix. You don't have free will, David. You have the appearance of free will."

A 0(Arc Zero) "Beginning Of The End" Story 0(zero) Origins

"2121; Army" United States Army (U.S.A.) @03 (Mar) – 30 (thirty), Planet Earth
"America, The Land Of Opportunity" "Peace To The Fallen"
Military genre character(s) = Maiandra GD ***"Green" for U.S.A. military branch.***
Human Population In The Solar System; "Let The Bodies Hit The Floor { Or In The Atmosphere }"
Military Belief Systems; "**United States (U.S.) Constitution,**
Uniform Code Military Justice (U.C.M.J.)"

"The Oath Of Enlistment (Enlisted, E)."
"I, ___________________(Full Name Here) do solemnly swear that I will support and defend the Constitution of the United States against all enemies, foreign and domestic; that I will bear true faith and allegiance to the same; and that I will obey the orders of the President of the United States and the orders of the officers appointed over me, according to regulations and the Uniform Code of Military Justice. So help me God."

"The Oath Of Office (Officer, O)."
"I, ___________________(S.S.A.N.), having been appointed an officer in the ___________________(Military Branch) of the United States, as indicated above in the grade of ___________________ do solemnly swear that I will support and defend the Constitution of the United States against all enemies, foreign or domestic, that I will bear true faith and allegiance to the same; that I take this obligation freely, without any mental reservation or purpose of evasion; and that I will well and faithfully discharge the duties of the office upon which I am about to enter; So help me God."

United States Army (U.S.A.) Military Ranks Without Any Visual Pictures

Enlisted (E) **Green and Gold**	**Company and Field Grade Officer (O)** **Gold and Silver { Commissioned }**
E-1 = Private	O-1 = 2nd (second) Lieutenant
E-2 = Private	O-2 = 1st (first) Lieutenant
E-3 = Private 1st (first) Class	O-3 = Captain
E-4 = Corporal or Specialist	O-4 = Major
E-5 = Sergeant	O-5 = Lieutenant Colonel
E-6 = Staff Sergeant	O-6 = Colonel
E-7 = Sergeant 1st (first) Class	
E-8 = Master Sergeant or 1st (first) Sergeant	
E-9 = Sergeant Major or Command Sergeant Major or Sergeant Major of the Army	

A 0(Arc Zero) "Beginning Of The End" Story 0(zero) Origins

"2121; Army"

United States Army (U.S.A.) Military Ranks Without Any Visual Pictures Continued

General Officer (O) Silver	Warrant Officer (O) Silver and Black
O-7 = Brigadier General	W-1 = Warrant Officer
O-8 = Major General	W-2 = Chief Warrant Officer
O-9 = Lieutenant General	W-3 = Chief Warrant Officer
O-10 = General	W-4 = Chief Warrant Officer
O-11 = General of the Army	W-5 = Master Warrant Officer

U.S.A. Military Branches;

Combat Arms; "Involved in actual fighting."

Infantry	Field Artillery	Corps of Engineers	Special Forces
Armor	Air Defense Artillery	Aviation	

Combat Support; "Provide operational assistance to combat arms to include combat missions as necessary."

Military Police (M.P.)	Aviation	Corps of Engineers
Chemical Corps	Military Intelligence	Signal Corps

Combat Service Support; "Provide logistical and administrative support to the army; personnel normally not directly involved in combat operations."

Adjutant General Corps	Finance Corps	Ordnance Corps Chaplain Corps	Transportation Corps	Quartermaster Corps
JAG Corps	Medical Corps	Dental Corps	Nurse Corps	Veterinary Corps

My Alt History, U.S. Department Of War (DOW) Military Branches, Aside from U.S.A.

Branch	Year Established	Servicemember Titles	Installations
Army	1775;	Soldiers	Fort, Garrison, Camp,
Navy	1775;	Sailors	Naval Station, Naval Base,
Marine Corps	1775;	Marines	Camp, Depots, Stations
Coast Guard	1790;	Coasties	
Air Force	1947;	Airmen	Air Force Base, Air Base,
Space Force	1958;	Spacemen, Spacers	Base, Stations,

A 0(Arc Zero) "Beginning Of The End" Story 0(zero) Origins

United States Armed Forces, Military (M) Services Branches Character(s)

United States Army (U.S.A.), Established 1775;

2121; Army	Model, Actor	Character(s) { Soldier Boys, Superman That Ho }
Rook	Michael Rooker	A.M.C.'s Walking Dead T.V. Show Merle Dixon, Guardians
Reed	Norman Reedus	A.M.C.'s Walking Dead T.V. Daryl Dixon, Death Stranding
Mikkelsen	Mads Mikkelsen	M.C.U.'s Doctor Strange K. "Mister Doctor", Hannibal
Jenkins	Tommie E. Jenkins	Hideo Kojima's Death Stranding Multiverse Die Hard Man
Grill	Frank Grillo	M.C.U. SHIELD Brock Rumlow Crossbones, Purge Movies
Mully	Callan Mulvey	M.C.U. SHIELD Jack Rollins, Batman V. Superman
Fukuhara	Karen Fukuhara	D.C.E.U. Katana, Kimiko Female Amazon's The Boys
Kinsman	Joel Kinnaman	RoboCop Remake, D.C.E.U. Colonel Rick Flag, Altered C
Harry	Woody Harrelson	War Planet Of The Apes Colonel, Venom 2 Carnage
Shannon	Michael Shannon	Groundhog Day, 99 Homes, D.C.E.U. General Dru-Zod
Traue	Antje Traue	D.C.E.U. General Dru-Zod's Right Hand Officer Faora-Ul
Renner	Jeremy Renner	Disney Marvel Agent Hawkeye, 28 Weeks Later Movie
Huston	Danny Huston	D.C. Wonder Woman (2017) Military, Wolverine Origins
Burton	Steve Burton	English Original Voice Actor Cloud Strife F.F. 7 (seven)
Gomez	Rick Gomez	English Original Voice Actor Zack Fair F.F. 7, Band Of Bro
Newbern	George Newbern	English Original Voice Actor Sephiroth F.F. 7 (seven)
Clarke	?	X (variable) ? (question-mark)

A 0(Arc Zero) "Beginning Of The End" Story 0(zero) Origins

United States Armed Forces, Military (M) Services Branches Character(s)

United States Navy (U.S.N.), Established 1775;

2136; Navy	Model, Actor	Character(s) { Sailor Suits, Sailor Moon That Ho, Yu! }
Lieu,	Ludi Lin	Power Rangers Re, Aquaman, Black Mirror, MK"2136-2139
Yellow Sisco, S.	Jacob Batalon	Disney Marvel Sony Spider-Man Home (O ;)"2136-2139"
Ogden,	Evangeline Lilly	Hobbit Movies, Marvel Ant-Man And Wasp "2136-2140"

(Comedy, Horror, Neo Noir) Blue Sailor Suit Student Trainees 'BladeMaster 刀大師' of 'BladeSociety 刀社會'. Established ????; Year Unknown.

Clones, Klones,	Model, Actor	Character(s) From According Universe, Mediums
Elle K. Mcgee	Rosa Salazar	Heathers' Blue Cosplay Shortcake Elle Mcgee, Maze Runner 2, 3 Movies, Alita Battle Angel
Kanon-Klones 0. Original-Copy	Kanon Miyahara	Real Action Pose Collection Photo-Book(s) Featuring Kanon Miyahara, Blood C Live Action Movies,
Fuka K. Fujimoto	Luna Fujimoto	2022's Blade Of 47 Ronin Movie To 47 Ronin Original

The Society; The **Immortal Qui Gon Jinn** And **Immoral Ra's Al Ghul** = The Boss

(Political) The Boss	Liam Neeson	Star Wars' Qui Gon Jinn, Chris Nolan Ra's Al Ghul

(Political) Black Eastern-Western Style Mao Fusion Business Suits 'BladeBoss 刀老闆'
Insane Inspiration By "The Tommy Wiseau Disaster Artist Room" So Bad It So Good Genre

"Oh Hi Mark"	Model, Actor	Character(s) From According Universes, Mediums
Mark D. eaf	Mark Dacascos	Double Dragon Movie, John Wick Chapter 3 Zero
Mark S. ubstitute	Mark Strong	Brothers Grimsby Sacha Baron Cohen, Shazam

(Political) Black Eastern-Western Style Mao Fusion Business Suits 'BladeMaster 刀大師'

Kung Fu, MA.	Model, Actor	Character(s) From According Universes, Mediums
No Name / Cecep Arif Rahman		Lionsgate's John Wick Chapter 3 Parabellum movie
No Name / Yayan Ruhian		Lionsgate's John Wick Chapter 3 Parabellum movie

A 0(Arc Zero) "Beginning Of The End" Story 0(zero) Origins

United States Armed Forces, Military (M) Services Branches Character(s)

United States Marine Corps (U.S.M.C.), Established 1775;

2121; USMC	Model, Actor	Character(s) { Marine Boys, Superboy That Ho, Yu! }
Guevara	Jonny Bernthal	A.M.C. Walking Dead Shane, M.C.U. Netflix Daredevil Punisher
Barney	Ben Barnes	M.C.U. Netflix's Punisher T.V., WestWorld T.V. season 1, 2
Drive	Adam Driver	Disney Star Wars Sequel Trilogy Episodes 7, 8, 9, TED Talk
Jane	Thomas Jane	2004 The Punisher, Dirty Laundry, The Predator, The Expanse

A 0(Arc Zero) "Beginning Of The End" Story 0(zero) Origins

United States Armed Forces, Military (M) Services Branches Character(s)

United States Coast Guard (U.S.C.G.), Established 1790;

21XX; Coast	Model, Actor	Character(s) { Sailor Suit, Sailor Moon That Ho, Yu! }
Morales	Oscar Issac	Ex Machina, Annihilation, Disney Star Wars 7, 8, 9, Moon

A 0(Arc Zero) "Beginning Of The End" Story 0(zero) Origins

United States Armed Forces, Military (M) Services Branches Character(s)

United States Air Force (U.S.A.F.), Established 1947; Post U.S. Army Air Corps

2145; Air	Model, Actor	Character(s) { Air-Men, Raw-Men }
Drastal	Alex Drastal	@AlexDrastal Actor Model Cosplayer "Serves 2145-2149"

A 0(Arc Zero) "Beginning Of The End" Story 0(zero) Origins

United States Armed Forces, Military (M) Services Branches Character(s)

United States Space Force (U.S.S.F.), Established 1958; { Our History, Was NASA Founded }

2121; Space	Model, Actor	Character(s) { Space Superman That Ho, Yu! } 2150; Space
Lang	Stephen Lang	James Cameron's Avatar Colonel Miles Q., Don't Breathe
Williams	Keith David	E.A. Mass Effect 1, 2, 3 David Anderson, Halo, COD
Drastal	Alex Drastal	@AlexDrastal Actor Model Cosplayer "Serves 2150-2153"
Wright	Gunner Wright	"Moved To Contemporary genre U.S.S.F. civ contracts"
Rockwell	Sam Rockwell	"Moved To Contemporary genre U.S.S.F. civ contracts"
Hale	Jennifer Hale	E.A. Mass Effect 1 2 3 Trilogy Female Shepherd FemShep
Meer	Mark Meer	E.A. Mass Effect 1 2 3 Trilogy Male Shepherd MaleShep
Richards	Denise Richards	Director Paul Verhoeven's Starship Troopers One Movie
Busey	Jake Busey	Director Paul Verhoeven's Starship Troopers One Movie
Brown	Clancy Brown	Director Paul Verhoeven's Starship Troopers One Movie
Meyer	Dina Meyer	Director Paul Verhoeven's Starship Troopers One Movie
Harries	Neil P. Harris	Director Paul Verhoeven's Starship Troopers One Movie
Vandien	Casper Vandien	Director Paul Verhoeven's Starship Troopers One Movie
Ironside	Michael Ironside	Director Paul Verhoeven's Starship Troopers One Movie
Judge	Chris Judge	Stargate S.G. One, God Of War; 4 2018, 5 Ragnarok
Weaver	Sigourney Weaver	Alien 1(one) Original, Aliens 2(two) Movie, Avatar Movie
Hamilton	Linda Hamilton	Terminator 1(one) Original, Terminator 2(two) Movie
Characters' Of The Expanse T.V. Universe Earthling Spacer Belter Public Sec Military Civilian(s)		
2121; Space	Model, Actor	Character Universe – 2150-2153; Mars
Chatham	Wes Chatham	Amos Burton, Earthling U.S.S.F. Military Police (M.P.)
Doyle	Shawn Doyle	Sadavir Errinwright, Earthling Civilian
Aghdashloo	Shohreh Aghashloo	Chrisjen Avasarala, Earthling Civilian Politician
Coleman	Chad L. Coleman	Fred Johnson, Earthling Military **Terrorist**, A.M.C. T.W.D.
Roach	Martin Roach	Admiral Souther, Earthling U.S.S.F. Military Officer
George	Brian George	Michael Benyaer, Earthling Civilian
Toufexis	Elias Toufexis	Kenzo Gabriel, Earthling Civilian Spy / Deus Ex H.E. M.D.

A 0(Arc Zero) "Beginning Of The End" Story 0(zero) Origins

United States Armed Forces, Military (M) Services Branches Character(s)

United States Space Force (U.S.S.F.), Established 1958; { Our History, Was NASA Founded }

Characters' Of The Expanse T.V. Universe Earthling Spacer Belter Public Sec Military Civilian(s)		
2121; Space	Model, Actor	Character Universe – 2150-2153; Mars
Tarabay	Nick E. Tarabay	Cotyar Ghazi, Earthling Civilian Public Sec
Whittaker	Jonathan Whittaker	Esteban Sorrento Gillis, Earthling Civilian
Mann	Byron Mann	Augusto Nguyen, Earthling Military Officer
Whittall	Ted Whittall	Michael Iturbi, Earthling Civilian Scientist
Mitchell	Elizabeth Mitchell	Anna Volovodov, Earthling Civilian Pastor= Parody, Contemporary C
Owens	Chris Owens	Scientist Kolvourd, Earthling Civilian
Williams	Genelle Williams	Tilly Fagan, Earthling Civilian Socialite
Hopkins	Anna Hopkins	Monica Stuart, Earthling Civilian Journalist
Dauda	Raven Dauda	Nono Volovodov, Earthling Civilian Spouse To Anna V.
Mcgibbon	Brandon Mcgibbon	Cohen Casti, Earthling Civilian Journalist
Millen	Ari Millen	Stanni Kulp, Earthling Civilian Electrician
Nunes	Paulino Nunes	Priest Cortez, Earthling Civilian Priest
Lisinska	Natalle Lisinska	Lieutenant Shaffer, Earthling Military Officer
Kelly	Morgan Kelly	Lieutenant Mancuso, Earthling Military Officer
Kapelos	John Kapelos	Ren Hazuki, Earthling Civilian Electrician
Gao	Lily Gao	Nancy Gao, Earthling Civilian Politician / RE 4 RE Ada Wong

A 0(Arc Zero) "Beginning Of The End" Story 0(zero) Origins

United States Armed Forces, Military (M) Services Branches Character(s)

United States Space Force (U.S.S.F.), Established 1958; { Our History, Was NASA Founded }

Contemporary genre character(s) = Century Schoolbook

Civilian Contractors Within The Military (M) genre character(s) = Maiandra GD

2121; Space	Model, Actor	Character(s) { Space Superman That Ho, Yu! }
Wright	Gunner Wright	E.A.'s Dead Space 1 Remake, 2, 3 Issac Clarke voice actor
Rockwell	Sam Rockwell	Sam In Moon Movie, Iron Man 2 Hammer, The Bad Guys
Alabi	Anthony Alabi	E.A.'s Dead Space 1 Remake Zach Hammond perform cap
Tahir	Faran Tahir	E.A.'s Dead Space 1 Remake Doktor Mercer, Mad Science
Canales	Brigitte K. Canales	E.A.'s Dead Space 1 Remake Kendra Daniels, Secret Agent
Characters' Of The Expanse T.V. Series Universe Earthling Spacer Belter		
Strait	Steven Strait	James Holden, Earthling Spacer Civilian "Belter"
Tipper	Dominique J. Tipper	Noami Nagata, "United States Merchant Ship (U.S.M.S.)"
Costanzo	Paulo Costanzo	Shed Garvey, Earthling Spacer Medical Technician
Faivre	Florence Faivre	Julie Mao, Spacer Civilian Private Sec
Gee	Cara Gee	Camina Drummer, Spacer Civilian Public Sec
Alexander	Keon Alexander	Marco Inaros, Spacer Terrorist Operative
Owens	Jasai C. Owens	Filip Inaros, Spacer Terrorist Operative
Nicole	Nadine Nicole	Clarissa Mao, Spacer Augmented Civilian And Terrorist
Rotilio	Andrew Rotilio	Diogo Harai, Spacer Civilian Public Sec
Chau	Francois Chau	Jules-Pierre Mao, Spacer Civilian Private Sec
Karkanis	Athena Karkanis	Tavi Muss, Spacer Civilian Public Sec
Harris	Jared Harris	Anderson Dawes, Spacer
Hernandez	Jay Hernadez	Dmitri Havelock, Spacer Civilian Public Sec
Glaudini	Lola Glaudini	Captain Shaddid, Spacer Civilian Public Sec
Hanchard	Kevin Hanchard	Detective Sematimba, Spacer Civilian Public Sec
Chen	Terry Chen	Doktor Meng, Spacer Civilian
Jung	Leah Jung	Mei Meng "Baby Girl", Spacer Civilian
Strathairn	David Strathairn	Klaes Ashford, Spacer Pirate Officer
Johnson	Brock Johnson	Grigui, Spacer Civilian
Noel	Jaeden Noel	Katoa Merton, Space Civilian

A 0(Arc Zero) "Beginning Of The End" Story 0(zero) Origins

United States Armed Forces, Military (M) Services Branches Character(s)

United States Space Force (U.S.S.F.), Established 1958; { Our History, Was NASA Founded }

Contemporary genre character(s) = Century Schoolbook

Civilian Contractors Within The Military (M) genre character(s) = Maiandra GD

Characters' Of The Expanse T.V. Series Universe Earthling Spacer Belter		
2121; Space	**Model, Actor**	**Character(s) { Space Superman That Ho, Yu! }**
Gorman	Burn Gorman	Adol Murtry, Spacer Military Private Sec, Dark Knight R
Greenwood	Lyndle Greenwood	Doktor Okoye, Spacer Civilian Biologist
Gilmore	Rosa Gilmore	Lucia Mazor, Spacer Terrorist
Kim	Patti Kim	Carol Chimene, Spacer
Madeira	Kyle Madeira	Felcia Mazor, Spacer Refugee
Mccarthy	Steven Mccarthy	Jakob Mazor, Spacer Refugee
Salgueiro	Jess Salgueiro	Chandra Wei, Spacer Military Private Sec
Smadu	Zach Smadu	Scientist Sarkis, Spacer Civilian
2014's Call Of Duty Advanced Warfare (COD AW) Set In Alternative History 2054 to 2061.		
Mitchell	Troy Baker	U.S. Marines Private 1st C., Player Char, Voices In V.G.
Gideon	Gideon Emery	Atlas' Private Military Contractors, Biggs Eng F.F. 7 R.
Irons	Kevin Spacey	Atlas' Private Military Contractors (P.M.C.) Mastermind
2016's Call Of Duty Infinite Warfare (COD I.W.) U.N.S.A. AKA "Earth's Space Force"		
Lt. Reyes	Brian Bloom	Acting Commanding Officer (C.O.) Captain, Retribution
Adam. Raines	John Marshall Jones	Admiral (Adm.) In I.W. "Earth's Space Force"
Lt. Salter	Jamie Gray Hyder	Lieutenant (Lt.) "**Peace To The Fallen**" COD I.W.
E3N "Ethan"	Jeffrey Nording	Artificial Intelligence (A.I.) Automation COD I.W.
Sgt. Omar	David Harewood	Sergeant (Sgt.), *Also In D.C.T.V. Supergirl Martian M*.
Navigator	Omid Abtahi	Navigator (Gator) Victor Diallo, COD I.W.
Cap. Ferran	Claudia Christian	Tigris warship Captain Maureen Ferran COD I.W.
ChEng Mac	Claudia Black	Chief Engineer (ChEng) Maureen Ferran, Retribution
Adm. Salen Kotch	Kit Harington (*Mars S.D.F. Leader*)	Admiral (Adm.) "Mars Aeternum = Mars Forever", W.B. H.B.O.'s Game Of Thrones, George R.R. Martin Story.
Riah	Jim Pirri	Akeel Min Riah, Senior Commander In COD I.W.

A 0(Arc Zero) "Beginning Of The End" Story 0(zero) Origins

The Militia, "Soy Mexico (I am Mexico)": Mexican-Men Mythology, Mexican Cartels

Characters Of The Narcos T.V. Series Columbia 1st (first), Mexico 2nd (second) Universe		
2121;→2154;	Model, Actor	Characters In Narcos T.V. Verse
Moura	Wagner Moura	(Not Mexican), Pablo Escobar, Narco T
Jimenez	@ Jorge A. Jimenez Actor	"Poison", Roberto Ramos.
Catano	Diego Catano	"La Quica", Juan Diego Diaz
Ruiz	Pau Gaitan	Spouse Wife Of Pablo Escobar
Najera	Bruno Bichir	Attorney Lawyer Of Pablo Escobar
Martinez	Raul Mijo Mendez	28th Colombian President
Alcazar	Damian Alcazar Castello	Gilberto Rodriguez, Colum Cali Cartel
Varela	Matias Varela (Swedish)	(Not Mexican), Salcedo, Surveillance Cali
Rodriguez	Javier Camara 1 (Spanish)	(Not Mexican), Pollomari, Accountant C
Reguera	Ana De La Reguera	Elisa Alvarez, M-19
Garza	Gabriela De La Garza	Diana Turbay, Columbia Journalist
Diaz	@ Julian Diaz Actor (Colombian)	(Not Mexican), "Blackie" "El Nergo"
Gomez	@ Leynar Gomez Cr (P. Rico)	(Not Mexican), "Limon" Jhon Burgos.
Canto	Adan Canto	Rodrigo Lara Columbia Lawyer, Politican
Big Ben., R.	@ Diego Luna	Miguel Angel Felix Gallardo, Narcos M
Huerta	Tenoch Huerta	"Rafa" Rafael Coro Quintero
Pena	@ Michael Pena "mvegapena"	Kiki Camerena
Diaz	Alyssa "Aly Diaz"	(Not Mexican), Mika Camarena
Cosio	Joaquin Cosio Oficial"	"Don Neto" Ernesto Fonseca Carillo
Yazpik	Jose Maria Yazpik "Jmyazik"	Amado Carrillo Fuentes
Edda	Alejandro Edda "Aljo-eddo"	"El Chapo" Joaquin Guzman
Ruiz	Teresa Ruiz " _teresaru"	Isabella Bautista
Taracena	Gerardo Taracena "Eltaracena"	Pablo Acosta Gonzales
Cedillo	Julio Cesar Cedillo	Commander Guillermo Calderoni
Not In Narcos T.V. Series Universe		
Cabral	Richard Cabral Official"	Mexican-American. Mexico Flag Waver
Baby Ben., J.	@ Gabriel Luna	Agents Of SHIELD Ghost Rider, Term.

A 0(Arc Zero) "Beginning Of The End" Story 0(zero) Origins

"2121; Collapse" X @ 05 (May) – 28 (twenty eight), Planet Earth
"Holocene, The 6th (sixth) Extinction Event Of Epoch"
"Mythology" genre character(s) =
Human Population In The Solar System; "Let The Bodies Hit The Floor
{ Or In The Atmosphere }"
Mythology Belief System; "***It's all biblical...***"

Insane Inspiration;

1990s Rodney King Trials And The Los Angeles (L.A.) City Riots

"An empire toppled by its enemies can rise again. But one that crumbles from within? That is dead. Forever.".
-Daniel Bruhl { Alter Ego, Ze Mo } in
Disney, Marvel Studios' Captain America 3 (three) Civil War movie

"The League Of Shadows has been a check against human corruption for thousands of years. We sacked Rome, loaded trade ship with plague rats, burn London to the ground. Every time a civilization reaches its pinnacle of decadence, we return to restore the balance.".
-Liam Neeson { Alter Ego, Da Mentor }
Warner Brothers (W.B.), Detective Comics (D.C.) Christopher Nolan "Chris-N"
Batman story 1(one), Batman Begins The League Of Shadows

The collapse of Afghanistan on 2021; 08(Aug) and its takeover by the Taliban

The Economist, 2018; 08 (Aug) – 04 (four), page 09 (nine)
"The world is losing the war against climate change
Planet Earth and the ever changing crisis of "climate change".

The Economist, 2018; 11 (Nov) - 17 (seventeen), page 89 (eighty nine)
"Burning out, Climate change makes fires worse, but agricultural development limits them".

Homefront: The Revolution, Homefront alternative history violet video games

2022; 02(Feb)-24(twenty four) The Invasion Of Ukraine By Russian Military Forces

A 0(Arc Zero) "Beginning Of The End" Story 0(zero) Origins

Kay's Private Security Services Limited Liability Company (L.L.C.)
Owner Operator Manual Guide genre = "OCR A Extended"
Standard Operating Procedure (S.O.P.), Established ? (Question-Mark)
Agent(s); K=Kool E=Enforce A=Assert N=Negotiate U=Unity

Model, Actor	Character(s) { Private Security Agent(s) }
Keanu Reeves	Bill And Ted, Matrix, John Wick, Cyberpunk 2077
@TheJackColeman	Universal N.B.C. Kim Kring's Heroes T.V. H.R.G.
@JimmyJeanLouis	Universal N.B.C. Kim Kring's Heroes T.V. The Haitian
@GregGrunberg	Universal N.B.C. Kim Kring's Heroes T.V. Mind Reader
@AnthonyMackie	Phil K. Dick's Adjustment Bureau, M.C.U. Falcon
John Slattery	Phil K. Dick's Adjustment Bureau, M.C.U. Old Stark
@ClarkGregg	Marvel Cinematic Universe (M.C.U.) Agent Coulson
@MeaganTandy	W.B. D.C. C.W.'s Batwoman T.V. Crow Private Sec
@RubyRose	W.B. D.C. C.W.'s Batwoman, John Wick, 'Ruby Rose'
Stephen D. Scott	W.B. D.C. C.W.'s Batwoman T.V. Crow P.S. J. Kane
@MazinAkar	Marvel Netflix's Jessica Jones Private Sec Hi Hank
@LeeMajdoub	Paramount Sega's Sonic Hedgehog 1, 2 Agent Stone
@FreddieWong	YouTube Rocket Jump Firearms Gun Fu Videos
@GerardButler	"This Is Sparta!" Gamer Movie, Has Fallen Movies

A 0(Arc Zero) + "A ? (Arc Question-Mark) "? ? ? ?;" and or "XXXX;"

"Super-Hero (Western Society) genre,

Vigilante (Revenge Thriller genre, AKA Judge, Jury, Xecutioner) genre,

Super Sentai (Eastern Society) genre,

Parody (Spoof, Satire, Mockery) genre character(s),

Parody {Pussy Cat} Contemporary Christian Catholic Church genre character(s) = Papyrus Sister Nun(s), Father Priest(s), Lord Jesus Christ, God, The Lord, He, Himself, The Church

Conscientiousness Cat/Kat Orange Artificial Intelligence (A.I.) Ai 愛 Love

Psychology Personality = OCEAN 5 (five)

Yellow White "HeartBladeEndureLove 心刀忍愛"

Daoriginal (D And Or The Original) First Vigilant Volunteer Non-Paid Worker"

XXXX;	Model, Actor	Characters' Portrayed Universe
Kiss-Hiss, AKA Sister K. W. White	@ Briana White "ItsBrianaWhite"	Square Soft Enix's Final Fantasy 7 (seven; VII) Remake, Crisis Core Remaster Prequel, Occupants.
Sister M. Mitchell	@ Elizabeth Mitchell	Universal's Purge, Expanse T.V. Pastor Anna V
Sister A. Adams	@ Janeshia Adams-Ginyard	Disney Marvel Black Panther Dora Milaje
Sister S. Shaye	@ Lin Shaye	Insidious Universe Movies as Elise, The Quarry
Father Greg Ford	@ John Eric Bentley	Final Fantasy 7 Remake, Rebirth, Re 3 Barret
Sister P. Poesy	@ Clemence Poesy	Chris Nolan's Tenet, Walking Dead Daryl Nun
Sister S. Sweeney	@ Sydney Sweeney	Sony Marvel Madame Web Spider Woman Julia

Next "**Arc Two Story Three Book Two (A 2 S 3 B 2)**" AKA "**2154; Incident**"

Lieu, Michael W.'s Science-Fiction Character Story Universe Blueprint (ver. 1(one), or
Lieu, Michael W.'s Hard Soft Science-Fiction Parody Fan Art Alternative History (ver. 2(two).

A 2 (Arc Two) "A Garden Of Vigilantes"
S 3 (Story Three)
B # (Book Number) Book Chapter Like Breakdowns… "2154; X (Variable)"

Since Arc Two Story One (A 2 S 1) "Blue 藍" "2154;_Arriving"
Was Almost 400 (four hundred) pages.

And Arc Two Story Two (A 2 S 2) "Red 紅" "2154;_Arcadia"
Was Past 500 (five hundred) pages.

Now **A 2 S 3 B 1 To X # (Book One To Variable Number),**
Sectioning To About 200 (two hundred) + (plus) pages.

A 2 (Arc Two) S 3 (Story Three) B 1 (Book One) 'This one'	**2154; Lieu, 廖,**
A 2 (Arc Two) S 3 (Story Three) B 2 (Book Two) 'Next one'	**2154; Incident**
A 2 (Arc Two) S 3 (Story Three) B 3 (Book Three)	**2154; Expo**
A 2 (Arc Two) S 3 (Story Three) B 4 (Book Four)	**2154; Ascension**

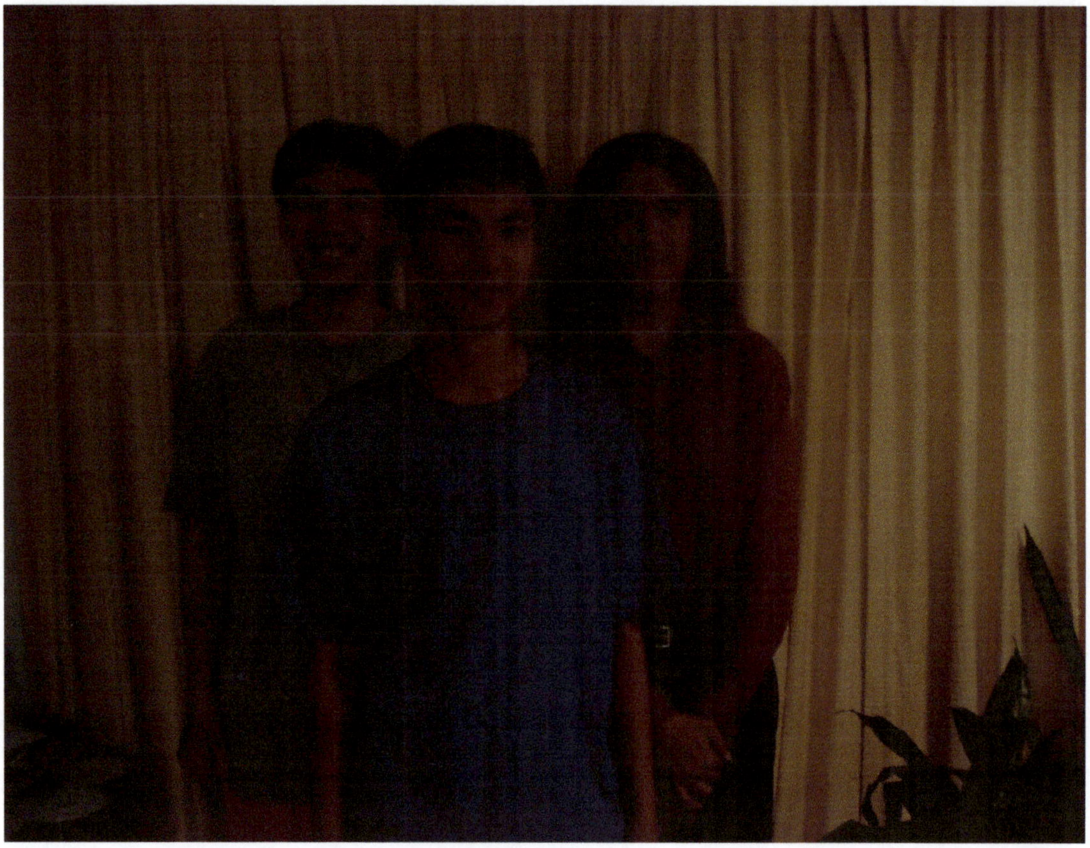

220 \ **2154;** Lieu, 廖, 'WhiteWolf 白狼' Left to Right (L→R) 220

Next "**Arc Two Story Three Book Two (A 2 S 3 B 2)**" AKA "**2154; Incident**"
Subject to change due to Murphy's Law and X Variables...

Act 1 (one) back story
creating composition
2154; 10 (Oct) – 30 (thirty)
{ "**BladeMaster 刀大師**, Sailor Suit Student Trainee Elle K. Mcgee" Arriving }
@ XX;XX p.m.
page #XXX

Act 2 (two) story forward
inciting incident
2154; 10 (Oct) – 31 (thirty one)
{ Terrorists' Attack In **Elysium Corp** Chicago Animal Sanctuary **Dome** Illinois? }
@ 22;31 p.m.
page #XXX

Act 3 (three) final fight
boss battles
2154; 11 (Nov) – 11 (eleven)
{ The Old World United **States** of North America's Military Veterans' Day }
@ 12;00 a.m. and or 00;00 a.m.
page #XXX

Act 4 (four) conclusion wrap-up
closing circles
2154; 11 (Nov) – 11 (eleven)
{ Prelude To 2154 Fall Entertainers' Expo Day 0 (Zero) }
@ 12;00 a.m. and or 00;00 a.m.
page #XXX

Arc Two Story Three Book Two (A 2 S 3 B 2) AKA 2154; Incident	
Act 1 (one)	2154; 10 (Oct) – 30 (thirty)
Act 1 (one)	2154; 10 (Oct) – 31 (thirty one)
Act 2 (two)	2154; 10 (Oct) – 31 (thirty one)
Act 2 (two)	2154; 11 (Nov) – 01 (one) to 11 (eleven)
Act 3 (three)	2154; 11 (Nov) – 11 (eleven)

Lieu, Michael W.'s Artist Author Info Dump.

A Hard Soft Science-Fiction
Parody Fan Art Alternative
History story...
Chicago, Illinois in **2154**,
formerly known as the
"ChiRaq: City Of Death"
in the 21st century,
by the 22nd century
of the new world
2121 Utopian Sovereignty movement,
Chicago City is now
an Animal Sanctuary utopia.
Within Elysium Corporations' engineering
bio**tech domes**, the temperature
is around +024° Celsius.
Meanwhile, outside the bio**tech**
domes, the temperature exposure,
varies according to the
apocalyptic unconventional warfare agents
affecting Planet Earth's atmosphere.
Retired veteran **Lieutenant Lieu,廖,**
AKA WhiteWolf 白狼 in suit,
is making plans to
celebrate the anniversary of:
Sir Arthur C. Clarke's
Space Elevator launch and,
meet up with **his**
Elementary, Middle, High School,
Friends from County Base.
Just another year in
a retired veteran's existence...

A 2 (Arc Two), "A Garden Of Vigilantes"

4 (four) letter stanza with sentence break indent.

during warm season outside:+150° Fahrenheit day time, +100° Fahrenheit night time. Rising sea water levels within 15 (fifteen) miles inbound to all United States (U.S.) North America sea shore, According to Warner Brothers' Alternative History **Blade Runner 2049 (twenty forty nine) story movie** and maybe assume world wide on Planet Earth by mid 21st (century first) century. Low Planet Earth natural resources inbound to Down Town L.A. (D.T.L.A.) South California (SCA) state. Sony's **Elysium 2154 (twenty one fifty four) story movie** = off planet space station.
Disney Fox? **Avatar, James Cameron, 2154 (twenty one fifty four) story movie** = off planet space mining.
Science-Fiction \ Hard Science-Fiction \ Soft Science-Fiction \ Alternative History \ Apocalyptic, Neo-Western { Macro Big } \ Utopia, Parody { Micro Small } Semi-Autobiography \ Traditional Art Character Designer \ Coloring-Book Line Art Style \ Photo-Book { Vigilant Sketches Journal Expansions = Verdana } 200 / 590

Lieu, Michael W.'s Artist Author Info Dump.
Updated And Abstract From "2154;_Arcadia"

(Elementary, Mid, High School Friends) = Eastern-Western SitCom genre

"Remy Lin / ***d. Chicago b. Basketball-Man***"	Jeremy Lin
@ JLin7.	書豪 林

An Asian-American that used to play in N.B.A. (National Basketball Association) Professional Sports Basketball-Man Player

2011;_06(Jun)-28 Jeremy Lin Academy from 2013's Linsanity Documentary, paraphrase "I remember in the D-League, um, I kept a little diary… Um… I know that's a little girly for some of you guys, but I like to write down some of my thoughts every once and while, so I can look back and remember. And, uh, I remember on December twenty nine, I wrote about how actually wished I never signed with the Warriors. I wish, you know, I wish I could quit basketball because, um, I wasn't having fun. And what got me through those times was… Obviously I don't feel that way now… But, what got me through that time, was the, uh… the number one reason why all the leaders are here today, and that's because we want to let you know that God loves me, and he, he has a perfect plan for me, and his plan will take you through a lot of ups and downs, but if I stay faithful to him. In the end, I know he's gonna work everything for my good. Now when I play basketball, I don't play for anybody else anymore. I only play for God, and that's the type of purpose he gave me. Once he gave me that purpose, that's when I have my peace. Once I got my peace, that is when I got my joy." -**Lin, Jeremy (林, 書豪) { Lin, Remy }**

"{ **Jia Wong** } / B.4-Burma3(three)III-M.M.B." = @ **The Chin Han** .
Warner Brothers + D.C.'s Nolan, Christopher's 2008 The Dark Knight, Batman movie
Asian Accountant Lau, "(confident) I'm good with calculation."
Warner Brothers, D.C. + C.W.'s 2012 The Arrow { Green Arrow } show series
Warner Brothers' 2011 Contagion movie, directed by Steven Soderbergh

"{ **Sen Wang** } / WackyJoker.exe" = @ **Masi Oka** / "Yatta! Heroes T.V. show series"

"Lieutenant { **Lieu, Michael W.** }" = @ **Ludi Lin** 2017 Power Rangers

Better Luck Tomorrow movie, directed by Justin Lin \ Ragnarok Online \ C.S. S.C.

Entertainer, **Mister Steven Z. J.**, Plastination Mister, Mistress = Thriller genre(s) **A super animated 6 rainbow tone spectrum** A.R. Holo HUD projection plays the music in the background of "**Henry Mancini's The Entertainer**" talks to her directly, as she looks at **it** with a curious cat look. **A rainbow color curtains displays itself first with English on top, Kanji on the bottom reading**: **Red "紅", Orange "橙", Yellow "黄", Green "綠", Blue "藍", Violet "紫". Purple Prince's Rain.**

www.ingramcontent.com/pod-product-compliance
Lightning Source LLC
LaVergne TN
LVHW070831220926
848486LV00030B/272